Storey's Guide to
RAISING SHEEP

Storey's Guide to RAISING SHEEP

Paula Simmons & Carol Ekarius

Schoolhouse Road
Pownal, Vermont 05261

The mission of Storey Communications is to serve our customers by publishing practical information that encourages personal independence in harmony with the environment.

Edited by Marie Salter and Deborah Burns
Copyedited by Mary Boylan
Cover design by Renelle Moser
Front cover photograph © Grant Heilman, from Grant Heilman Photography, Inc.
Back cover photographs by PhotoDisc
Illustrations by Elayne Sears, except those on pages 17, 19, 247, 286 by Brigita Fuhrmann; 88, 91, 92, 122 by Chuck Galey; 97 by Carl Kirkpatrick; and 54, 159, 190, 194, 216, 220, 314

Series design by Mark Tomasi
Text production by Susan Bernier, Leslie Noyes, and Deborah Daly
Indexed by Nan Badgett/Word•a•bil•i•ty

Storey's Guide to Raising Sheep was previously published under the title *Raising Sheep the Modern Way*. This new edition has been expanded by 112 pages. Chapter 4, Herding Dogs, was written by Beverly Lambert. All of the information in the previous edition was reviewed and revised for this new text, which offers the most comprehensive and up-to-date information available on raising sheep.

Printed in the United States by Versa Press
10 9 8 7 6 5 4 3 2 1

Library of Congress Cataloging-in-Publication Data

Simmons, Paula.
Storey's guide to raising sheep / Paula Simmons and Carol Ekarius.
p. cm.
Includes index.
ISBN 1-58017-262-8 (alk. paper)
1. Sheep. I. Ekarius, Carol. II. Title.
SF375 .S563 2000
636.3—dc21 00-058804

Contents

Foreword

Raising Sheep the Modern Way has been used by more than 100,000 sheep lovers since its publication in 1976, but times have changed, laws have changed, technology has advanced, and the resources (which change so rapidly) all required updating.

Carol Ekarius, author of *Small-Scale Livestock Farming* (also published by Storey Books) undertook the update for this new edition, *Storey's Guide to Raising Sheep*. She is particularly knowledgeable regarding genetics, sheep dogs, guardian dogs, sheep showing, and ecological concerns, which were perhaps not sufficiently addressed in my original book. Much appreciation to Carol for her good work in updating and adding to a book that has been relied upon by so many sheep and their owners.

A book is never the work of just one or two people. Many contributed to my meager knowledge from the beginning. The Lunds, original publishers of *Shepherd* magazine, were most helpful; *sheep!* magazine, edited by Dave Thompson, has been a constant source of useful articles and veterinary columns. In my opinion, this magazine is a "must have" in any shepherd's household, for success with sheep is more certain with a regular supply of current information. And, the more you know, the more you will enjoy your sheep.

Paula Simmons

PREFACE

This project has been a privilege and pleasure for me. I first heard of Paula Simmons in the early eighties (long before I had my first sheep, or even thought about having any). I'd taken up weaving and spinning, and her book *Spinning and Weaving with Wool* was always near at hand — a patient mentor, waiting to walk me through some technique or problem. I was honored when Deborah Burns, one of the editors at Storey Books, dangled the carrot of working on the revisions to Paula's *Raising Sheep the Modern Way* in front of my nose.

Paula's original manuscript was wonderful, and had provided the same type of mentorship to aspiring and new shepherds for decades. But as is apt to happen, even with the finest books in this realm, the times had changed, leaving some parts of the book outdated. Laws have changed, technology has changed, and even consumer's tastes in book design have changed. The book needed some freshening up, and Paula decided she didn't have the time or inclination to undertake the project. I was drafted.

The bulk of this book is still Paula's, with her wonderful wit, compassion, and knowledge shining through. I appreciate the opportunity she and Deborah gave me to work on the update.

A book is never the work of just one person — and this one isn't just mine, or just Paula's. We both benefited from the input of many knowledgeable and wonderful people. The shepherds who agreed to have their stories told deserve a big round of applause. The technical experts from various groups, who answered my questions, and the secretaries for the various breed associations, who filled in a questionnaire, all deserve special thanks. For preparing the chapter on herding dogs, thanks to Beverly Lambert. For reviewing part, or all of the revised version, thanks Dr. Ann Wells, DVM, Jeff Green of USDA's Wildlife Services, and Dave Thompson of *sheep!* magazine. Thanks also to Deborah Burns and Marie Salter, editors at Storey Books, and to the rest of the Storey staff for their help and support.

Carol Ekarius

The following poem was prepared by Paula's good friend, Dr. Darrell Salsbury, DVM, for the book's previous edition, but its wisdom hasn't changed at all:

The Shepherd's Lament

Now I lay me down to sleep
 Exhausted by those doggone sheep;

My only wish is that I might
 Cause them not to lamb at night.

I wouldn't mind the occasional ewe,
 But lately it's more than just a few:

Back into bed, then up again,
 At two o'clock and four A.M. . . .

They grunt and groan with noses high,
 And in between a mournful sigh,

We stand there watching nature work,
 Hoping there won't be a quirk:

A leg turned back, or even worse,
 A lamb that's coming in reverse.

But once they've lambed we're glad to see
 That their efforts didn't end in tragedy.

There's no emotion so sublime
 As a ewe and lamb that's doing fine.

I'm often asked why I raise sheep,
 With all the work and loss of sleep;

The gratification gained at three A.M.,
 From the birth of another baby lamb —

How can you explain, or even show?
 'Cause only a shepherd will ever know!

D. L. Salsbury, DVM

ONE

STARTING WITH SHEEP

"Sheep are the dumbest animals on God's green earth," our neighbor avowed with a vigorous shake of his head when he saw the newest additions to our farmstead. His belief is not uncommon. In fact, sheep are a love–hate animal — people either really love them or really hate them. And the people who really hate them, love nothing more than to malign them.

But sheep don't deserve the bad rap they've received. They fill a niche that needs filling: They provide economically efficient food and fiber, they eat many kinds of weeds that other livestock species won't touch, they're relatively inexpensive to begin raising, and they reproduce quickly so that a minimal capital outlay can become a respectable flock in short order.

Plus, sheep are simply nice, gentle animals. Watching a group of young lambs charging wildly around the pasture or playing king of the hill on any mound of dirt, downed tree, or other object that happens to occupy space in their world has to be one of life's greatest joys.

Admittedly, there are some difficulties to raising sheep: They think fences are puzzles that you've placed there for them to figure a way out of. Their flocking nature can sometimes make handling a challenge. Although they're less susceptible to many diseases than other critters, they're more troubled by parasites. They're also vulnerable to predators. But with the help of this book, even a novice can learn to manage the negative aspects of raising sheep while enjoying the benefits.

Some Background on Sheep

Scientists consider sheep to be members of the family Bovidae, which includes mammals that have hollow horns and four stomachs (ruminants). All sheep are in the genera Ovis, and domestic sheep are named *Ovis ammon aries*.

The human need for animals isn't new: Food, fiber, traction (or the ability to do work, such as pulling, pushing, and carrying), and companionship led humans to domesticate animals more than 15,000 years ago. Dogs were the first animals to be domesticated, but humans bonded with sheep and goats early on as they settled into agriculturally based communities. Both sheep and goats were domesticated about 10,000 years ago, according to the latest theories.

Biologists believe that modern sheep were primarily descended from the wild Mouflon sheep of western Asia, although other wild sheep (such as the Mouflon sheep of Europe) have been mixed in since domestication took place. Some breeds, like the Soay of Europe, still retain many of the characteristics of their wild ancestors, but most modern breeds have changed substantially. Traits of wild sheep include naturally short and fat tails, coarse and hairy outer coats, short and wooly undercoats, and great curling horns on the rams. Wild sheep are endangered or threatened throughout the world

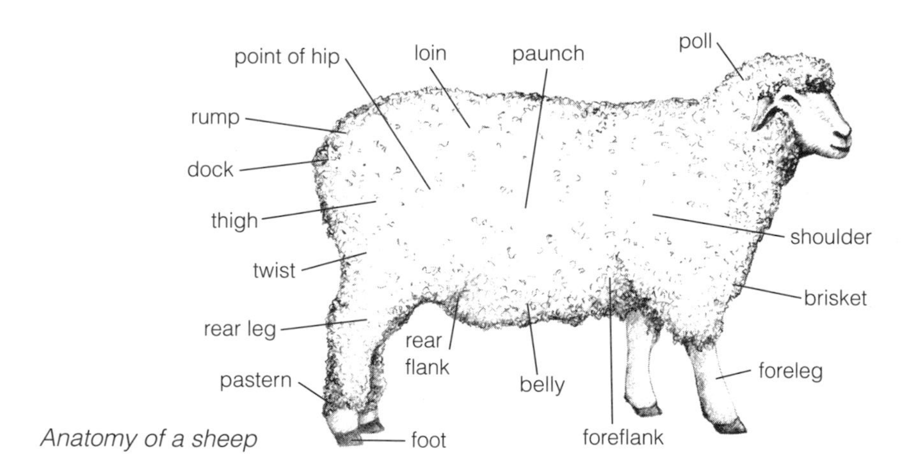

Anatomy of a sheep

Sheep Farming Today

Sheep numbers continue to fall in the United States and Canada. During the 1990s, the total number of sheep in the United States dropped below 9 million. At the middle of the twentieth century, there had been more than 50 million.

Worldwide, sheep numbers are still high, but because synthetic fibers have largely replaced natural fibers, warehouses are clogged with wool surpluses. In North America, the lamb dollar has also shrunk: Consumers purchase only about one-third of the lamb that they purchased 30 years ago. However, these changes have not been devastating to the sheep industry. This is due, at least in part, to its lack of vertical integration.

SHEPHERD STORY

Karl and Jane North's primary enterprise is a seasonal sheep dairy. Forty-five ewes lamb on pasture in early May and are milked in a six-stanchion parlor through the end of September. The milk is then made into cheese in an on-farm cheese plant.

"Although we only milk for 4 or 5 months, we are able to carry an inventory of cheese year-round, because the types of cheese we produce require long aging," Karl says.

Marketing is critical to the success of an operation like the North's. They market cheese, lamb, yarn, and tanned hides.

"About 60 percent of our cheese is marketed at the Ithaca, New York, farmer's market. The market has a cosmopolitan clientele of people who can afford $12 per pound for cheese." The rest of the cheese is marketed via mail order and an in-town outlet. The mail-order business grew out of write-ups in various magazines and books about cheese and from taste-testings they did in New York City. "For us, mail order works. Our product is unique, and the customers are willing to pay not only for our cheese, but also for the shipping."

Ethnic markets also help Karl and Jane with marketing their lambs. The lambs come in May and are sold or butchered when the grass runs out in December. "With this short window, most of our lambs weigh less than 80 pounds when we sell them. One-third of our lamb crop we direct market, but the other two-thirds are sold to a packer. Luckily, we have an ethnic packer who likes lambs this size." Conventional packers want lambs that weigh up to 120 pounds.

Vertical integration occurs when large multinational companies begin controlling all facets of production and marketing. When a market segment becomes vertically integrated, it's very hard for small producers to exist in that segment. The poultry and pork industries are good examples. The sheep industry, on the other hand, hasn't been taken over by corporate giants, so small producers who can produce by using low-cost methods can remain in the black. In fact, if you're willing to market your own product, you can do quite well.

Homestead Flocks

Sheep are especially good animals for small-property owners who don't have the space to raise cattle but want some kind of livestock. Five to seven ewes and their offspring can typically be run on the same amount of land as only one cow and a calf. Sheep can graze lawns, ditches, wood lots, and orchards (with full-size trees only — the sheep will eat dwarf trees if you plant them).

Starting small gives you the opportunity to gain low-cost experience. If you start with fewer sheep than your land will support (see chapter 3), you will be able to keep your best ewe lamb each year, for a few years at least. After a while, as your purchased ewes become unproductive, they can be replaced with some of your best lambs.

Although a homesteader may occasionally sell a few lambs or fleece, the flock is raised primarily for personal use. Providing your own meat and some fleece for hand-spinning or a 4-H project for the kids are all reasons homesteaders choose to keep a few sheep. Typically, these flocks are small — usually no more than a dozen ewes and a ram.

Commercial Flocks

Commercial flocks vary in size from fairly small flocks of twenty to fifty ewes to vast flocks that number in the thousands. Today, more than 80 percent of the sheep raised in the United States are raised in large "range bands" in the western half of the country. These bands typically have 1,000 to 1,500 ewes and are tended by one or two full-time shepherds and their dogs.

The main thing to consider is that for commercial flocks — even relatively small ones — marketing must be vigorous. This can be direct marketing or marketing through the conventional agricultural system of sale barns and middlemen, but to do it profitably it's going to take time, energy, and thought (see chapter 11).

More than one commercial flock has grown out of a homestead flock. Suddenly, a flock that began with one or two ewes grows to twenty or thirty, and the homesteader is looking for a larger piece of land or some additional places to graze the sheep on other people's land.

Then, there are folks who jump from virtually no experience with sheep to acquiring a commercial flock in one step. Perhaps they've inherited a farm or have decided to purchase their dream farm. These folks face a greater challenge than those who take the "grow-your-own-flock" approach, but the rest of this book should help either type of new shepherd.

Intensive Versus Extensive Management

There are two approaches to any type of agricultural enterprise: the high-input, intensive system or the low-input, extensive system. The high-input system is the one that currently dominates U.S. agriculture. This system requires tremendous inputs of labor and cash for fertilizers, pesticides, harvested feeds, veterinary services, extensive lines of machinery, and specialized buildings. Farmers practicing high-input, intensive agriculture hope to produce enough product to meet those costs and make a profit regardless of what "the markets" are doing. In the intensive system, there is an expectation that more lambs means more money, but that isn't always the case. Although the intensive approach works for some folks, there are far more who are drowning in worthless products and piles of bills.

The low-input, extensive management system places far less emphasis on production volume and more on profitability. This is also the system that's been tagged "sustainable agriculture" in recent years. Sustainable practitioners look to maximize profit while protecting the environment and the social structure of their rural communities. They consider quality of life to be as important as gross income, but they would probably agree that *net income* plays a big role in having a good quality of life.

In this system, farmers try to mimic nature — for example, by lambing in the spring when the grass is coming on (and wild animals are having their young). They look to their animals to carry a fair share of the workload, harvesting their own feed and spreading their own manure for a large portion of the year. Successful practitioners of low-input, extensive agriculture find that both labor and costs are dramatically reduced. The time they save allows them to maximize profits by working on direct marketing. This book emphasizes the low-input, extensive system because this type of management is especially well suited to homestead flocks and small commercial producers.

Sheep Production Systems

The sheep production systems currently in use are:

- **Accelerated lambing.** The most intensive approach to sheep production, accelerated lambing calls for each ewe to lamb at least three times every 2 years. This system requires a high outlay of capital for lambing barns, and feedlots or barns (finishing facilities). It requires sheep that have the genetic capability of lambing more than once a year and phenomenally good management to keep the ewes healthy enough to do so.
- **Winter-confinement lambing.** In this intensive system, lambing occurs in January and February in lambing barns. Lambs are able to nurse and self-feed in creep feeders. Creep feeders allow the lambs free-choice access to extra feed but prevent the ewes from getting at the self-feeders. After weaning, usually around 2 months of age, the lambs are kept on free-choice feed in finishing facilities until marketing.

- **Phase lambing.** Another highly intensive approach, phase lambing seeks to have the ewes lamb only once a year, but the flock is broken into three or four groups. This allows lambing throughout the year, so lambs can be marketed throughout the year. It requires capital outlay for building and feeding facilities, but these can be somewhat smaller than those required for accelerated lambing or winter-confinement lambing, because only part of the flock is lambing in any given season of the year.
- **Early spring–confinement lambing.** With March and April lambing, the lambs must be lambed in a barn but are often finished after weaning on high-quality pastures instead of in finishing facilities. This system is an intensive–extensive hybrid.
- **Fall lambing.** Like early spring–confinement lambing, this system is a hybrid of intensive and extensive systems, but capital outlay is for finishing facilities instead of lambing barns.
- **Late-spring pasture lambing.** This is the most extensive system. Few facilities are required, and less labor is required than in the other approaches because lambs drop and finish on pasture.

Organic Production

Some readers may be interested in organic production. There are some successful organic producers in the sheep business, although internal parasites can challenge those producers who want to be "certified organic," because such certification prohibits the use of most commercial worming medications. Organic certification also requires some expenses and lots of extra record keeping. Chapter 11 talks about products and marketing and will help you to evaluate the costs and benefits of becoming certifiably organic.

Behavior

When you decide to get sheep, it helps if you understand their behavior — in other words, what makes them tick. The more you understand about their behavior, the easier it will be for you to spot problems (for example, is that ewe in the corner sick or about to lamb?). Understanding behavior also makes handling animals much easier, on both you and them.

Behavior can simply be thought of as the way an individual animal, or a group of animals, responds to its environment. Behavior falls into three main categories: normal, abnormal, and learned. Remember, in the case of sheep, most of their behavior stems from their position in the food chain: They are prey animals — as such, they are rather small and vulnerable.

Sheep that are behaving normally are content and alert. They have good appetites and bright eyes. They are gregarious animals, which contributes to their flocking nature. Youngsters, like those of other species, love to play and roughhouse. Groups of lambs will run, jump, and climb for hours when they are healthy and happy. Then they'll fall asleep so deeply, you may think they're dead.

Sheep learn to adapt to new environments, or new conditions within their environment. With patience and the right cues, you can teach them to move into handling facilities or through gates and into new pastures. Through "punishments" they can quickly learn to avoid certain things, like an electric fence.

Abnormal behavior is usually either related to stress or disease and can take many forms, such as wool eating, fighting and other aggressive actions, lack of appetite, excessive "talking," or sexual nonperformance. Stress-related abnormal behavior most often occurs in the close confinement of intensive production systems; these abnormal behaviors rarely occur in animals that are raised in an extensive, pasture-based system. If left unchecked, the stress that contributes to abnormal behaviors creates an environment in which disease can get a strong foothold. It's best to eliminate or minimize stress-causing agents on your farm. See chapter 3 for what you can do to relieve stress through pasture, fence, and facility design.

Sheep Behavior

Understanding your sheep's behavior will be easier if you remember the following:

- Sheep fear noise, unfamiliar surroundings, or unfamiliar items in their surroundings (that jacket hanging on the shovel against the wall, for example), strange people, dogs, and water.
- Sheep move readily from a dark area to a light area, from a confined area to an open area, from a lower area to a higher area, and toward food.
- Sheep like to follow one another and move away from people, dogs, and buildings.
- Sheep will cling to a wall in a pen, and if there are sharp corners will bunch up in the corners and stay there.

Social Structure

Although all sheep are generally considered to be gregarious, there are always differences among breeds and differences among individuals. Relationships are generally strongest between animals of the same species, although animals can learn to have a relationship with animals of other species (for example, you, your horses or cows, your dogs). Relationships also tend to be strongest between members of the same breed. Oftentimes, if two distinct breeds are run in the same large pasture, they'll group together in two distinct flocks that avoid each other's space. Within a flock, relationships tend to be strongest among family units. An older ewe, her children, her children's children, and so on will behave as a unit.

Like most prey species that join together as a group, a sheep flock has a *pecking order*, or dominance hierarchy. On pasture, dominant and subordinate relationships don't tend to have much impact on any members of the flock, but they can be important if the sheep are fed in a confinement system. Animals at the top of the pecking order will "hog" the feed at a trough, and animals at the bottom will starve.

Dominant animals are simply the more aggressive members of the group. This aggressiveness may be the result of age, size, sex, or early experiences. The dominant ewe within a flock will have dominant daughters, but I don't know if this is as much an inherited trait as a learned position.

Dominance among rams just before and during the breeding season is easy to observe, as they actively fight for the top position in the pecking order. Rams fight by backing up, and then, with heads down, charge forward at a full run to butt heads. This type of fighting generally ends when one ram backs down, but rams can be seriously injured or killed during these fights.

Leader–follower relationships are strong in sheep. Interestingly, the leader of the flock may not be its most dominant member. Because of the stronger relationships among family members, the oldest ewe with the largest number of offspring often becomes the leader of the flock.

Emotions and Senses

If behavior is thought of as being the way animals react to their environment, then senses are the tools they use to investigate their environment and emotions are the outward manifestations of this reaction. Let's talk about emotions first.

Sheep, like other mammals, are capable of displaying a full array of emotions, from anger to happiness, to the most common emotion we humans usually see when dealing with animals: fear. Scientists have discovered that fear memories are stored in a primitive part of the brain, and as such, these memories stay with an animal for long periods. If an animal has an especially bad fright, for example, upon entering a barn, then it will continue to fear entering that building. (See chapter 3 for more about handling and providing guidance for overcoming fear reactions.)

As much as fear reactions can be a pain in the neck for shepherds, they must remember that those reactions are genetically programmed for survival. Sheep are prey animals, and the speed with which they have a fear reaction is part of their defense against predators — make no mistake about it, when sheep see you, they see a predator. Though with patience and training, you can win their trust.

Like humans, sheep count on their senses for understanding the world around them.

Sight

As prey animals, sheep rely heavily on their sense of sight. A sheep's eyes are set off to the sides of its head, creating a wide field of vision (between 270 and 320 degrees, depending on how much wool it's wearing). This wide field of vision allows it to see predators at great distances — and more important, in almost any direction that a predator may approach. Sheep do have a blind spot directly behind them (so if you are approaching from the back, make sure to let the sheep know you're there by talking to them).

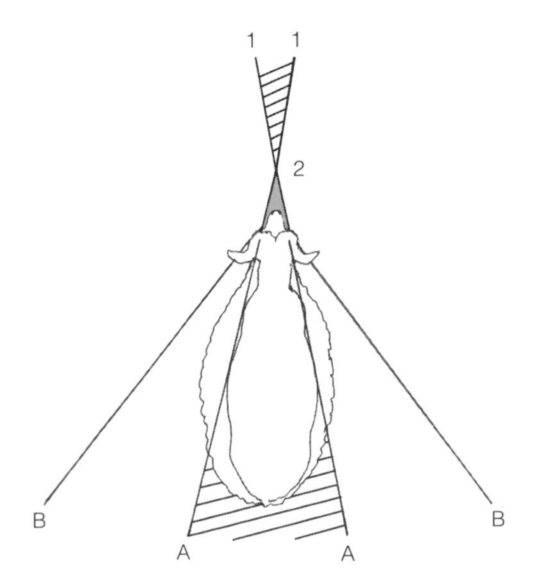

Sheep have a fairly wide field of monofocal vision and a small area of bifocal vision (shaded area; *between point 2 and the sheep's eyes). They also have blind spots directly in front and in back of their bodies* (hash marks). *When sheep are sheared, they can see everything between point 1 and point A on either side of their bodies. As their fleece grows out, they see less toward the back but can still see everything between point 1 and point B.*

They also have a small area of bifocal vision directly in front of their noses. Bifocal vision allows them to focus on an item with both eyes at the same time and greatly enhances depth perception. This is why you'll sometimes spot a sheep staring at something right in front of its nose with great intensity: It's bringing the item into focus with both eyes. Scientists believe that sheep have keen vision and that they see and can differentiate between colors.

Hearing

After sight, hearing is probably the second most important of the sheep's senses. Sheep hear at a much higher frequency of sound than you do. Low-pitched rumbling sounds won't really disturb them, but any kind of high-pitched sound, like that of a human yelling, will send them off the deep end. Very loud or novel sounds will also cause stress. On the other hand, with training, your sheep can learn that certain loud sounds have meaning, like a whistle that means it's time to come for food. Ewes and lambs can hear, and differentiate between, the voices of their offspring or mother over fairly long distances. An animal that is separated from the flock will call incessantly to try to locate its buddies. One interesting fact about the sheep's auditory system is that it can pinpoint where subtle sounds are coming from by "tuning in" its two ears separately.

Smell

The sense of smell is extremely important to sheep. A sheep's sense of smell is much stronger than that of a human's. Smell is the first sense that ewes use to identify their lambs at birth and while they are nursing. The ewe recognizes her own scent from the amniotic fluid that is coating the lamb and later from the milk odor. One method of grafting an orphan lamb to a ewe is to fool her sense of smell by rubbing the orphan down with her amniotic fluid. Rams also use the sense of smell to detect ewes that are coming into heat or are ready to mate; at that time, females release a chemical pheromone, which is like a perfume, that the male smells.

Taste and Touch

Taste and touch are the least important senses. Sheep use taste the same way humans use it: to decide if something is good to eat. Touch is used in courting, in parental bonding, and sometimes to become more familiar with something. Unlike humans, however, a sheep does most of its touching with its nose. You'll often see this behavior if something new has been introduced to the animal's environment: First, it approaches with its neck stretched way out while it sniffs the air, then it touches the item with its nose.

Working Sheep

When you are working a flock of sheep (for example, for moving or shearing), it helps to have patience, to move slowly, and to work quietly. Working your animals is always stressful on them, so your efforts to reduce their stress will pay dividends in better production, less illness, and fewer injuries. If you're working sheep within the first month after they've been bred, stress can actually cause abortion. Your patience and slow, quiet approach will not only reduce the stress on them but on you as well.

Moving and controlling flocks in large areas are often best accomplished with the assistance of a herding dog. Chapter 4 discusses the use of these working companions. Working your flock also becomes easier if you use good handling facilities such as catch pens, chutes, and gates. Read more about these in chapter 3.

Breeds

A breed of animals is a group that has been raised to exhibit similar, inheritable traits. Most breeds have a breed association or registry that establishes the standards for the breed and maintains records of "registered" breeding stock. A purebred possesses the distinct characteristics of the breed and is registered, or eligible for registration, with the breed association.

The advantages of purebreds are greater uniformity in appearance and production and a chance of income from the sale of breeding stock, although in most cases the additional cost associated with maintaining and marketing purebred animals isn't offset by the extra income. If you or your children are interested in showing sheep, then purebreds offer a much wider array of show opportunities. The disadvantages are higher initial cost and the costs of registering lambs, with no better price for wool or meat.

Different breeds were developed in response to market needs and the conditions under which the animals were to be raised. For example, some breeds were raised to flourish in hotter climates and others in cooler climates. Some breeds have a higher incidence of multiple births (which is fine if you are able to give them sufficient attention to ensure survival and good growth), and some breeds are able to lamb more than once a year (known as "out-of-season" lambing).

Crossbred Sheep

Crossbred sheep are those that have blood from one or more breeds in their lineage. Crossbreeds often produce as well, if not better, than purebreds as a result of a phenomenon known as hybrid vigor. Although purebreds usually exhibit certain desirable traits, inbreeding can also bring out some undesirable traits; when sheep of two different breeds are bred to each other, the most desirable traits of each breed tend to come out and the less desirable ones don't. This makes for hardier, more vigorous, and more productive offspring; hence the term *hybrid vigor*.

Most commercial flock owners run a crossbred flock for their production animals, though many also maintain smaller, registered flocks. A typical cross in commercial circles is a ewe with ½ Finn and ½ Rambouillet blood; these crossbred ewes are typically bred to a Dorset ram, yielding ¼ Finn, ¼ Rambouillet, and ½ Dorset lambs. See chapter 2 for help deciding which breed, or crossbreed, might be best for you.

Native and Western Ewes

In the areas of the country where sheep are raised most commonly, some sheep are classified as *native* and some as *western*, or *range*, sheep. In reality, these terms have little significance to you as a shepherd, but you may hear people use them from time to time. The terms don't necessarily refer to a specific breed — or even a specific cross — but they refer to a "type." Native sheep are primarily raised for meat and are large, prolific, and usually blackfaced. Western sheep are usually fine-wool sheep or are a cross of fine-wool and long-wool breeds. Fine-wool sheep were often preferred on the western ranges not for their wool but for their strong flocking instinct.

Buying Sheep

If you are new to sheep, then *read the rest of this book* before purchasing your first sheep. Studying before buying will save you money, time, aggravation, and possibly the lives of your sheep! But if you are ready to buy, here are some things to keep in mind.

Unless you plan to have only a few sheep, try to obtain ewes with similar breeding for your first foray into shepherding. Not only will these ewes share traits such as temperament, breeding period, gestation period, and maturity dates, they'll also produce lambs of similar quality that mature at about the same time, which will enhance the marketability of your lambs.

If you don't have a preference for a particular breed, consider the predominant one in your area. It's likely to be well suited to the climate, and buying close to home cuts down on shipping costs and a stressful ride for the animals.

You can get replacement rams more easily, even trading with other breeders nearby, when you have used yours for a while and want to avoid inbreeding.

Until you become an experienced shepherd, it may be best to seek a mentor who can help evaluate the animals before you buy. Another shepherd, or a veterinarian, can help you evaluate the conformation and general health of the animals you're considering, and paying for their consultation can actually save lots of money down the road.

If you're considering buying sheep, think seriously about the timing of your purchase. It's best to buy at a period when the animals aren't going to have to do anything too significant right after they arrive on your farm. Moving to a new home is as unpleasant for them as it is for you, and they take some time to settle in.

Buyer's Guidelines

- Purchase through a private treaty sale.
- Expect some kind of production and health records.
- Look for animals of similar breeding.
- Demand healthy-looking coats.
- Demand good conformation.
- Look for alert and active ewes but with relatively calm dispositions.
- Look for bright-eyed animals, with no discharge around the eyes, no squinting, and no eye damage.
- Examine the teeth.
- Make sure the manure is solid and pelleted.
- Consider your breeding goals, such as whether you want multiple births, to produce a particular type of wool or meat, or to raise a particular breed.

Where to Buy

Your first purchases should be directly from a farmer or rancher who raises sheep. (These sales between individuals are called *private treaty sales*.) If you're buying purebred stock, the breed association can help identify shepherds in your region who raise the breed that interests you. If you don't have your heart set on a specific breed, ask around for references to a reputable farmer. Veterinarians, county Extension agents, and other small-flock owners may be able to give you some names of shepherds to talk to.

Don't buy from the first farmer you visit. Try to check out two or three farms. Look around at each, but don't judge on the "fanciness" of the facilities. Some excellent shepherds (especially if they're full-time farmers) have old, unpainted buildings but still have excellent, healthy animals, and that's what you're there for! Although the facilities may be old and in need of a coat of paint, they should be fairly clean. This doesn't mean that there will be no manure piles around or any equipment stuck in a corner, but it does mean that bottles of medicine, bags of chemicals, used needles, and just plain trash shouldn't be in evidence anywhere you look. If it's been raining or snowing for a while, the ground may be muddy, but the sheep should never be chest deep in wet manure or mud.

At each farm, ask the shepherds about their breeding plans:

- What are they trying to accomplish with their flocks?
- Do they have production records and health records on the flock?
- Will they provide a 5-day health warranty? (Some farmers won't do this, and with good reason: They don't know how you will take care of the animals. But many will stand behind their animal's health for a short period.)
- Will they deliver your animals? (Within a reasonable distance, this may be part of the sale price, but for long distances, expect to pay the farmer for trucking.)
- Will they provide some technical support after purchase, like answering phone questions?

If a seller seems to be unwilling to answer your questions or is impatient with you, go somewhere else.

Don't purchase your first sheep at the sale barn or livestock auction house. Although some good ewes may go through there from time to time, it's the most dangerous way for beginners to purchase their animals. First, even if the animals are healthy when they get there, they're exposed to all kinds of other animals that are there specifically because they aren't healthy. Second, as a neophyte, you probably don't have the ability to distinguish good, healthy animals from those that aren't, especially at a distance as they run through the ring. If you do have a sale barn nearby, though, go there for educational purposes. Talk with the farmers, and study the pricing of terminal market animals (those that are going for butcher). If you see some sheep that look good to your untrained eye, ask whose farm they come from, and by all means, give that person a call.

Also try to move animals during mild weather, if possible, and avoid rough handling and overcrowding in transport. All animals become stressed by moving, but the worse the stress, the more likely they'll come down with shipping fever, which can run the gamut, from a small nuisance to a major calamity.

Sheep Age Versus Price

The age of the sheep is important in relation to the asking price. Fine, young ewes that have already lambed once or twice usually bring the most money; they're already proven breeders but they still have lots of years, and lambs, in front of them. But don't rule out older sheep if you're on a tight budget. You can get started with the least outlay of capital by purchasing someone else's culls.

Commercial shepherds often cull ewes at 7 or 8 years of age, although their expected productive life is 10 or 12 years. And older ewes are often the previous owner's better ewes to have remained in the flock for a long time. Their years may be numbered, but with good care, older ewes can be even better for you than they were for their former owner because they don't have to compete with younger ewes. By keeping the very best ewe lambs produced by these old ladies, you'll soon have a nice young flock at a reasonable price.

When trying to decide on a fair price for someone else's culls, ask yourself:

- Just how much more fleece and how many lambs could this ewe be expected to produce?
- If she is quite old, how much additional and higher-quality feed will she need to compensate for her poor teeth?
- Does she have a history of twins and triplets?

Let your offer reflect these conditions.

The opposite age extreme — baby lambs — may also provide a cheaper route to starting out with sheep. Oftentimes, shepherds who have a bunch of *bummer lambs* (orphans or rejected lambs that have to be hand-raised on a bottle) are glad to get rid of some. But before you think of traveling down this path, you need to understand that bummers got that name for a reason: Until they are weaned, feeding them is very time consuming. But if you have the time, it can also be very rewarding, and your hand-raised lambs will be close pets for life, running to greet you whenever you enter the pasture or barn. For more about feeding bummers, see chapter 10.

Teeth

Sheep have no teeth in the top jaw. They have a hard palate, or dental pad, on top. Their bottom teeth consist of 8 incisors in the front and 24 molars, or cheek teeth, in the rear of the mouth. Up to a certain age, the incisors can help you figure out the sheep's age.

A lamb has eight small incisor teeth until it reaches approximately 1 year of age. Each year thereafter, one pair of lamb teeth is replaced by two permanent teeth that are noticeably larger. By the time the sheep is about 4 years old, all of its lamb teeth have been replaced by permanent teeth. After this point, it is no longer possible to *accurately* determine the animal's age by its teeth, although you can make estimates based on the condition of the teeth.

All that grinding action begins to wear the sheep's teeth, shortening their useful life and thereby the life of the sheep. As the incisors wear down, the amount of tooth below the gumline (about ½ inch [1.3 cm])is gradually pushed out to help compensate for the wear. This is partly why the teeth of an older ewe look so much narrower — the wider part at the top of the tooth is being

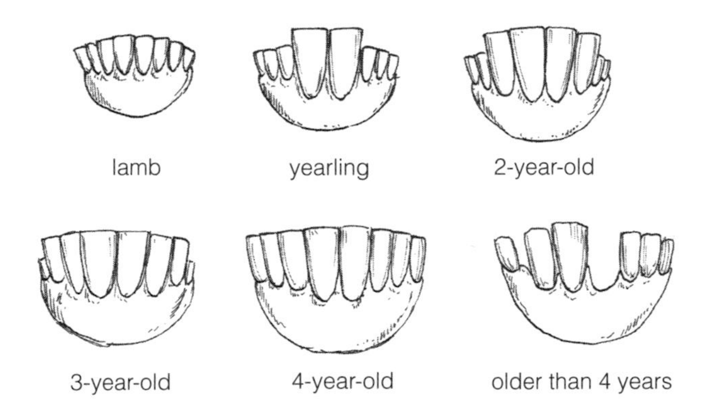

To some extent, you can determine the age of sheep by their teeth. Sheep have no teeth in their upper jaw. In the front bottom jaw, they have four pairs of incisor teeth that change with age. Lambs have four pairs of incisors — all small baby teeth that, like human baby teeth, fall out to make way for permanent teeth. At about 1 year, the center pair fall out, and the first pair of larger, permanent incisors appear. For each year until the sheep is 4 years old, it loses one pair of baby teeth and gains one pair of permanent teeth. After a sheep turns 4 years old, you can't really tell its age by looking at the teeth, but it will begin losing permanent teeth at this point, hence the name broken mouth.

worn back toward the narrower center part of the tooth while the even narrower part below the gumline is being pushed up. With narrowing, gaps that reduce the efficiency of the ewe's bite occur between the teeth. If you listen to an old ewe grazing, you can hear sound as the grass slips between her teeth.

On very short or overstocked pasture, the wear is faster, because sand and soil particles are picked up as the animal grazes and act like sandpaper on the teeth. The closer to the soil sheep graze, the more dirt and sand they ingest. On short pasture, ewes must take more bites to get a pound of grass, and each bite contributes to the wear of their teeth.

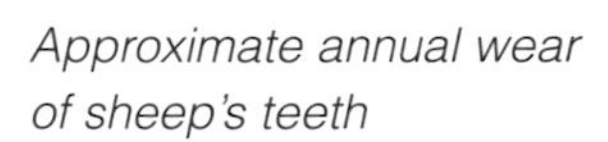

Approximate annual wear of sheep's teeth

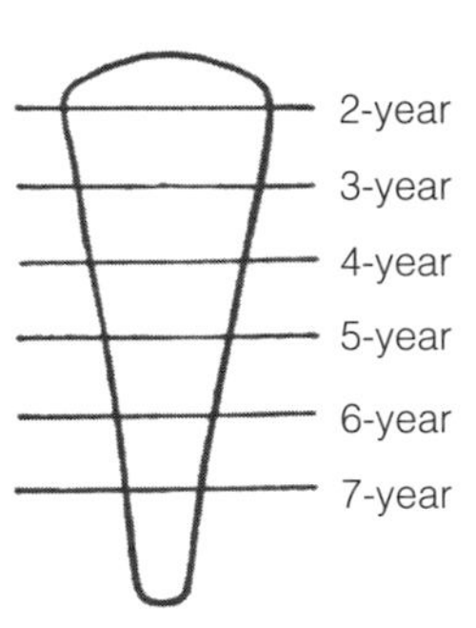

Terms to Define the Condition of a Sheep's Teeth

- **Solid mouth.** Sheep up to about 4 years of age having all adult teeth in place.
- **Spreaders.** Older animals with teeth showing wear, and the narrower parts of the teeth that are under the gums are moving up into position.
- **Broken mouth.** Sheep with some teeth missing. Ewes may have one or two seasons of lambs left.
- **Gummers.** Sheep that have lost all of their front teeth. These sheep are a very poor buy, although at times a gummer will do better than a sheep with a badly broken mouth, as the gums harden and enable them to still chomp off grass. If you have an old ewe with only one or two teeth left that you are determined to keep, get a pair of pliers and pull the remaining teeth.

Conformation

In livestock terminology, the word *conformation* means the shape and size of an individual animal compared with the ideal. Animals with good conformation are more likely to produce well, although we've had a few critters over the years that were pretty ugly by conformation standards but still did fine, so don't obsess about perfect conformation.

When you're shopping for sheep, choose animals that have good conformation. The sheep on the left shows good conformation, with a nice straight back, a strong chest, legs well placed under the body, and so on. The sheep on the right is far less desirable, with a sway back and belly, hock-kneed legs, and a weak chest and neck.

Special Considerations: Ram Conformation

When looking at a ram, remember the old saying, "The ram is half of the flock." The choice of the ram will most rapidly affect the character of your flock, for good or bad, because he breeds all (or a large number) of your ewes. So inspect him more closely than you would the ewes.

- Not only should a good ram display the basic conformation points, he should also show a heavy, muscular neck and a deep, wide body.
- His genitals should be well developed, which, for all but the smallest breeds, means that he should have a scrotal circumference of at least 12 inches (30 cm) by the time he's a yearling.
- He must have good, healthy feet, as bad feet can render him useless.
- His head can't be too large for his type because big-headed lambs often cause problems for the mom during lambing.

Since the ram has such a major influence on your entire flock, buy or lease the best one you can afford.

Teeth and Shape of Head

First, study the teeth and the shape of the head. Not only should the teeth be in good shape, but the bite itself is also important. As Paula's friend Darrel Salsbury, DVM, says, "They can't shear grass if the blades don't match." In a well-conformed animal, the upper jaw is the same length, or just a hair longer, than the lower jaw. In other words, the teeth of the lower jaw have to line up with the dental pad of the top jaw.

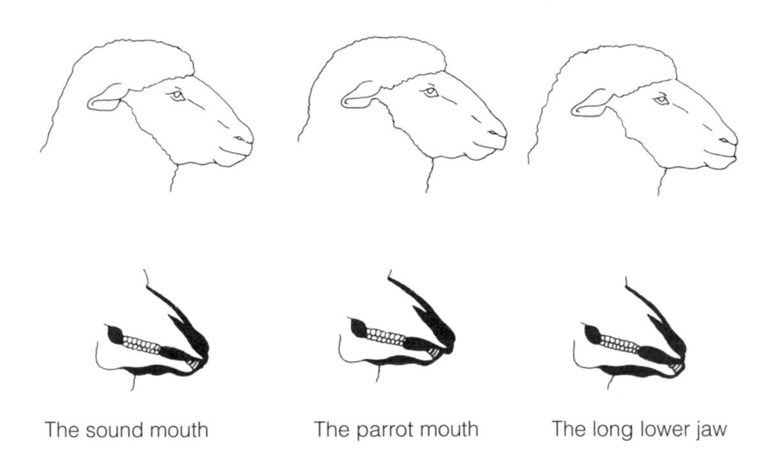

Types of mouth conformation

Body

Next, look at the body. The back should be long and straight, and the belly should also be fairly straight. Both the chest and the pelvic area should be broad and firm. The legs should be widely set, fairly straight, and forward-facing, with feet well placed on the ground. The rump should be rounded with a slight downward curve but should not look like a slope that you could ski-jump off of. Sheep being raised for meat should be large with strong muscles and trim features. Sheep being raised for wool should have a slightly more angular body with dense, clean, bright fleece.

Udder

In a mature ewe, look at the udder next. A healthy udder is soft and pliable, warm (but not hot) to the touch, and symmetrical with two good teats widely spaced on each side. The teats should not show signs of chapping or hardness.

General Health

The final thing to think about when planning your purchase is the general health of the animal. Determining a sheep's general health should involve a close physical examination. If you're considering purchasing a large flock from one seller, you may decide to closely inspect only a portion of the animals, but if you're buying a small number of animals from one seller (say, fewer than twenty), then take the time to give them all a complete examination. (See chapters 7 and 8 for an in-depth discussion of health.)

Is This Sheep Healthy?

Healthy animals are, as the saying goes, "Bright-eyed and bushy-tailed." Their ears and eyes are alert to their environment. They have good appetites and drink plenty of water. An animal that is listless and does not eat or show much interest in the world around it isn't a good bet. A sunken appearance around the eyes is a sign of dehydration, which often accompanies illness. The skin under the eyelid and the gums should have a nice, bright-pink color; pale color indicates anemia, which is often associated with internal parasites.

Mucous Membranes

The animal should have no suspect discharges from the eyes, ears, or nose. Just like in people, if the weather has been cold and windy, a little clear fluid may discharge from the eyes or nose and not indicate anything of consequence, but if the discharge is crusty or pussy or if there is excessive slobbering or frothiness around the mouth, beware. There should never be a runny discharge from the ears, period.

Respiration

Respiration should be easy and steady. Unless the animals had to be chased for their examination, they shouldn't be panting or breathing heavily. If they were chased, let them rest for a few minutes. Their respiration should return to normal within about 20 minutes. Coughing and wheezing should be considered a warning of a real problem.

Coat

The wool or hair should be shiny and even. Clumpy fleece and bald spots may be a sign of poor nutrition, illness, or most often external parasites. Separate the fleece with your hands around the neck, and look for signs of sheep ticks, or keds. (These are not a true tick, like what the dog picks up in the woods, but a wingless fly that passes its whole life cycle on the body of the sheep.) There may be a little mud around the ankles if the weather's been wet, but there shouldn't be caked manure in the wool. Manure around the rump and on the backs of the rear legs indicates scours (diarrhea) and is a definite problem.

Feet

Pick up the feet, and look for signs of foot rot. The hooves shouldn't be too overgrown. If any of the hooves look long, ask the owner to trim one while you watch. This is an easy way to learn how it's done and how the feet should look after they're properly trimmed. The legs should move fluidly with no signs of lameness or stiffness.

Skin

Closely inspect the whole body for rashes or for wounds that haven't healed. Turn the animal up into the sitting position used for shearing (see chapter 11) to inspect the belly and the scrotum or udder areas. Several diseases manifest with skin lesions, and sheep with these disorders are best avoided. If there are wounds, are there signs of infection, like a hot, red area around the wound or draining pus? During fly season, make sure there is no sign of flystrike, which is eggs in the wool or maggots or screwworms at the wound site. (Flystrike can also happen in the hooves of sheep with footrot.) Minor wounds that appear to be healing correctly shouldn't rule out an animal.

Health Records

After you've inspected the animals, inspect the health records. Check the vaccination record. If you're buying a ram, the enzyme-linked immunosorbent assay (ELISA) should be negative for epididymitis. Some shepherds have had their flocks monitored for certain diseases, like scrapie, ovine progressive pneumonia, or Johne's disease. If the flock owner has not done this type of testing, ask your local veterinarian which tests are recommended. The decision should be based in part on where you live, how many animals you're purchasing, and whether the seller is willing to provide you with a healthy-animal warranty. (Also, while you're talking to the vet, find out if there are any recommended changes to the vaccination program for the flock.)

Home at Last

So, congratulations — you're the proud owner of some sheep. Now what? Before you even bring the flock home, make sure your facilities are ready (see chapter 3). A small holding pen or drylot that is *well fenced* should be the flock's first stop.

Feed the same type of feed as the farm where your sheep came from. Before leaving the farm, ask the owner what kinds of forage or grain the sheep have been eating. (If the flock was fed something that isn't readily available at any feed store, buy some from the farmer.) Then gradually change the sheep from their accustomed diet to whatever you intend to feed. Never change abruptly! (See chapter 6 for specifics on feeds and feeding.) To avoid scours or bloat, sheep should be given their fill of dry hay before being turned out onto a pasture that's more lush than what they've had before.

As you unload the sheep, you can get a head start on future health problems. If they need to be vaccinated, now is a good time to do it. And absolutely deworm them. Hold them in a drylot or small pen for 24 hours after you've given worming medicine; after they've passed any viable eggs, move them to a clean lot or pasture. Treat again 14 to 21 days later to kill any worms that hatched from eggs left after the first worming. (This is the only time I recommend worming without bothering to take a fecal sample.)

Quarantine

When you bring new sheep home, quarantine them for at least 3 weeks if you have any other sheep on site. This period gives you time to watch closely for illnesses that didn't show up during your examination.

TWO

Breeding and Breeds

Any discussion of sheep raising needs to begin with breeding and breeds, whether you are purchasing your first sheep or adding to an existing flock. Remember, although it may be tempting to bring in an exotic breed from a far-off place, it may not be practical or it may simply not be affordable. The best route may be to use a more available breed, or crossbred sheep that are locally available, knowing that a careful and patient breeding program can upgrade your flock and may even provide many of the desired qualities of the less available breed. The addition of one special ram at a later date might just accomplish your breeding goals.

Breeding and Genetics

For anyone interested in breeding sheep, a primer in genetics is helpful. The idea of selecting for desired traits has been around since humans began domesticating animals. Statues and artwork from ancient cultures in the Middle East and Africa show that selection for wool was taking place by 4000 to 3000 B.C. But early breeding was somewhat haphazard.

In the mid-1700s, Robert Bakewell elevated the art of breeding to a higher level and began to establish what we know today as true breeds. Bakewell was influenced by Charles Darwin's work in evolution. He began keeping extensive breeding records to help select animals for breeding, and began using line-breeding. Much of Bakewell's work was with sheep, with his most important contribution being the development of the Leicester Longwool.

In the mid-1800s, an Austrian monk named Gregor Mendel was the first person to begin moving breeding from an art to a science. Mendel experimented with peas to demonstrate that observable traits could be passed from parents to offspring with predictable results. In his experiments, Mendel used green peas and yellow peas. But what he discovered about the genetic process with those peas applies equally to all living organisms.

The essence of Mendel's work demonstrated that inheritance is controlled by a hereditary unit now called a *gene;* that genes come in pairs (one gene from the mother and one from the father); and that each gene maintains its function from generation to generation. (Mutations are the rare exception to that last part, and we'll discuss that in a minute.) Mendel also hypothesized that each gene could come in different forms. These forms are called *alleles*. In peas, there is an allele for yellow pods and an allele for green pods.

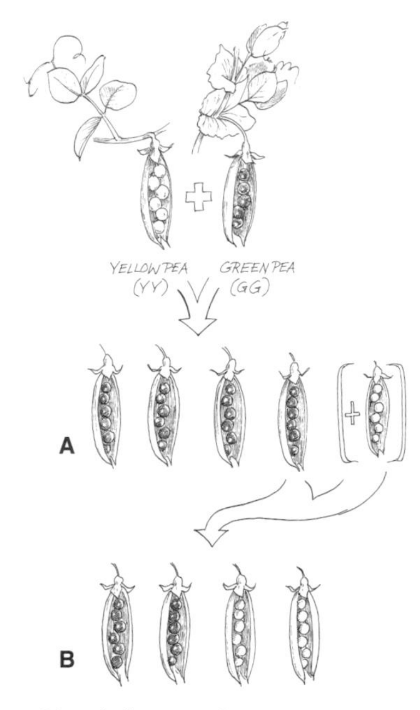

Mendel's experiment.
A. *Green trait.* **B.** *One-half green trait; one-half yellow.*

Some alleles are dominant, and some are recessive. When a dominant allele is present in a gene pair, the trait represented by that dominant allele is observable. For peas, the green pod is dominant. For a recessive trait, or the yellow pod, to show up, both halves of the gene pair have to have that allele. In actuality, most traits in most animals are the result of not one gene pair, but many gene pairs working together. These are referred to as *polygenic* traits. Some recessive alleles are extremely undesirable. *Lethals* are one type of recessive allele. Like their name implies, an animal that receives two lethal alleles dies, often before birth. Fortunately, lethal alleles by their nature are rare.

Mutations

A mutated gene makes a fairly sudden change in how it exerts itself in the next generation. For the most part, mutations are rare — however, as with everything in nature, there are exceptions to that rule. For example, the AIDS virus mutates fairly quickly, making research on a cure challenging. Usually (thanks to television and movies) we think of mutations as all being bad, like the radiation-mutated, three-headed, 2,000-pound, killer frog kind of thing. But mutations don't have to be bad. Several of the breeds discussed later in this chapter, such as the Booroola Merino and California Variegated Mutant, are the result of a mutation that a farmer, rancher, or researcher spotted in a flock and began purposefully breeding for.

Undesirable Recessive Traits

ABNORMALITY	DESCRIPTION
Lethal defects	
"Daft lamb"	Brain isn't fully developed. Lambs are born alive, but are usually unable to walk due to poor balance.
Dwarfism	Short legs, thick shoulders, bulging forehead. Lambs usually only live a week or two.
Earless and cleft palate	Affected lambs are born alive but die quickly.
Lethal gray	Lambs die early due to digestive disorder.
Muscle contracture	Lambs are usually stillborn. Limbs are rigid.
Paralysis	Hind limbs are paralyzed. Lambs die within a few days.
Nonlethal defects	
Cryptorchidism	One or both testicles do not drop down into scrotum (associated with polled trait in Merino and Rambouillets).
Earless	In Karakuls, some lambs may be born with no ears or some with short ears.
Naked	Lambs born with a few hairs, and black shiny skin. Lambs grow okay, but they are sensitive to temperature changes.
Skin folds	Most often occur in fine-wool breeds.
Yellow fat	Yellow carcass fat, which is objectionable in some markets.

Inheritance

So, how do these genes actually get passed along? Well, in every cell of a body — except the sex cells (eggs in the female and sperm cells in the male) — there is a full complement of the genetic code that defines who and what that animal is. This code is complex and consists of many gene pairs that are strung together like two strands of string, twisted around each other. In humans, for example, there are more than 100,000 gene pairs that make up the code. Each egg or sperm cell carries one-half of the parent's genetic string. Remember, there are lots of sex cells in each parent's reproductive tract, with half the cells carrying one string of code, and the other half carrying the other string.

At conception, a sperm cell fertilizes an egg, and the resulting cell (called a *zygote*) has a full complement of genetic material for its species. It is a matter of chance that determines which half of each parent's genetic code is brought to the mating, but there are just four possible combinations of alleles for each gene pair from any pair of parents.

Geneticists use letters to represent all possible alleles on each gene pair. A capital letter is used for the dominant allele, and a small letter is used for the recessive allele. Let's look at a simplified example. For domestic sheep, white is a dominant color and black is a recessive color. We'll use the capital letter *W* for the white allele and the small letter *w* for the black allele. In this example, both the ram and the ewe are white, and both have one dominant allele (represented by W) and one recessive allele (w). On average, these two animals would produce three white lambs out of every four lambs they had as offspring.

Remember, this example is a very simple example; the final color of the fleece over the whole body is actually controlled by as many as sixteen gene pairs, depending on the breed of sheep. This is why, in colored flocks, there are many shades of black, gray, and brown. However, in dark sheep, age can play a role in the color of fleece, as brown may turn to tan and black may gradually turn gray, just like in humans.

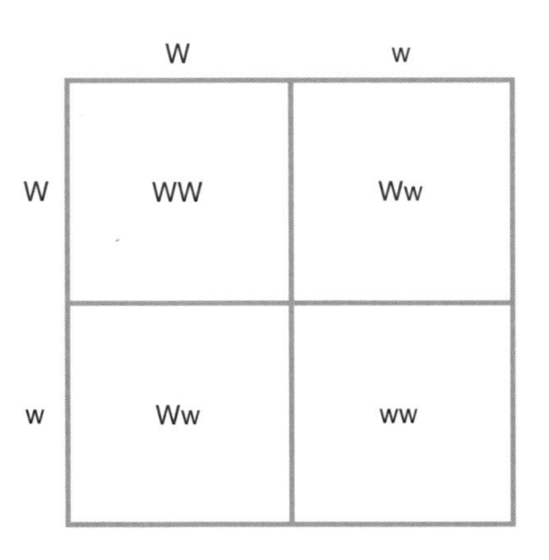

Color inheritance in sheep (W= white allele; w= black allele)

Heritability

Some traits, such as eye color, are fully the result of genetics, but most traits are influenced by both genetics and environment. A particular breed may have a high incidence of multiple births, but if the ewes aren't fed adequately, their conception rate won't be as high despite their genetic potential. Or, a breed may be known for producing really fine fleece, but if an individual animal has been sick, its fleece may be of poor quality. Since many traits are at least partially heritable, they can be taken into account when you're making breeding program decisions.

Many breeds of sheep are only able to breed during the late fall and early winter because their estrus cycles are controlled by hours of daylight. But some breeds are able to breed out of season, or during other times of the year. Out-of-season lambs can command good prices, and this trait may be desirable for shepherds interested in accelerated lambing. (*Accelerated lambing* refers to a program in which ewes give birth more than once per year on average, but accelerated lambing requires exceptional skill and management and may not provide "accelerated profits").

SHEPHERD STORY

People choose different breeds of sheep for different reasons. In Bill White's case, the choice of Shropshires stemmed from some research coming out of Europe.

"I've been raising Christmas trees for years," Bill told me, "and I spent most of my spare time in the summer out in the tree lot keeping the weeds and grass trimmed around the trees. I saw some research reports from the European Christmas tree industry, where they'd experimented with vegetation control using different species of domestic animals, and different breeds within those species. The Shropshire breed of sheep consistently performed the best in these applications, so I decided to do my own experiment."

Bill obtained eight ewes and a ram. This turned out to be no small feat. "I wanted to get all the animals from one flock. Shropshire flocks aren't that common, and most are pretty small flocks. I finally found a guy across the state that had a larger flock and could sell me nine head."

The main trees grown in Europe for the Christmas tree industry are spruces and firs, but on Bill's Missouri farm, he also grows several species of pines. At first, he was leery of placing the sheep right in with his trees regardless of what the European research showed, so he fenced them into a windbreak just to see what they'd do. Things seemed to go fine, so he began fencing out ½-acre paddocks in the Christmas trees with temporary electric fences, and moving the sheep every 3 to 5 days.

The sheep have proven themselves as effective tools for Bill, saving him time in the summer. Though after his years of working with them, he's learned a few lessons the hard way. "If I put the flock into the White Pine in early spring, they'll eat the shoots like candy. And they really have a taste for White Pine seedlings. They can be rotated through the other kinds of trees with no problems, but rotation is a key. If they stay too long in one paddock they'll start chewing on branches."

Multiple Births

Although multiple births require more attention and care, the profits seem well worth the effort. A 1987 University of Wisconsin analysis stated that 5,721 ewes producing 1 lamb each generate the same profit as 353 ewes producing 2 lambs each. This may not sound feasible, but remember that the amount of feed for the smaller number of ewes would be radically less, and each ewe would produce twice as many lambs.

Of course, if it isn't possible for someone to be at home during the day during lambing season, then ease of lambing is a much more important trait to select for than high growth rate or even multiple births. In Minnesota, we raised Karakul sheep, which have a fairly low rate of multiple births (the flock average was about 120 percent), but we were also milking cows at the time so we had little time to deal with the sheep. Despite their relatively low lambing percentage, they were great, self-sufficient mothers that never required any assistance with lambing.

Undesirable Characteristics

When you are evaluating breed characteristics (and making culling decisions), keep in mind that wool on the sheep's legs should be considered a disadvantage because it is unusable and makes shearing more time consuming. Wool on the face, which is very heritable, is another disadvantage. Tests have shown that an open-faced ewe (with little or no wool on the face) will raise more and heavier lambs than sheep that have wool on the face. Also, open-face sheep don't suffer from wool blindness or collect burrs on the face. But, if you live in a cold climate and expect your sheep to brave the great outdoors in January, those same traits suddenly become less of a disadvantage. Wool on the legs and head, like socks and a hat on people, help the sheep maintain body temperature.

Skin folds, in general, are also undesirable. They do produce a higher grease weight of fleece, but they also cause more shrinkage. For hand-spinners, excessively greasy fleeces are harder to wash. Folds make shearing more tedious and cause more second cuts, and since maggots can hatch and thrive in moisture-retaining folds, the folds predispose sheep to flystrike. According to the U.S. Sheep Industry Development program, skin folds usually indicate somewhat lower fertility and productivity.

Marketing Considerations

If you intend to spin your wool, your needs differ from those of people who sell large quantities of wool to a dealer or who are primarily interested in selling market lambs. If you plan on selling hand-spun fleece or direct marketing lambs, remember that some research indicates that hair sheep and sheep with coarser wool have less muttony flavored meat. However, many of the heritage breeds (see page 30), though they grow slower than the commercial breeds, can provide both delicious meat and desirable fleeces.

What Is a Breed?

A breed is a group of domesticated animals that have been bred to predictably pass on recognizable traits to their offspring. Some breeds show traits that were specifically selected by farmers and ranchers — such as color, wool quality, or high milk production. Other breeds demonstrate traits that are the result of environmental pressures, such as heat tolerance or resistance to parasites. Most breeds have a combination of traits selected by both humans and the environment, and at times it's hard to say to which category a particular trait belongs.

Genetic Diversity and Sheep Breeds

Although humans have only successfully domesticated about fifty species of animals, they have developed many thousands of distinct breeds from those species. The United Nations' Food and Agriculture Organization (FAO) estimates that there are currently about one thousand breeds of domestic sheep throughout the world and about fifty breeds in North America.

The one thousand breeds of sheep located throughout the world provide diversity for our gene pool. But this diversity is being lost at an alarming rate. The FAO, the American Livestock Breeds Conservancy (ALBC), and the Canadian Rare Breeds Conservancy estimate that worldwide 30 percent of all domestic animal breeds are at risk for extinction and that six breeds are lost each month.

The problem isn't limited to other areas of the world. In North America, nineteen sheep breeds are considered at risk for extinction, whereas just four breeds account for more than two-thirds of the total sheep population. The Suffolk breed alone accounts for about 40 percent of the North American sheep population.

The development of any new breed is slow, even with modern technology — which includes frozen embryos, frozen semen, ova transfers, and computerized record keeping combined with vigorous selection and extensive culling. Within the past several decades, new breeds have been developed, like the Booroola, the California Red, and the Cormo. *Composite* is a term often used for new breeds during development, indicating that they are being bred from a mixture of other, older breeds. It takes years for a composite to develop the genetic integrity of the breeds that have been around for a long time, though undoubtedly, some will.

Breeds that were uncommon in North America, but have been bred in other areas of the world for long periods, are now being imported. For example, the Perendale was first imported from New Zealand in 1977, the Romanovs were imported from Russia to Canada in the 1980s, and the Dorpers and Icelandic sheep have been imported into the United States in the 1990s.

Rare and Heritage Breeds

The breeds that have fallen out of favor with high-input, industrialized agriculture are referred to as *rare, heritage,* or *minor breeds.*

Many of these breeds were major breeds just a generation or two ago. For instance, in 1900, there were 71,000 registered Cotswolds; at the turn of this century, there were fewer than 1,000.

Some heritage breeds are now getting attention, and for a few, their numbers are actually improving. For example, the Jacob sheep, which was once an endangered species, is making a comeback. There were no Jacobs registered in the 1970s and early 1980s. In 1986, ALBC helped start a registry with one hundred Jacobs; and in 1990, the last year a census was completed for Jacobs, there were 665 registered.

Traits of the Heritage Breeds

The loss of heritage breeds can have especially grave impacts on homesteaders and small commercial operators who are interested in low-input agriculture. These breeds, although not the most productive in an industrialized system, have traits that make them well suited to low-input systems: Some are dual purpose, able to produce both meat and fiber well. Others are acclimatized to regional environments, like hot and humid, or cool and dry conditions. Many perform well on pasture with little or no supplemental feeding. Others resist diseases and parasites. Some have such strong mothering skills that the farmer doesn't have to do much work during lambing season.

As Don Bixby, Executive Director of the ALBC says, "It does seem as though the heritage breeds of sheep are doing better — in large part, due to hand-spinners who appreciate wool diversity. Many heritage breeds of sheep have moved up in status on our conservation priority list, and some have been removed from this list altogether. Sheep are amenable to conservationists and accessible to a wide range of people since they are small and easy to work with. Many commercial breeds need full intervention, but you can select for self-sufficiency in breeding populations."

Population Status of Rare and Heritage Breeds

The American Livestock Breeds Conservancy has established a conservation priority list for rare breeds of livestock in the United States. The categories are:

- **Critical.** Fewer than 200 North American annual registrations and fewer than 2,000 in the global population.
- **Rare.** Fewer than 1,000 North American registrations and fewer than 5,000 in the global population.
- **Watch.** Fewer than 2,500 North American registrations and fewer than 10,000 in the global population.
- **Recovering.** Breeds that were once listed in one of the other categories and have exceeded "Watch" category numbers but are still in need of monitoring.

Conservation Priority for Rare Breeds

Critical	Rare	Watch	Recovering
California Variegated/Mutant	American Jacob	Barbados Blackbelly	Black Welsh Mountain
Hog Island	American Karakul	Dorset Horn	Lincoln
Romeldale	Cotswold	Gulf Coast Native	Oxford
Santa Cruz	Leicester Longwool	Katahdin	Shetland
		Navajo-Churro	Shropshire
		St. Croix	Southdown
		Tunis	
		Wiltshire Horn	

Individual Breeds

Currently, there are about fifty breeds in North America. The following descriptions come from a variety of sources.

Barbados Blackbelly

The Barbados Blackbelly is a dark, tropical, hair sheep that came originally from Barbados in the West Indies. These sheep were developed from those brought to the island by African slave traders during the 1600s. Although the U.S. Department of Agriculture (USDA) imported some Barbados in 1904, they were used for crossbreeding and their pure bloodlines were soon diluted. But several years ago, some breeders began selecting again for the Barbados traits and the Ministry of Agriculture in Barbados gave some purebred sheep to the North Carolina State University for research.

Some of the recent interest in Barbados sheep centered on their ability to lamb almost twice a year. Not only are they prolific, but they're also hardy, early breeders. The ewes are nervous around strangers and protective of their young. Mature ewes have between 1.5 and 2.3 lambs per lambing on average. The record is 8 lambs that were born to a ewe on Barbados.

Two other points make the Barbados an interesting alternative (especially for shepherds in the hot, humid Southeast): They exhibit great resistance to internal parasites and heat stress (in fact, university trials have confirmed that they carry only about 10 percent of the parasite burden of wool breeds in the Southeast), and they suffer a low incidence of heat stress.

The pure-blooded Barbados Blackbellies that live on Barbados are naturally *polled* (hornless), but the American strain often shows impressive horns due to crossbreeding. The breed's body color varies from dark brown, to almost red, to tan, with distinct black markings on the legs, belly, and inside of the pointed ears, and on the chin and face. Face markings include black bars down the front of the face, as well as a line of black that goes across the top of the head.

Black Welsh Mountain

The Black Welsh Mountain is a small, extremely hardy sheep that originated in the high mountain valleys of South Wales. The breed was introduced into the United States in 1972.

The ewes are known for easy lambing, high fertility, and good milk production (which means fast-growing lambs). They are very active and independent, making them somewhat hard to fence.

Barbados Blackbelly (Courtesy Dr. Lemuel Goode, North Carolina Experiment Station)

Black Welsh Mountain (Courtesy American Livestock Breeds Conservancy)

Bluefaced Leicester

There are three Leicester breeds: the Longwool (or English) Leicester (see page 56), the Border Leicester (see page 36), and the Bluefaced Leicester.

Developed in northern England during the beginning of the 1900s, Bluefaced Leicesters are quite prolific, with good maternal instincts. These sheep are a medium to large size and get their name from the dark blue skin on their heads, which shows through white face hair. They have one of the finest longwool fleeces (56s–60s count), which is semilustrous, with a silky handle and pencil locks.

Booroola Merino

Booroolas are a strain of sheep that started with a single prolific ewe at the Booroola Merino farm in Australia during the 1960s. Booroolas are noted for being prolific and for their high-quality, fine wool that has a long *staple*. Booroola wool is typical of all Merinos. Booroolas have such a high lambing rate because a single gene (the F gene) affects ovulation; ovulation in other prolific sheep is controlled by a large number of genes. The Booroola strain is also capable of breeding out of season.

J. Sloan, a Canadian breeder of Booroolas, notes, "Booroola rams may be crossed on most medium-sized, maternal breeds with a history of excellent milk production." There aren't very many Booroola breeders in North America, and no breed association exists at this time.

The F Gene and Fertility

The unique F gene is epistatic, or dominant over other genes, so it enhances ovulation rates and the fertility of lambs born to ewes that are a crossbreed of a Booroola ram and other breeds of ewes. Purebred Booroola ewes average almost 300 percent lamb crops, and crossbred ewes average 120 percent more lambs than they would without the Booroola crossed in. While lambing percentages can increase with Booroola crossbreeding, some of the undesirable traits that are often associated with other prolific breeds may not increase. But bear in mind that increased lambing rates don't guarantee an increase in the total lamb weight at weaning per ewe.

Bluefaced Leicester (Courtesy Diane Kelly, Firesong Farm, Cooksville, MD)

Booroola Merino (Courtesy Agriculture Canada)

Border Cheviot (Southern Cheviot)

The Border Cheviot is one of two distinct types of Cheviot sheep raised in North America. The smaller Border type was improved by selection from the original stock, rather than by crossbreeding, and is the predominant type of Cheviot in the United States.

Cheviots started as a mountain breed, native to the Cheviot Hills between Scotland and England. These sheep are extremely hardy and can withstand hard winters. They graze well over hilly pastures at high altitude. They lack the herding instinct needed for open range but do well in a small farm flock.

They are active and high strung, being alert both in appearance and behavior. They are good mothers despite their nervousness, with a high percentage of twins, and their newborn lambs are hardy. Because of their small head, they experience few lambing difficulties. They also raise good meat lambs.

These sheep are short and blocky and have no wool on their faces or legs. They're recognizable by their black nostrils and lips and their erect, sharp ears. They have a light-weight, medium-wool fleece that is easy to hand-spin.

Border Leicester

Developed in 1767 by the Cully brothers of Northumberland, England, the Border Leicester was quite popular in England by the mid-1800s. This breed is thought to be a cross of the Leicester Longwool (see page 56) and the Teeswater, though some folks believe that Cheviots were also crossed in.

Border Leicester sheep are prolific, good mothers, and are known for rapidly growing market lambs. They are medium sized and known for their docile dispositions. They have bare legs and an open face, making them easy to shear. The Border Leicester is a white wool breed, but there are beautiful colored Border Leicesters, with locks that grow from 8 to 12 inches (20–30 cm) after 12 months. Fleeces yield between 65 and 80 percent and weigh anywhere from 8 to 12 pounds (3.6–5.4 kg). The numeric count of the fiber is 36 to 48. (See chapter 11 for more information on interpreting counts.) Border Leicester wool dyes beautifully and has a mohairlike sheen.

Border Cheviot (Courtesy American Sheep Industry Association 2000)

Border Leicester (Courtesy New Zealand Farmer *magazine)*

California Red

In the early 1970s, Dr. Glen Spurlock of the University of California at Davis began crossing Tunis (see page 76) and Barbados sheep in order to establish a new breed with superior qualities of both wool and meat production. The result is the California Red sheep, which developed rapidly and are quite consistent in maintaining the desirable characteristics of the breeds from which it was crossed.

California Reds are medium-sized, dual-purpose sheep. Rams weigh from 200 to 250 pounds (90–113 kg), and ewes weigh from 130 to 160 pounds (59–73 kg) at maturity. The fleece yields about 7 to 8 pounds (3.0–3.6 kg) annually. The texture of the wool is silky and contains reddish hair, which makes it desirable to spinners and weavers. It also makes good-quality felt.

Rams are active and aggressive, even in hot weather, and ewes are good milking mothers that tend to be free of lambing problems. Like the Barbados and Tunis from which they are derived, California Reds breed out of season, and many breeders aim for three lamb crops every 2 years. Lambs are red at birth, but the wool lightens to an oatmeal color with age, though the legs and face retain a reddish tinge.

California Variegated Mutant

The California Variegated Mutant breed was developed by Glen Eidman in the early 1960s when one of his purebred Romeldale (see page 68) ewes gave birth to a multicolored ewe lamb with a badger-patterned face (dark stripes along the sides of the face and a lighter-colored center.). Several years later, a ram lamb was born with the same coloring. Eidman crossed these two half-siblings, and the same pattern came through. For the next 15 years, Eidman continued breeding for the trait, concentrating also on characteristics like spinability of the fleece, twinning, and ease of lambing.

These sheep are medium sized. The rams are aggressive and virile breeders and are reported to be able to breed larger numbers of ewes than other rams can. The breed is known for longevity and prolific lamb production. They breed out of season. Fleeces are long stapled and fine and average 8 pounds (3.6 kg) per animal per year. Wool colors run from white to gray to black, with some spotting, and unlike other colored breeds that lighten with age, these sheep get darker.

A breed registry has formed for the California Variegated Mutant, so the number of these sheep should begin to increase.

California Red (Raised by Paulette Soulier)

California Variegated Mutant (Courtesy American Sheep Industry Association 2000)

Clun Forest

This breed was developed in the Clun Valley in southwestern Shropshire, England, in the 1800s. Early breed selection was for hardy, fertile sheep that could thrive on grass and whatever they could forage. The first six Clun Forest ewes were imported into North America from Ireland in 1959, but the first large importation didn't take place until the 1970s.

Clun Forest ewes are prolific and almost always have twins. With narrow, sleek heads and wide pelvic structures, they lamb very easily and without assistance. Even yearlings show strong mothering instincts; ewe lambs breed at 8 or 9 months old and lamb as yearlings. The ewe's milk has a higher protein and fat content than that of other breeds, contributing to quick-growing lambs, and ewes are also capable of producing a high volume of milk. As a result, Cluns are garnering interest among sheep-dairy operators.

Clun Forest sheep are adaptable to all kinds of climates and all kinds of grazing conditions. Another quality that makes them valuable is their longevity and that they have good fleeces until 10 or 12 years of age. Their medium wool is 58s count. (The *s* has to do with spinning count and means the number of hanks, or 560 yards [511.8 m], that the wool can spin. See chapter 11 for more on spinning counts.)

Columbia

The Columbia is the first breed to come out of U.S. government and university research. Developed by the USDA in 1912, it was intended as an improved, true breeding type for the western range. It is the result of a Lincoln (see page 58) ram and Rambouillet (see page 66) ewe cross, with interbreeding of the resulting lambs and their descendants without back-crossing to either parent stock. The object of the cross was to produce more pounds of wool and lamb under range conditions, but they have also adapted well to the lush grasses of small farms in other parts of the country. Heavy wool clips, hardy and fast-growing lambs, open faces, and ease of handling are characteristics for which the breed is known.

They have medium wool in the 50s to 60s range but predominantly about 56s. It has light shrinkage and makes excellent, all-white fleece for hand-spinning.

Clun Forest (Breeder is Mrs. Warren G. Menhennett, Cochranville, PA)

Columbia (Courtesy American Sheep Industry Association 2000)

Coopworth

Coopworths were developed in the 1950s at Lincoln College in New Zealand by crossing Border Leicesters with Romneys (see page 68), and the breed has unusually strict registration requirements. Performance recording is mandatory, with emphasis placed on multiple births and high weaning weights. The breed is very docile, but this makes them more susceptible to predators.

Coopworths were first imported to North America in the 1970s and have proven to be excellent foragers on lush pastures. The wool is very strong, soft, curly, and thick like carpet wool. It is well suited to felting.

Cormo

Cormos are smaller than Columbias and Rambouillets (see page 66) but yield 70 to 73 percent clean-weight, fine fleece under range conditions and have a high-yielding carcass. The fleece is uniform and therefore valuable to industry. Hand-spinners find it the most exciting of the fine-wool breeds.

Traditional pedigrees aren't kept. Instead, the sheep are numbered and allowed into the registry based on performance. Computer management makes the Cormo the most strictly scientific genetic improvement scheme in sheep history. The criteria for selection are clean fleece weight, fiber diameter between 17 and 23 microns, fast body growth, and high fertility.

Origins and Characteristics of the Cormo Breed

Cormos were developed in Tasmania, guided by principles originated by Dr. Helen Newton Turner who is believed to be one of the world's leading sheep geneticists. A group of Australian scientists selected Tasmanian–Corriedale rams to cross with superfine Saxon Merino ewes, resulting in the Cormo breed. The Cormo's outstanding qualities are fine, well-crimped wool, excellent conformation, fast growth, high fertility, and the ability to thrive in areas of heavy snowfall, severe climatic conditions, and rough terrain.

Coopworth (Courtesy American Sheep Industry Association 2000)

Cormo (Courtesy This and That Farm, Danby, VT. Photograph by Jim McRae, Vermont Camera Works, Pittsford, VT.)

Corriedale

The Corriedale is a Merino–Lincoln–Leicester cross that was developed in Australia and New Zealand during the late 1800s and first brought to Wyoming in 1914. The breed is now distributed worldwide, with its greatest population found in South America.

The fleece is dense and medium fine, 56s grade, and soft and has good length and light shrinkage. It falls somewhere between a medium wool and long wool and is a favorite of hand-spinners in many areas of the United States. The Corriedale's face is clean of wool below the eyes and is naturally polled. It's a large-framed breed that has been developed as a dual-purpose sheep: It has good wool and good meat for greater profits and is noted for a long productive life, which means a greater return on investment. Because of a strong herding instinct, it does well as a range animal.

Cotswold

The Cotswold is a large sheep, known for its very long, coarse, and lustrous wool that is 8 to 12 inches (20–30 cm) long and wavy; it hangs in pronounced ringlets. Thought to have originally been introduced to England by the Romans, more than 2,000 years ago, today's Cotswolds were developed between 1780 and 1820 through the introduction of Leicester Longwool (see page 56) genetics to the sheep of the Cotswold Hills in Gloucester. The first Cotswolds were imported to North America in the early 1800s, and the breed's 1878 registry was also the first U.S. sheep registry.

The breed was popular early on for crossbreeding with western range sheep, but it fell out of favor as selection moved toward meat breeds. Today, the breed finds favor with smaller farm flocks, where their docile personalities make handling easy.

The fleece weighs from 13 to 15 pounds (6.0–6.5 kg), and there is very little shrinkage. It is quite lustrous, with a count in the 40s range, and grows at the rate of about 1 inch (2.5 cm) per month. Because of this fast growth, many shepherds shear twice a year. One unique characteristic of the Cotswold is that the locks fall over its forehead in cords. The breed is white to silvery gray and sometimes has excessive wool on the thighs.

Corriedale (Courtesy Shepherd *magazine)*

Cotswold (Courtesy Sheep Breeder and Sheepman *magazine)*

Debouillet

Development of the Debouillet was begun by Amos Dee Jones in New Mexico in the 1920s. The result of breeding Ohio Delaine Merino (below) rams with Rambouillet (see page 66) ewes, successful crosses of these sheep show the length of staple and character of the Delaine fleece and the large body of the Rambouillet. By 1927, the ideal type was attained and a line breeding program began. The breed was registered in 1954, starting with 231 rams and 1,587 ewes.

Debouillets are open faced below the eyes and over the nose, have a good wool covering over the belly, and shear a heavy fleece of long-staple, fine wool. Rams can be horned or polled. Even under adverse conditions, ewes produce desirable market lambs of excellent weight.

Debouillet lambs that are eligible for registration by bloodline must be 1 year of age and in full fleece when inspected by an association inspector. Wool must be 64s grade or finer, with 3-inch (7 cm) minimum staple and deep, close crimp.

Delaine Merino

The Merino sheep, which are so famous for their fine wool, originated in Spain. They are descended from a strain of sheep developed during the reign of Claudius (A.D. 14–37). Spaniards crossed the Tarentine sheep of Rome with the Laodicean sheep of Asia Minor, resulting in one of the world's most popular sheep breeds; most modern wool breeds have some Merino in their background. Merinos were first imported to North America during the late 1700s.

Since the lambs are small and mature slowly, the main income is from the sale of fleece and breeding stock. The Merino fleece is heavy in oil and, like the Rambouillet (see page 66), loses much of its weight in washing.

Not too long ago, the Merinos were classified into types A, B, and C, depending on the amount of wrinkling in their skin. Type A had excessive wrinkling and is now considered extinct, type B had fewer folds, and type C had the fewest folds. Delaines are a type C Merino, which were bred in North America starting in the nineteenth century, and are the most common type of Merinos in North America today.

Delaines have good herding instincts and can travel far for feed and water, so they work well on open range. They are medium sized and hardy. They breed year-round and are excellent mothers, but twinning is not the norm.

Debouillet (Courtesy Dr. Tim Ross, New Mexico State University, Las Cruces, NM)

Delaine Merino (Courtesy Morehouse Farm, Red Hook, NY. Photograph by Francis J. Twomey.)

Dorper

The Dorper is a hair sheep that was bred in South Africa by crossing Blackhead Persian sheep with Dorset Horns (below). Their solid-white bodies are often accompanied by a solid-black head, giving them an unusual appearance.

Dorpers are highly fertile, have a long breeding season, and are quite docile. They are adaptable to a wide range of climates, from hot and dry to humid and cold. Although they put on wool in cold climates, they shed in warm weather and don't require shearing.

Dorset

Dorset sheep originated in England and although their history is not well known, it is believed that they developed more by selection than by crossbreeding. There are two types of Dorsets, Dorset Horns and Polled Dorsets. The first Polled Dorsets were developed at North Carolina State College and first registered in 1956.

The Dorset has very little wool on the face, legs, and belly; its fleece is lightweight and good for hand-spinning. They have a large, course frame and white hooves and skin. The ewes are prolific and often have twins. They are good milkers, having even been kept in dairies at one time in England, and are good mothers. (A *Shepherd's Guide* from 1749 described them as "being especially more careful of their young than any other.") Ewes breed early, allowing for fall lambing, and it's possible for them to lamb twice per year.

East Friesian

A German breed, East Friesian are raised primarily for milk used particularly for cheese production. These sheep produce more milk than any of the other European breeds, and they are very prolific. In southeastern France, this milk breed is one of three that are crossed for production of Roquefort cheese. Its high milk production accompanied by prolific lambing make it valuable for crossbreeding as well as for sheep dairying.

Although the East Friesian is a large sheep, and the lambs have good growth rates, it is not considered an especially good sheep for meat production. It shears a heavy fleece of 48s to 50s wool.

Dorper (Courtesy Cynthia M. Brasfield, Robertsdale, AL)

Dorset Horn (Courtesy American Livestock Breeds Conservancy. Photograph by Joe Greene.)

East Friesian (Courtesy American Sheep Industry Association 2000)

Finnsheep

Finnsheep, or the Finnish Landrace, were first brought to the United States from Finland in 1968. Since then, their numbers have grown rapidly. They are said to "lamb in litters" because they are known to have up to six lambs per lambing, with three to four being normal for the mature Finn ewe.

Finn ewes can breed at 6 or 7 months. Because they are so prolific, Finnsheep are being widely used for crossbreeding. The lambs that result from crossbreeding a Finn with a meat breed are indistinguishable from those of the meat breed, while the lambing percentage is greatly increased.

Finns are known to be good mothers and easy lambers, but they require exceptionally good care during gestation to meet the nutritional needs required to support multiple lambs. When litters greater than three occur, the bonus lambs are either left with the dam and supplemented or taken away and hand-raised.

Fleeces are generally very soft with a lustrous quality, appealing to handspinners. Although white is predominant in Finns, natural-colored flocks have also been developed that include black, gray, brown, and spotted patterning. The tails of Finn's are naturally short and don't require docking. The friendly disposition of Finn's makes them especially popular with small-flock raisers.

Gulf Coast Native

As the name implies, Gulf Coast Native sheep hale from the southern coast of the United States along the Gulf of Mexico. They developed from early Spanish sheep introductions in the 1500s in Florida and are one of the oldest breeds in the United States. Before World War II, hundreds of thousands of these sheep roamed free on unimproved pastures throughout the subtropical regions along the Gulf, but after the war, the Southern sheep industry turned to commercial, high-input, improved breeds.

Through hundreds of years of natural selection for withstanding the hot and humid conditions under which parasites thrive, Gulf Coast Natives are one of the most resistant breeds to internal parasites. This trait is helping to stimulate renewed interest in these sheep. In fact, their tolerance to parasites is generating interest outside their traditional subtropical range and they are now being raised as far north as Minnesota. This breed also tends to be resistant to foot troubles.

The Gulf Coast breed is small, has clean legs, and an open face. Though usually white, they can also be brown, black, or spotted. They grow slowly and have a low lambing percentage, but in subtropical conditions, the percentage of lambs finished per ewe mated is higher than other breeds because of excellent lamb survivability. The lambs mature early, and the ewes can lamb out of season.

.*Finnsheep (Courtesy Finnsheep Breeder's Association, DeRuyter, NY. Photograph by Elizabeth K. Luke.)*

Gulf Coast Native (Courtesy American Livestock Breeds Conservancy)

Hampshire

The Hampshire is one of the largest of the medium-wool, meat sheep. While they don't do well on rough or scanty pasture because of their size and weight, they do nicely on good pastures, and the lambs can usually be marketed directly from the grass of high-quality pasture.

Hampshires are another British breed, from Southdowns (see page 72) crossed with a Wiltshire Horn (see page 76)–Berkshire Knot cross. They were first imported in the early 1800s.

The ewes are good milkers and are fairly prolific, but they do not lamb easily, probably because of the large head and shoulders of the lambs and heavy weight at birth. The lambs grow rapidly and are known for good carcass cutability.

The Hampshires have large heads and ears and are polled. The dark face and legs are a rich, dark, chocolate brown. Their fleece is lightweight, and they have fairly short, medium wool.

Hog Island

Two hundred years ago, Hog Island, a barrier island off the coast of Virginia near the mouth of the James River, became home to a flock of sheep that was established from locally available British breeds. These animals have since evolved into a unique breed of feral sheep.

Feral sheep are rare worldwide because sheep do not easily adapt to unmanaged habitats. Feral sheep are most often found on islands, where predators don't exist. Under those conditions, natural selection yields a hardy sheep with excellent foraging ability (they are often able to utilize feeds that other sheep couldn't hope to survive on) and reproductive efficiency.

Hog Island was purchased by The Nature Conservancy during the 1970s, and the entire flock of sheep was removed from the island to improve survival of native vegetation. The breed is now kept primarily at historic sites in Virginia, such as Gunston Hall Plantation and Mt. Vernon, to portray eighteenth-century sheep raising.

Hog Island sheep have medium-weight wool, and mature animals weigh 125 to 200 pounds (56.7–90.7 kg). Most are white with spotted legs and faces, though about 10 percent are black.

Hampshire (Courtesy John D. Wibbels, Jeffersonville, IN)

Hog Island (Courtesy American Livestock Breeds Conservancy)

Icelandic Sheep

Viking settlers brought sheep to Iceland, and few sheep have been imported since settlement ended there about 900 years ago. As a result, Icelandic sheep are one of the purest breeds in the world today. In Iceland, these sheep account for about 25 percent of the island's total agricultural output. Its first North American importation was into Ontario, Canada, in 1985.

Icelandic sheep are of northern European descent, have short tails, and are distantly related to Finnsheep, Romanovs, and Shetlands. But Icelandic sheep are the biggest of these short-tail types. These sheep have good conformation for meat production, and while they are raised for meat, milk, and wool in Iceland, they are well known internationally for their wool, which is mostly marketed as Lopi yarn. The fleece is dual coated, with an outer coat that can reach 15 inches (38 cm) and a shorter, softer inner coat, and comes in a wide range of colors.

The breed is well suited to small farms, as their herding instincts are poor. They are alert and aggressive, showing great determination in going after their feed. The lambs, though they are born small, are eager nursers and can reach finishing weight in 3 to 4 months if raised on good pasture. Both ewe and ram lambs mature early and begin breeding at about 8 months. The meat has a fine texture and a delicious flavor, the wool is sought after by hand-spinners, and the skins make beautiful rugs.

Jacob Sheep

The mottled fleece of Jacob sheep is light, with 4- to 7-inch staple (10–18 cm). Its medium-fine texture and high luster have great appeal for hand-spinners. Tanned pelts bring premium prices.

The First Recognized Breeding Program

Jacob sheep get their name from the book of Genesis, where it is recorded that Jacob's father-in-law paid Jacob for his labors with all the spotted and speckled sheep in his flock. Then, in a dream, God told Jacob to use only spotted rams, creating a spotted flock that all became Jacob's. Although the origin of modern Jacob sheep is unknown, this small sheep with random spots all over its body is known to have been bred in England for at least 350 years. One of its most unusual characteristics is that it often sports up to six horns, but four is more common. It is thought that parks and zoos were the first to import these sheep into North America in the early 1900s.

Icelandic (Courtesy Mongolds Tongue River Farm, Miles City, MT)

Jacob (Courtesy Prairie Mary's Acres)

Karakul

Native to the Central Asiatic region of Russia, the Karakul, a fat-tailed sheep, is thought to be one of the oldest breeds in the world. They were first introduced into North America in the early 1900s.

Karakuls are small, fine-boned sheep with long, droopy ears. They are quite hardy, adaptable to a wide range of climatic conditions, and known for their longevity. Their breeding season is fairly long and allows for out-of-season breeding. Single lambs are the most common (though we had one Karakul ewe that had triplets every year, which she raised without any assistance from us).

Most lambs are born black and lighten with age, though there are strains of blue and red as well. The lamb's pelt, which is tightly curled, was traditionally prized for its lustrous "fur." Although the fleece grades as carpet wool, it is long stapled and good for hand-spinning. It also has excellent felting quality. Karakul meat is less muttony tasting than some other sheep breeds, and the fat from its tail provides good tallow for soap or candle making.

Katahdin

The Katahdin is a hair sheep breed and originated in Maine. In 1957, Michael Piel, an amateur geneticist who enjoyed raising livestock, read an article in *National Geographic* about West African hair sheep. He eventually imported three African hair sheep from St. Croix. Piel began experimenting with various crosses, trying to develop a hair breed with a good conformation for meat production, high fertility, and good flocking instincts. By the 1970s, Piel felt he had the sheep he was looking for and named the breed Katahdin after Mt. Katahdin, the highest peak in Maine.

Katahdins are hardy, adaptable, and low maintenance. Docile and easy to handle, these medium-sized sheep produce good lamb crops with lean, meaty carcasses. Ewes have good mothering ability and lamb easily, and the breed is considered to be ideal for an extensive pasture-lambing system.

Leicester (or English) Longwool

The Leicester Longwool (or English Leicester), was originally bred by Robert Bakewell in the 1700s for early maturity and improved mutton quality and quantity. Bakewell was an early leader in the development of selective breeding practices for livestock and is said to have been influenced by the work of both Mendel and Darwin.

Karakul (Courtesy American Sheep Industry Association 2000)

Katahdin (Courtesy American Livestock Breeds Conservancy)

Leicester Longwool (Courtesy Colonial Williamsburg Foundation)

Leicester Longwools were imported into the United States early, with references made to the breed in some of George Washington's correspondence. Although it was once very popular in both England and the Americas, by the 1980s the breed was almost extinct in both areas. Although still rare, these sheep escaped that doom in the United States, largely through the work of the Colonial Williamsburg Foundation Coach and Livestock Department. In the late 1980s, Colonial Williamsburg imported some purebred Leicester Longwools from Australia, where the breed's numbers hadn't dipped quite so low.

The Leicester Longwool has a mop of wool over the crown of its head. The breed is hardy and adapts to a wide variety of environmental conditions. They have large frames. The fleece is generally white and falls in long, lustrous ringlets, with a 15-inch (38 cm) staple.

Lincoln Longwools

The Lincoln Longwools are from Lincolnshire, England. Although they are the largest of the sheep breeds, they mature slowly. Their long fleece is dense, strong, and heavy, and they have forehead tufts. The breed is fairly hardy and prolific, but lambs need protective penning for the first few days.

The Lincoln is not an active forager and is best adapted to an abundance of good pasture and supplements. They don't stand cold and rainy weather too well because their fleece parts down the middle of the back, allowing cold air to hit their backbone (a sensitive area on sheep). However, the fleece is resistant to the deterioration shown in the wool of other breeds that parts along the back. This lustrous fleece is sought by hand-spinners for its long-wearing qualities. When spun alone, the wool makes an almost indestructible sock yarn; when blended with other wools, it makes a strong weaving warp (or threads strung through a loom to create the foundation for weaving) and has an attractive sheen.

Montadale

The Montadale is an American breed that originated around 1932 in the St. Louis area. Montadale sheep are a cross between Cheviot rams and Columbia ewes. The small head eliminates many lambing problems; the ewes are prolific lambers and good mothers.

Heavy fleeced with little shrinkage, the wool grades medium. The breed has a beautiful face, and alert, Cheviot-style ears. These sheep are open faced with clean legs.

Lincoln Longwool

Montadale (Courtesy Larry E. Mead, Sheep Breeder and Sheepman magazine)

Navajo-Churro

Much like the Gulf Coast Native sheep, the Navajo-Churro breed developed from sheep imported by the Spanish more than 400 years ago. Connie Taylor, secretary of the breed association, wrote, "The Navajo-Churro endured primitive, ocean transport and the rigors of trailing from Mexico to the Southwest. They survived the pressures of providing food and fiber to the early mining settlements of Mexico, California, Arizona, and New Mexico."

Native Americans acquired flocks of sheep from the Spanish ranches and villages (either through raids or trading) in the early seventeenth century. Within 100 years, herding and weaving became the main enterprise of the Navajos. Their sheep became so important that the Navajo name for sheep is *bee'iin' á át'é*, which means "that by which we live."

During the late 1800s, the U.S. Army decimated flocks of Navajo sheep in an effort to subjugate the Navajos, and then in the 1900s U.S. agencies slaughtered large numbers of the flock in an effort to control "overgrazing and erosion." Only a few small flocks remained. But today, conservationists and Navajo Native Americans are once again breeding the Navajo-Churros for their important characteristics, such as hardiness, longevity, and high lamb survivability on range.

The fleece is double coated — the inner coat has fine wool and the outer coat is long, course, and lustrous. It is the fleece of these sheep that gives the classic Rio Grande and Navajo weavings their strong and lustrous traits, and these fleeces are again inspiring fiber artists and weavers.

North Country Cheviot

This general-purpose breed originated in northern Scotland and is well adapted to northern climates and hilly, rough terrain. They were first imported to North America in 1944. North Country Cheviots are larger than their kin, the Border Cheviots, and probably show more of the traits of the breed's early ancestors in Scotland.

The breed is polled and has open faces and bare legs. They produce a medium-wool fleece with good staple length. In Scotland, their wool, which is free from hair or kempy fibers, is used to make the famous Scottish tweeds. Ewes are good milkers, easy lambers, and fairly prolific.

Navajo-Churro (Courtesy American Livestock Breeds Conservancy. Photograph by Tanya Charter.)

North Country Cheviot (Courtesy American Sheep Industry Association 2000)

Oxford

Oxfords are an English breed that was named after the county of Oxford. These sheep were bred primarily from Cotswold and Hampshire foundations, which makes them large and heavy. Breeders were successful in combining the hardiness, muscle, and wool quality of the Hampshire with the great size, rapid growth, and wool characteristics of the Cotswolds. Oxfords were first recognized as a true breed in 1862 and were imported into the United States as early as 1864.

Oxfords have a good fleece weight and medium wool of reasonable length. Their faces and legs are usually light brown, but anything from light gray to dark brown is acceptable. A white spot on the end of the nose is common. Because their faces are partly open, there is no tendency toward wool blindness.

These sheep are most valuable as a sire breed. Rams weigh up to 300 pounds (136.1 kg); they contribute size and muscle to the offspring. Because they are easily handled in small pastures, Oxfords are well suited to farm raising and thrive when given good feed. The ewes are docile, heavy milkers, and since the breed has a small head, lambs are born easily.

Panama

As a crossbreed of Lincoln ewes and Rambouillet (see page 66) rams, the original Panama stock has the reverse parentage of the Columbia, which are Lincoln rams crossed with Rambouillet ewes. Breeder James Laidlaw wanted to replace the small Merinos that were common in Idaho. His goal was to develop more rugged sheep with finer wool and better herding instincts than those of Columbia sheep. He felt that the ram had more influence on the offspring than the ewe, although this view is controversial.

Laidlaw made the first cross in 1912, starting with 50 rams and 1,600 ewes. With this large number of animals, he was able to avoid the inbreeding problems that often arise in the attempt to start a new breed. After 3 years, he switched from Rambouillet to Panama rams. After 5 years, Laidlaw sold the remaining Lincoln ewes. The registry started in 1951, and all registered Panamas must be direct descendents of the original Laidlaw flock.

Panamas are good-sized, hardy, polled sheep. Ewes are good mothers that produce plenty of milk. Their heavy fleece weighs 9 to 14 pounds (4.1–6.4 kg) and is medium to fine.

Oxford (Courtesy Larry E. Mead, Sheep Breeder and Sheepman *magazine)*

Panama (Courtesy Troy L. Ott, PhD, PAS, University of Idaho Sheep Research and Teaching Center, Moscow, Idaho. Photograph by Dave Casebolt, MS.)

Perendale

The Perendale is a cross of Cheviot rams on Romney (see page 68) ewes. Developed in New Zealand, they were first imported to the United States in 1977 and are growing in favor.

Perendales have clean faces and legs and dense, usually white, wool of a 4- to 5-inch (10–12 cm) staple. Perendale wool is prized by hand-spinners who dye their own wool, is easy to spin, and makes good garments.

The breed is well suited to hilly areas. They are easy to care for and lamb unassisted, though they inherit a bit of nervousness from their Cheviot ancestry and need gentle handling.

Polypay

The Polypay was developed in the U.S. Sheep Experiment Station in Idaho. Announced as a new breed in 1976, it started with initial crosses of Targhee (see page 74)–Dorset, and Rambouillet–Finnsheep breeds. These crosses were then recrossed, resulting in a breed that is one-quarter of each type of sheep.

Polypays are a superior lamb-production breed. Not only are they outstanding in twice-a-year lambing operations, their lambs have a quality carcass.

The Rambouillet and Targhee breeding is included to retain hardiness and flocking instincts. Dorset blood contributes to carcass quality, milking ability, and a long breeding season. The Finnsheep contribute early puberty, early postpartum fertility, and high lambing rate.

The fleece of the Polypay is medium to fine and weighs about 8 pounds (3.6 kg). Wool weight is higher in flocks that are bred less than twice each year.

Perendale (Owned by Norlaine Schultz)

Polypay (Courtesy American Sheep Industry Association 2000)

Rambouillet

The Rambouillet is the French version of the Merino. Louis XVI imported 359 Spanish Merinos for his estate at Rambouillet in 1786, and the Merinos were crossed with the resident sheep. The resultant Rambouillet sheep were first brought to the United States in 1840.

Rambouillets are very large and have strong bodies and little wrinkling, except perhaps some across the brisket. They are hardy and possess remarkable herding instincts. They spread out to graze during the day and gather closely together to sleep at night, making them excellent for open range. They also adapt well to a wide range of climates and feeds, making them equally suitable for farm flocks.

They are considered to be a dual-purpose breed, with a desirable carcass and good wool production. The ewes can be bred early to lamb in November and December, and the lambs give good yield in boneless, trimmed meat cuts. The fleece is less oily than that of the Merino, so it also shrinks less. The rams have horns, and both sexes have white feet and open faces. They show relatively strong resistance to internal parasites.

Rambouillets have been widely used as foundation stock for the development of new breeds.

Romanov

Like the Finnsheep, Romanovs are a northern European "rat-tailed" breed. These sheep first arrived in North America in 1980 when Agriculture Canada imported fourteen ewes and four rams from France.

It appears that the Romanov's fertility, body size, growth, and carcass characteristics are similar to those of the Finnsheep. The lambs are born black, with a silky hair coat over their wool. As they mature, the hair is shed and replaced by wool. The breed was traditionally raised for its fur (or pelts) in Russia.

One Romanov advantage is early sexual maturity, which occurs at 6 months of age. Ewe lambs first give birth when they are only 11 or 12 months old. They also have the ability to breed out of season; in Canada, they produce lambs every 8 months.

Rambouillet (Courtesy American Sheep Industry Association 2000)

Romanov (Courtesy Agriculture Canada Research Station, Lethbridge)

Romeldale

An American breed that dates back to the 1920s, the Romeldale is a cross of Romney (see below) rams and Rambouillet ewes. They produce medium-to-fine wool that shrinks very little, making more pounds of clean wool than is normally produced from the fleeces of fine-wool breeds. The wool of the Romeldale is finding favor with hand-spinners. The purebred Romeldale lamb is very marketable and the females can be saved for replacement ewes. Ewes are excellent mothers, prolific, and long lived. Twinning and ease of lambing are two traits for which the breed is known.

The breed is found primarily in California, and its popularity has not spread. This is due in part because, unlike the California Variegated Mutant, the breed has never had an active association and registry.

Romney

The Romney is an English breed, which is called the Romney Marsh in their native region after the low, marshy area where they are thought to have originated. Romneys are said to be somewhat resistant to foot rot, liver flukes, and other problems that often plague sheep in damp pastures.

This breed has a quiet temperament and does well on a good pasture. They are not suitable for hilly country or hot, dry climates. They have little herding instinct, but can be managed easily in a farm flock.

Although they are a long-wool breed, the wool of Romney sheep is much finer and more lustrous than that of other long-wool sheep. Their fleece doesn't have the tendency to part along the back, so they do well in rainy climates. Except for a tuft of wool on the forehead and short wool on the lower chin, the rest of the head is clean. Romneys produce an excellent hand-spinning fleece. Their meat is of good quality and has a delicate taste.

Romeldale (Hank Saxton flock)

Romney (Courtesy American Sheep Industry Association 2000)

Santa Cruz

Another island feral breed, Santa Cruz sheep hale from Santa Cruz Island, one of the Channel Islands off the coast of California. The Santa Cruz breed has lived on the island for about 200 years and has been feral for at least 70 years.

Like other island sheep, Santa Cruzes are small. Most are white, but black, brown, and spotted sheep are also found in the breed. They have medium to fine fleece, which is especially soft.

Sheep and Conservation Efforts

Like the Hog Island breed, Santa Cruz sheep were removed from their island after it was acquired by The Nature Conservancy during the 1980s. Individual breeders, primarily in California, now maintain the population. Don Bixby of the ALBC says, "Though this type of conservation effort is not for everyone, it does have the satisfaction of protecting a truly unique genetic resource from disappearance."

Scottish Blackface (Black Face Highland)

Another mountain sheep from Scotland, the Scottish Blackface is a hardy, quick-maturing meat animal. The breed is one of the most numerous in the British Isles and is raised where conditions are hard. Monastery records from the twelfth century refer to this breed.

Scottish Blackface sheep are adapted to cool, damp conditions. They do well on sparse forage. The coat has excellent water-shedding properties.

The ewes are excellent mothers. Although they are not prolific under "hill conditions," they are fairly prolific on good pastures.

These sheep have an attractive and stylish lightweight fleece of long, coarse wool. In addition to the attractive coat, the Roman nose and unusual black-and-white face markings give these sheep a unique appearance. The mottled faces are preferred over the solid-color black face in England, where the markings are said to indicate greater resistance to disease. Both ewes and rams have horns.

Santa Cruz (Courtesy American Livestock Breeds Conservancy. Photograph by Tanya Charter.)

Scottish Blackface (Courtesy American Sheep Industry Association 2000)

Shetland

An ancient breed raised on the Shetland Islands (located north of Scotland and west of Norway in the Atlantic Ocean), these sheep are quite hardy. Shetland rams have two horns, while the ewes are polled. Both sexes have short tails that don't require docking. Their wool is fine, more durable than Merino wool, and less likely to pill. A great range of colors adds to its value, especially for hand-spinners. The many fleece colors include sparkling white, shades of gray, lustrous black, tan, and shades of a deep, dark chocolate brown. Many of the colors are referred to by their traditional names, such as sholmit, shaela, eesit, mooskit, mogit, and moorit. Shetland wool is used for traditional wedding shawls that can be pulled through the bride's ring.

Shetlands were imported into Canada in the 1980s and into the United States later in the decade. The value of these sheep to hand-spinners resulted in a huge expansion of the breed, both in North America and in Britain. Considered endangered only a short time ago, the breed has now been removed from conservation lists on both sides of the Atlantic.

Shropshire

Shropshire sheep are one of the "down" breeds, developed in southern England in the low hills called downs. First known as a fixed breed in 1848, it was imported into the United States in 1855 and became a well-established, popular breed.

It is a medium-sized sheep that produces good meat lambs, but it needs abundant feed.

Southdown

The Southdown, one of the oldest meat breeds in the world, still ranks among the top four purebred sheep breeds today. The medium-sized lambs produce excellent smaller-type lamb carcasses. They are popular with small-flock producers and make great youth projects.

Southdowns originated in the South Downs, a hilly area of southeastern England. This heritage makes them especially adaptable to varied climates and terrains.

Southdown fleeces produce a medium wool fiber, 56s–60s. The short staple is suitable for hand-spinning into fine yarns.

Shetland (Courtesy Julie Guilette)

Shropshire (Courtesy Bill White, Missouri Department of Conservation, St. Joseph, MO)

Southdown (Courtesy American Southdown Breeders' Association. Photograph © Wendy Hall.)

St. Croix

The St. Croix is a hair breed, with a long breeding season, that developed in the hot, island climates of the Virgin Islands. Like the Gulf Coast Native, years of selection in a tropical climate has provided the St. Croix with strong tolerance for internal parasites and good heat tolerance. The breed will typically breed back in 30 to 40 days after lambing.

St. Croixs are a medium-sized breed. They are quite docile, and because they lack predators and have scanty forage in their native islands, they have weak flocking skills, so they require good predator control.

Suffolk

The Suffolk is the most common breed in the United States. It is a handsome sheep with a black face, ears, and legs that are free of wool.

The ewes are prolific and good milkers. Lambs grow rapidly; they have more edible meat and less fat than many other breeds.

Suffolks are active grazers and are able to rustle for feed on dry range. When raised on high-quality feeds, they have one of the fastest growth rates of any breed, and are considered to have excellent feed-conversion characteristics.

Another English contribution, the Suffolks were developed by crossing Southdown sheep with old Norfolk sheep. This latter breed had a black face and horns, was hardy and prolific, and produced meat of superior texture, but its conformation was poor. The resulting breed combined the best characteristics of its parents and became popular for use in both purebred flocks and in crossbreeding.

Targhee

The Targhee is a hardy American breed, developed by mating outstanding Rambouillet rams to either ewes of Corriedale and Lincoln–Rambouillet stock or ewes of only Lincoln–Rambouillet stock, followed by interbreeding the resulting lambs. This work was done in 1926 at the Sheep Experiment Station in Dubois, Idaho, to meet the demand for a breed of sheep that had thick muscles, was prolific, produced high-quality, apparel-type wool, and was adapted to both farm and range conditions. These sheep are named after the Targhee National Forest where the station flock grazes in the summer.

Targhees are large-framed, dual-purpose sheep that produce good meat and heavy fleece (11 to 16 pounds [5–7 kg]) of good, medium wool.

St. Croix (Courtesy American Livestock Breeds Conservancy. Photograph by USDA-ARS.)

Suffolk (Courtesy American Sheep Industry Association 2000)

Targhee (Courtesy American Sheep Industry Association 2000)

Texel

Texel sheep have been bred in Holland, Finland, and Denmark for more than 160 years. In 1985, the USDA was the first to import the breed. The breed resulted from crosses of native "polder sheep" (graziers on "polderland," or lands reclaimed from the sea) with British breeds, such as Border Leicesters and Lincolns.

Texels are hardy animals that can adapt to many climates and conditions. They do very well as a foraging breed. They are lean, medium-sized sheep and have a high muscle-to-bone ratio. These sheep lamb only once a year, but in farm flocks they have a high percentage of twins and triplets. Their lack of herding instinct makes them a poor choice for ranges. Texels have a white fleece of medium wool, with no wool on the face or legs.

Tunis

The Tunis is an American breed developed from the Tunisian Barbary sheep. The foundation stock was first imported into the United States in 1799, and they spread throughout the Southeast. A Tunis ram was used by George Washington to rebuild his flock, which had declined in number and vigor while he was serving as president.

Tunis are medium sized, hardy, and docile. The ewes are very good mothers and are known for breeding out of season — with proper management, they can be bred almost any month of the year. The lambs are a reddish color when they're born and gradually lighten to white, though they retain an unusual color of reddish tan hair on their faces, their legs, and their long, pendulous ears.

As their African heritage would suggest, the Tunis does very well in warm climates, and the rams remain active in very hot summer weather. Although they are a superior breed for a hot climate, they are raised successfully almost anywhere.

Wiltshire Horn

The Wiltshire Horn is an ancient British breed of hair sheep, once known as the Western. Unlike some breeds of hair sheep, the Wiltshire Horn grows wool as well as hair. It sheds its wool each spring. The breed is considered endangered globally but has a brighter future now than in the recent past.

Wiltshire Horn sheep are large — rams weigh up to 300 pounds (136.1 kg). Both sexes are horned, with impressive, curling horns being the norm. The breed is hardy, doing well on marginal pastures.

Texel (Courtesy Larry E. Mead, Sheep Breeder and Sheepman *magazine)*

Tunis (Courtesy American Sheep Industry Association 2000)

Wiltshire Horn (Courtesy American Livestock Breeds Conservancy)

THREE

Pasture, Fences, and Facilities

More than 2,255 million acres (913 million ha) of North America can't be used efficiently to grow anything other than grass, and all farms have areas that aren't really suitable for crops. We can't eat this grass, but our sheep sure can.

Sheep are really efficient at converting grass into meat. For the shepherd who is interested in low input and high profit, grass is the key to success. In the low-input, pasture-based system, sheep have their lambs on the pasture in late spring, and the lambs grow to market age on the abundance of grass during the summer. The lambs can then be sold in the late summer or early fall, about the time the pastures begin to give out. This means that you don't need to carry the animals through the winter on hay and grain.

Sheep are among the best grazing animals in the world. Even breeds with "poor" foraging abilities are still good grazers, but they need high-quality, tame pastures — they won't do well on rough, native pastures without supplemental feed. Breeds that are excellent foragers produce nicely on those rougher pastures.

For all shepherds, regardless of the breed they choose to raise, pasture should be the cornerstone of their operation. In fact, we think of ourselves as grass farmers, capturing solar energy in the grass of our pastures and converting it to a product (food and fiber) with our sheep. To be successful, a grass farmer must learn to manage pasture for both the plant's and the animal's needs. Even on very small parcels of land, a shepherd can use managed grazing to provide the bulk of feed for a small flock of sheep. By adopting the techniques discussed below, grass farmers can be stewards of their environment: Erosion on a well-sodded pasture is up to 300 times less than that of crop ground.

Foraging Capabilities

Excellent	Good	Poor
Barbados Blackbelly	Bluefaced Leicester	Booroola Merino
Black Welsh Mountain	Border Leicester	Dairy Breeds
Border Cheviot	California Variegated Mutant	Finnsheep
California Red	Clun Forest	Hampshire
Debouillet	Columbia	Oxford
Gulf Coast Native	Coopworth	Suffolk
Hog Island	Cormo	
Icelandic	Corriedale	
Karakul	Cotswold	
Katahdin	Delaine Merino	
Navajo-Churro	Dorper	
Panama	Dorset	
Polypay	Jacob	
Rambouillet	Lincoln	
Romanov	Longwool Leicester	
Romney	Montadale	
St. Croix	North Country Cheviot	
Santa Cruz	Perendale	
Scottish Blackface	Romeldale	
Shetland	Shropshire	
Targhee	Southdown	
Tunis	Texel	
Wiltshire Horn		

Note: *All sheep are, by nature, foraging animals, but those that are listed as* excellent *do well on poorer pasture or range, with little or no supplement. Those listed as* good *do well on good pasture with little or no supplement, and those listed as* poor *do well on a good pasture, with supplementation.*

Pasture

A pasture is simply an area of land where forage plants (grasses, legumes, and forbs) grow. Pastures may also include brush and trees and are generally classified as one of two types: tame pasture, which is an improved and seeded pasture, or native pasture, which consists of whatever plants naturally grow in the area.

Tame pastures, as a rule of thumb, are capable of much higher levels of production per acre, but some native pastures produce remarkably well without the cost of developing a tame pasture. Tame pastures are generally found in areas of high rainfall or in irrigated fields in arid areas. Native pastures run the gamut from the lowland to hilly; unimproved pastures found in the humid East to the dry rangelands of the arid West.

Reasons to Manage Grazing

There are lots of good reasons to manage your grazing!

- Good grass cover is aesthetically pleasing and increases property value.
- Even on a small piece of land, you can cut purchased feed bills.
- Many health problems, such as parasites, dust-related illnesses, and foot problems, are reduced.
- Erosion (both from wind and water) and pollution are reduced.
- Better grazing management improves profitability.

SHEPHERD STORY

Ken Kleinpeter manages the Old Chatham Sheep Company, the largest sheep dairy in the United States. Old Chatham produces cheese and yogurt, which they market nationally, from a thousand-ewe flock. The sheep are milked year-round, on a revolving schedule, with anywhere from two hundred to five hundred ewes producing at a given time, depending on the season. To meet the market demand for sheep cheese, they also purchase frozen sheep milk from other sheep-dairy farms and from a sheep-dairy coop in Wisconsin.

In Old Chatham's sheep-dairying operation, as in cow dairying, the lambs are taken from their moms after 24 hours and reared artificially. Raising hundreds of bottle lambs at a time presents challenges. As Ken explained to me, "We have nicely ventilated, conventional barns, but we were still having trouble with lamb death loss and illness. In the conventional barns, we had pneumonia running as high as 15 percent! This was completely unacceptable.

"Our first method of trying to deal with the problem was to move the lambs outdoors, using metal calf hutches to raise them in. We'd put about ten lambs to a hutch, and we did cut our losses, but although the lambs did better it still wasn't an ideal situation. For one thing, it could be miserable for the people taking care of the lambs. The lambs would be all snuggled back in the hut on wet and cold days, but the staff had to be right out in the middle of it. And during really wet periods, the pens in front of the hutches would become mud messes."

Ken began investigating the possibility of using a greenhouse barn, and the more he read, the more he thought he'd come to the solution. "For us, the greenhouse barn has been an unqualified success. We've cut total death loss of

Carrying Capacity

Carrying capacity (which is sometimes referred to as stocking rate) is a measure of how many animals a farm can support over the course of the year. Carrying capacity depends on many factors, including:

- Type of soil (rock, sand, clay, and so on)
- Plant species that are growing
- Amount and timing of annual precipitation
- Availability of irrigation water
- Temperature
- Fertility of the soil
- Lay of the land (hill, marsh, level)
- Whether lambs, ewes with lambs, or dry ewes will be run on it
- Whether any supplemental feed will be purchased (hay or grain)
- Other, more nebulous factors

our lamb crop to less than 4 percent, and we have virtually no pneumonia."

The greenhouse Ken settled on is manufactured by the Harnvois Company of St. Thomas, Quebec, Canada, and is specifically designed for livestock, as opposed to being designed for plants. Ken said the main difference is in the roof design. "Unlike a greenhouse made for plants, this greenhouse has an adjustable roof vent running the whole length of the greenhouse. This provides excellent ventilation, so the air in the greenhouse stays fresh. It's the buildup of ammonia fumes in barns that contribute to pneumonia, so this design is really key to eliminating that problem for our lambs.

"For what we're doing, this is really the least-cost building option. If I was thinking of starting a new operation from scratch, on a farm that didn't already have any barns, I'd certainly say it would be practical to go with greenhouse barns for all the buildings I needed."

Ken did point out that like any other kind of building, the greenhouse barn isn't problem free; it does require some maintenance. The plastic sheeting needs to be changed every 4 to 7 years, and the sides, which can be raised and lowered to adjust temperature and humidity in the barn, need to be manipulated regularly. It can take some time to learn how to manipulate the sides for the best airflow and temperature. But all in all, Ken says, "The benefits of the greenhouse far outweigh the problems.

"It's a nice atmosphere to work in for the people, and the lambs are comfortable and happy."

Sounds like a win-win situation. (See page 102 for a photo of the greenhouse.)

Some shepherds estimate that an acre of really good, tame pasture can support four sheep during the year. Rougher, native pasture may not even be capable of carrying one, so take a good look at the condition of your acreage before you bring your sheep home. It is better to keep too few for your first year until you see how your pasturage holds up.

Forage Plants

True grasses, legumes, and forbs are considered forage plants. True grasses, such as timothy, brome, bluestem, fescue, orchard grass, and even corn, are *monocotyledons* (monocots), or plants that initially grow from a single leaf. The legumes, such as alfalfa, clover, and bird's-foot trefoil, start growing from two leaves and are called *dicotyledons* (dicots). Another difference between monocots and dicots is the root system: Typically, monocots have a more fibrous and shallow root system, and dicots have a thicker taproot, which looks something like a carrot, and reaches much deeper into the soil.

Forbs are basically weeds. Most often they are dicots, though a few are monocots. One of the great things about keeping sheep is that they relish most forbs (when other livestock species avoid them). In fact, sheep are considered the ideal "biocontrol" for leafy spurge, one of the most serious noxious weeds spreading throughout North America.

Ideal pasture is a mixture of grasses, legumes, and forbs — not a monoculture of one kind of plant. The diversity of plants provides a more balanced diet for longer periods during the growing season under a wider variety of weather

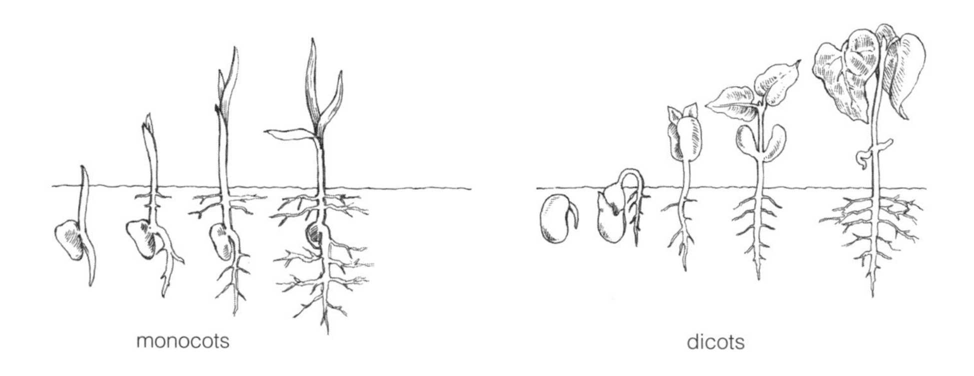

Plants that sprout with a single leaf are monocots; those that sprout with two leaves are dicots. Most grasses are monocots, and most legumes are dicots. The monocots tend to have a shallower and more fibrous root mass, whereas dicots have a deeper tap-rooted mass.

conditions. In tame pastures, the planting mix is usually 60 percent grass to 40 percent legumes. If you're planning to seed a new pasture or reseed an existing one, it's also a good idea to have a variety of both cool-season and warm-season plants. Cool-season plants perform well in spring and fall, and warm-season plants do well in the heat of summer. For sheep, the shorter sod-type grasses and legumes, such as bluegrass and white clover, are ideal because they don't trample down as much.

Deciding which kinds of grasses and legumes to plant if you're establishing pasture depends on the factors mentioned above. (But, it's usually better — and cheaper — to see what kinds of grasses your pasture will grow when the sheep are put on it. So many seeds are in the ground that when your grazing is managed they have a chance to germinate and grow). Your county Extension agent is an excellent resource for information on which forage plants grow best locally.

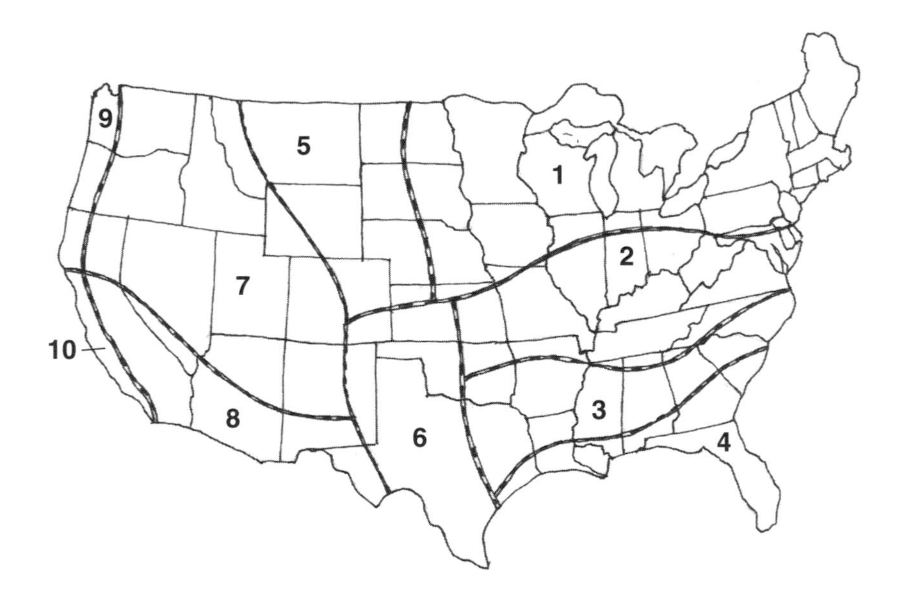

This map shows the ten generally recognized pasture areas in the United States; the numbers correspond to the information given in the table on pages 84–85. Remember when studying this table that some plants may be widely adapted to grow in a given area, or they may only grow there under limited circumstances, or when using selected varieties. For example, white clover shows up in all ten areas, but in arid environments it will only grow in subirrigated areas. (Redrawn from M. E. Ensminger and R. O. Parker. Sheep & Goat Science, *5th ed. Danville, IL: Interstate Printers & Publishers, 1986, p. 405)*

Plants and Their U.S. Pasture Areas

	U.S. Pasture Area									
Plants	1	2	3	4	5	6	7	8	9	10
Grasses, Shrubs, Forbs										
Bahia grass			✓	✓		✓				
Bermuda grass		✓	✓	✓		✓		✓		✓
Bluegrass	✓	✓			✓			✓		
Bluestem					✓	✓		✓		
Brome grass	✓				✓		✓	✓		
Buckwheat (wild)								✓		
Buffalo grass					✓	✓				
Canary grass	✓	✓			✓		✓		✓	
Chamiza (salt grass)								✓		
Dallis grass			✓	✓		✓		✓		✓
Fescue	✓	✓	✓		✓	✓	✓	✓	✓	✓
Foxtail							✓		✓	
Grama grass					✓			✓		
Indiangrass						✓				
Johnson grass			✓	✓		✓				
Junegrass					✓	✓		✓		
Love grass						✓		✓		
Millet	✓	✓	✓	✓						
Oat grass									✓	
Oats	✓	✓	✓	✓		✓		✓	✓	✓
Orchard grass	✓	✓	✓	✓	✓		✓	✓	✓	✓
Pearl millet		✓	✓	✓						
Redtop	✓						✓			
Rye	✓	✓	✓	✓	✓	✓		✓		✓

	U.S. PASTURE AREA									
PLANTS	1	2	3	4	5	6	7	8	9	10
Ryegrass (annual)			✓	✓					✓	
Sorghum-Sudan		✓	✓	✓		✓				
Sudan grass	✓	✓	✓	✓		✓	✓	✓		✓
Switchgrass					✓	✓				
Timothy	✓	✓					✓		✓	
Wheat	✓	✓	✓	✓	✓	✓	✓	✓	✓	✓
Wheatgrass					✓	✓	✓	✓		
Legumes										
Alfalfa	✓	✓	✓	✓	✓	✓	✓	✓	✓	✓
Black medic			✓			✓		✓		
Clover-Alsike	✓	✓	✓		✓		✓	✓	✓	
Clover-Bur		✓	✓							✓
Clover-Crimson	✓	✓	✓	✓						
Clover-Ladino	✓	✓	✓	✓	✓	✓	✓	✓	✓	✓
Clover-Prairie								✓		
Clover-Red	✓	✓	✓	✓	✓		✓	✓	✓	
Clover-Strawberry					✓		✓	✓		✓
Clover-White	✓	✓	✓	✓	✓	✓	✓	✓	✓	✓
Cowpeas			✓	✓						
Field peas			✓	✓					✓	
Lespedeza		✓	✓	✓						
Soybeans	✓	✓	✓	✓						
Trefoil	✓	✓	✓	✓	✓		✓	✓	✓	✓
Vetch			✓	✓	✓	✓		✓		

Feeding Your Pasture

Just like animals, plants need to eat and drink. Most of their "food" is absorbed through the roots, though a small amount can be absorbed through leaves. They require nutrients like nitrogen, phosphorus, potassium, calcium, and magnesium. The true grasses have to get all their nutrients from the soil, but the legumes can acquire part of their nitrogen from air molecules that are trapped in the soil through a process known as *nitrogen fixation.*

Nitrogen Fixation

The ability of legumes to trap nitrogen from the soil is an amazing process. Soil bacteria, known as *Rhizobium,* live in nodules on the roots of the legumes in a symbiotic, or mutually beneficial, relationship. *Rhizobium* actually convert the nitrogen that's trapped in soil air molecules into a form that the legumes can absorb through their roots. Nitrogen fixation not only provides nitrogen to the legume, it also provides extra soil nitrogen for the grass plants that are growing nearby (reducing the need for nitrogen fertilizers) and increases the protein content of the legumes. Protein is important for your sheep. (See chapter 6 for more about feeding your sheep.)

Nitrogen fixation by the *Rhizobium* depends on several factors. First, the soil has to have some of these beneficial bacteria living in it, or in the case of a new pasture seeding, the seeds have to be inoculated with bacteria. Second, the soil has to be healthy enough to support the bacteria. Soil that is low in organic matter may not support the bacteria very well; soil that has been treated with strong, chemical fertilizers, pesticides, and herbicides may not support them at all.

Deciding what kinds of fertilization program to implement for pasture optimization requires soil tests or plant tissue tests or both. Soil tests are a little less expensive than plant tissue testing (ask your county Extension agent about soil testing availability), but plant tissue testing is more accurate; see Forage Information System at Oregon State University in Resources. If you do opt to use chemical fertilizers (like ammonium sulfate), several light applications during the year are far better for the soil health and soil organisms than one large application. Natural soil amendments such as compost and green manure actually enhance soil health.

In humid environments, the fertilizer that usually provides the most bang for the buck is some kind of calcium supplement (lime being the most familiar).

Soil in areas of high rainfall suffer from *leaching*, or movement of nutrients down through the soil to depths that plants can't access, and calcium leaches quickly. In some areas of the United States, certain micronutrients are also in short supply in the soil, and you need to either boost their levels in the soil or make sure you're supplying them to your flock in a supplemental trace mineral product. Again, your county Extension agent should know about soil deficiencies in your area. (See chapter 6 for more about feeding minerals, and the resources for some books that provide additional information on composting, soil amendments, and so on.)

Controlling Overgrowth and Rejuvenating Pastures

If a pasture has become overgrown with brush and excess weeds or, on the other hand, if it has developed large bare spots and eroded places, it may require rejuvenating.

Overgrowth

When the problem is overgrowth of brush and weeds, you have several options.

- Get a breed of sheep that's considered an excellent forager and/or get some goats. (When it comes to taking out really overgrown brush, goats are the superior animal — but they are even more difficult to fence than the most challenging sheep.) By continuing to strip growth off brush, sheep will eventually kill it.
- If you're intending to go with a breed that doesn't rate as highly on foraging capabilities, you may need to do some mechanical clipping or mowing right away.
- If you opt to mechanically clip brush or weeds, keep in mind that by optimizing the timing of your clipping, you can have a big impact on the success of your labors.
- Clip weeds just as they begin to flower but before the seed head has opened.
- Clip brush in the early spring, while the sap is running.

Fixing Bare Spots

Carrying around some pasture seed in your jacket pocket whenever you walk through the pasture and simply tossing handfuls on the bare places can help rejuvenate bare spots. Another trick is to feed bales of hay right on the bare spots during the winter — the hay gets stomped into the ground and protects the soil surface, allowing seeds to germinate and grow better.

Frost Seeding

Another technique that can rejuvenate a pasture or just increase the diversity of plants found in it is frost seeding. As the name implies, frost seeding requires seed to be spread during the spring, when the nights are cold enough to frost but the days are warm enough to thaw the soil surface. This freeze–thaw action allows planting of the seeds at a shallow level where they are most likely to germinate.

For frost seeding, broadcast the seed thinly over the pasture. If the pastures are large, use a tractor-drawn broadcast seeder; for smaller pastures, use a handheld version. The technique works really well with legume seeds — it makes a great method for increasing the diversity and percentage of legumes in an older pasture where they have become too thinly populated.

Grass Growth

Understanding how forage plants grow helps you get the most out of your pasture. Growth primarily takes place at the plant's basal growth point, which is just above the soil surface. Initially, plant growth is kind of slow, but as the leaves reach above the basal growth point, things really speed up. Then, as the plant reaches maturity, growth slows down again because the energy that had been used for growth switches to flower and seed production.

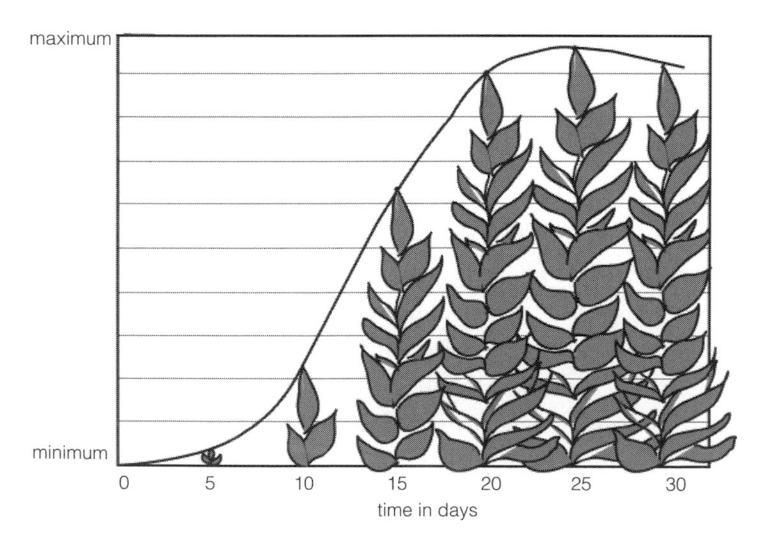

This is a typical S-type curve that represents growth of most living organisms. Note how growth begins somewhat slowly, then in the middle period it speeds up, and then finally slows down and tapers off. (Modified from André Voison, Grass Productivity, *Covelo, CA: Island Press, 1988, p.12.)*

Forage plants store extra energy in their roots when they're growing quickly (the steep part of the S curve). This stored energy can be used later to jump-start new growth after the leaves have been grazed down or to make the first spurt of spring growth after the winter dormancy. Their ability to recover quickly after grazing makes the forage plants valuable, but don't be deceived into thinking that leaves can be continuously removed without injury. In fact, if the leaves are grazed off repeatedly, the plant keeps drawing on the energy stored in its roots to grow new leaves until the energy supply is exhausted and the plant dies.

Grazing Approaches

The traditional approach to grazing has been to put animals in a pasture at the beginning of the season and let them stay there until they run out of feed. This approach is called *set stocking*, and it results in simultaneous overgrazing and undergrazing of plants in the same pasture.

The reason set stocking causes overgrazing and undergrazing at the same time is because critters are sort of like kids in a candy store. When they're set stocked, they eat what they like the most and ignore the "flavors" they don't like. Some plants are constantly being grazed, while others aren't touched. Interestingly, the result of both overgrazing and undergrazing is the same: The plants lose energy and don't perform at their peak potential. Both scenarios can ultimately kill the plants.

Managed Grazing

By using managed grazing (which you may also see referred to as *rotational grazing, management intensive grazing, or planned grazing*), you control the flock's access and grazing time, thereby obtaining peak plant performance; this, in turn, results in peak animal performance. The trick to this is to subdivide your pasture into smaller pieces, known as paddocks, and to *time your flock's movements through the paddocks according to how the grass is growing.*

Managed grazing is a complex skill that takes time to master, but the payoff is well worth the effort. Finding a mentor who uses managed grazing — even if it's a cattle grazier — will hasten the learning process. Many states now have organized grazing groups, which your county Extension agent can help you find.

Some forage plants, such as alfalfa, can't stand the pressure of the continuous grazing that occurs in a set-stocked pasture, but they can stand hard grazing for short periods. Alternating paddocks allows these plants to compete. In fact, all desirable forage plants grow much better when grazed hard and then given a period of rest.

The Advantage of Paddocks

Sheep graze smaller paddocks more evenly and have less tendency to pick and choose than they would in one larger pasture. When the area has been adequately grazed, the sheep are moved to a fresh paddock. When you move the sheep out of a paddock, the grass will receive the rest it needs to keep growing at its fastest growth rate. If the rest period is adequate — usually at least 28 days — then many of the parasite larvae will die before the flock returns. An old Scottish shepherd's saying goes, "Never let the church bell strike thrice on the same pasture."

You'll also find that managed grazing gets your sheep to eat better than they would on a set-stocked pasture. Sheep prefer not to feed continually in the same place. They like fresh pasture that hasn't been walked on, and managed grazing keeps them on fresh pasture regularly.

Timing the flock's movements. The key to managed grazing is time! There are ideal points for beginning and ending grazing (or mechanical clipping if the grass is getting too long but you're not ready to bring the flock back around). By controlling your flock's access to pasture through carefully timed movement, you can maintain growth between points C and B (see illustration, opposite).

Animals should be moved from a paddock before they've grazed off 50 to 60 percent of the forage in the paddock, because most forage plants reach their maximum vigor and growth when no more than 60 percent of their leaf surface is removed during any grazing period. For example, if the sheep enter a paddock when the forage is 6 inches (15 cm) long, then they should be removed while there's at least 2.5 inches (6 cm) left standing.

The last time-related issue to consider is the rest period. After you move a flock to another paddock, the one you're leaving needs enough time to grow back to the starting height. The rest period varies by season. In early spring, it may be as short as 7 to 10 days; in the dog days of summer, it may take 45 days.

During the spring when the grass is growing rapidly, you can move the flock through the paddocks quickly and don't need to worry about taking the full 50 to 60 percent of the forage each time. Just let the flock lightly graze 20 to 30 percent, and move them to the next paddock. Then in the latter part of the growing season, when the rest period is getting longer, slow down their movements between paddocks so they take the full 60 percent.

Paddock Management

When managing paddocks for sheep, try not to let the grass get too tall before you turn the sheep in, or they'll trample more than they eat. When the grass has reached a height of 6 to 8 inches (15–20 cm) is an ideal time for the sheep to enter the paddock. (If you happen to be grazing your sheep with cattle or horses, then begin grazing the paddock when it's 8 to 12 inches [20–30 cm] high.) Occasionally, especially in the spring, some paddocks may get too long; you can mechanically clip — or mow — these paddocks for hay or just leave the clippings on the field as green manure.

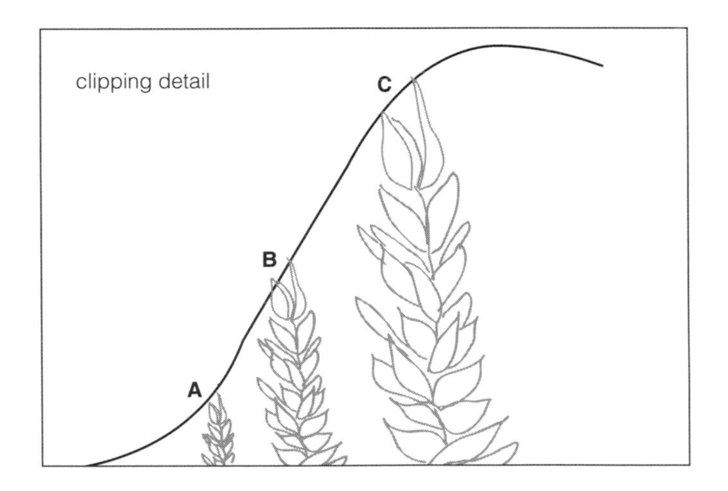

The objective of clipping is to take off growth in the high-growth part of the curve. Ideally, you would clip grass when it reaches point C, taking it down to point B. Never *clip below point A.*

How well you can control these time factors depends primarily on how many paddocks you have available.

Paddock numbers. So, just how many paddocks do you need? As paddock numbers increase, the time spent in each paddock decreases and the possible rest time before the paddock is regrazed increases. So, the answer is: as many as you can reasonably create. At the very minimum, shoot for four paddocks. Eight is even better, and twelve provides lots of flexibility and control through all kinds of conditions.

In the past, because of added expense and labor, fences for paddocks were often neglected, but with today's modern electric fencing, there's really no excuse not to create subdivided paddocks. These barriers don't have to be as heavy or as high as perimeter fences — they also don't need to be predator or dog proof. In fact, they can be created with temporary polywire and step-in posts. (See more about fencing on pages 94–99.)

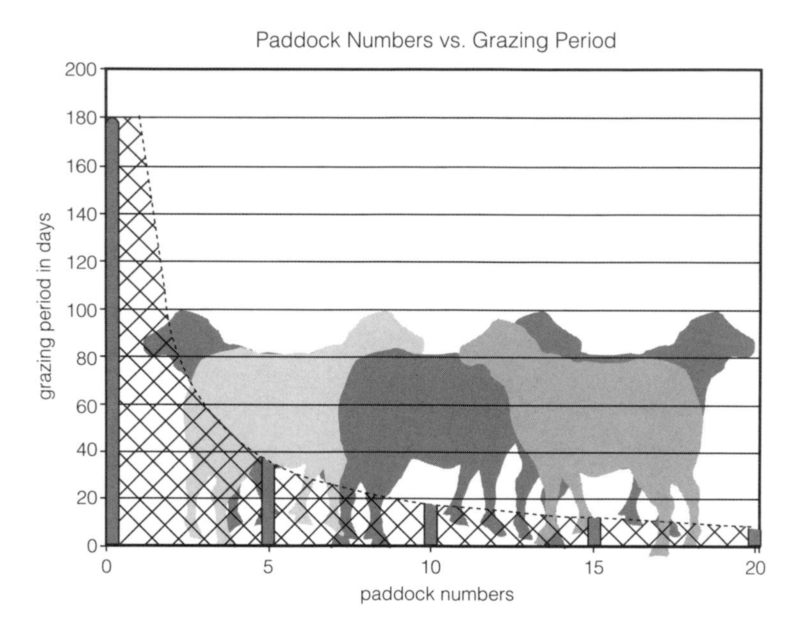

As the number of paddocks in a field increases, the time spent in each paddock, or grazing period *(in days), decreases. For example, if a field was grazed for 180 days with all animals placed in at the beginning and removed at the end of the 180-day period, the grazing period would be 180 days. Divide that field into two paddocks, and the grazing period drops to 90 days in each paddock.*

Orchards

Orchards are a special class of pasture and are one of the favorites on small farms in humid areas. The sheep can make use of the shade in summer, and if a little care is exercised to prevent them from getting too much windfall fruit at a time, they can also make good use of the fruit.

The flock should never be turned into an orchard with unlimited access to the fruit. The sudden change in their diet can cause bloat and other health problems, some of which may be deadly. As a rule of thumb, if there are fewer than half a dozen pieces of fruit lying around per animal in the flock, you'll see no problems. If the orchard floor is full of fruit, then pick some up and take it away. Gradually increase the amount you leave for the sheep over a week or so, because as they become accustomed to the fruit, they can eat quite a lot without adverse effects.

Subdividing the Orchard

By subdividing the orchard with temporary electric fencing like you would any other pasture, you can limit access to the amount of fruit the sheep can get at each helping and control the grazing like you would in any other paddock. On our Minnesota farm, we had a 1-acre (0.4 ha) apple and plum orchard that we subdivided into three paddocks with temporary fencing during the spring and fall.

Apples that are in good shape when they hit the ground can be stored and doled out well into the winter. If you opt to lamb in winter or early spring, you may be able to save them until lambing and give them to the ewes as treats in the lambing pens.

Protecting the Trees

If you decide to graze an orchard, you need to think about the trees as well as the sheep. Any newly planted trees, or dwarf trees, must be completely protected by a rigid fence or the sheep will inevitably eat the trees. Even with larger trees, you'll find an occasional sheep with goatlike habits, standing on its hind legs and nibbling the branches and leaves of fruit trees. Sheep that spend a long time in the orchard will start chewing on the bark of the trunks and can do a lot of damage if you don't protect the trees, but if you're treating the orchard as a paddock — or several paddocks — and moving the sheep through quickly, then this isn't a problem.

There are a couple of ways to protect your trees:

- Wrap the trunks with several layers of chicken wire or a single layer of rabbit wire. Use baling twine or wire to secure these temporary cages to the trees — don't permanently attach them, or you'll damage the tree. If you want to do the work only once, use three T posts formed in a triangle about 6 inches (15 cm) away from the trunk and construct a "fence" of wire mesh around the trees. These fences can stand for years.
- Another temporary solution is to make "manure tea" from sheep droppings and paint it on the trunks, but this needs to be repeated after it rains.

Fencing

There are two basic kinds of fence you need to consider — the perimeter fence and interior fences. As they serve different purposes, they are quite different beasts.

If you buy an old farm, the fences and buildings will probably need repairs. You can work on the buildings after you get the sheep, but the perimeter fences should be in good shape before you bring your first sheep home. Sheep quickly learn to jump sagging fences or to crawl through loose strands of barbed wire. One loose sheep in the neighborhood (and there's usually more than one) can be quite a problem, and sheep in a garden, especially if it happens to be the neighbor's garden, can be disastrous.

If you wait until sheep have the jumping habit, they may continue to jump the fence after it is repaired. One jumper can set a bad example and should either be sold or slowed downed by temporary *clogging*. To clog, attach a piece of wood to one front ankle with a strap — it gets in the way just enough to prevent most jumpers from doing their thing.

Again, the best policy is to have at least your perimeter fences up, tight, and ready to do their job before your sheep arrive. The investment in time, money, and effort will more than pay for itself in sound sleep.

Fencing is a unique skill that takes time to master and is a book-length subject in its own right. There isn't enough space in this book to do the topic justice, so if this is going to be your first fencing project, then I strongly recommend that you check out Gail Damerow's book *Fences for Pasture and Garden*. Gail has done a really great job on the topic in her book and includes all the

---

The Importance of Fencing

For any kind of livestock farmer, fencing is the most important asset on your farm, and this is even more true for shepherds than for those who are trying to raise cattle or horses. A well-constructed perimeter fence serves two crucial purposes: It keeps your sheep on your property, and it can help *reduce* the impacts of predators. (By themselves, fences generally can't be expected to completely eliminate predator problems. Chapter 5 talks more about predator control, including the use of guard animals.) Interior fences don't need to be nearly as substantial as the perimeter fence, because interior fences are used for your convenience and to enhance your management, so if the sheep get out of an interior fence they'll just graze another paddock a little sooner than you planned.

---

information you'll need on selecting materials, using the proper tools, construction tricks and techniques, and maintenance requirements. (See Resources.)

Types of Fencing

For perimeters, the fence should be at least 48 inches (1.2 m) tall and tight, with only small spaces between the fencing material. This type of an arrangement can be constructed from barbed wire, wooden rails, woven wire, and smooth electric wire (in either high-or low-tensile varieties), or a combination of several types.

Barbed-wire fencing. Invented in the mid-1800s, barbed wire had its place in history, but today, in the age of high-quality electric-fencing technology, barbed wire is the least desirable choice for containing sheep. For obvious reasons, you will find barbed wire difficult and unpleasant to use, and animals that get caught up in a barbed wire fence can hurt themselves badly. Another problem with barbed wire is that making an effective perimeter requires at least six, and preferably eight, strands of barbed wire, and that's a lot. By the time you build a barbed-wire sheep fence, you've spent a whole lot of money.

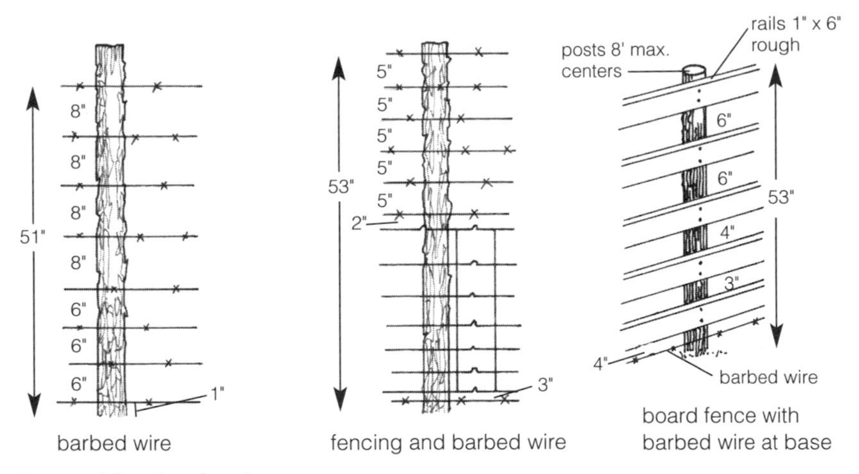

Three types of fencing for sheep

Wooden rail fences. An attractive option that can be constructed to keep sheep in, rail fences are unfortunately inefficient at keeping predators out. A barbed wire or electric wire can be run at the bottom of a rail fence to enhance predator control. Rail and other styles of wooden fencing should be constructed with the boards on the inside of the fence, so the sheep won't loosen them. Posts also need to be close together, which is part of the reason that rail fences are the most expensive type of fence.

Woven-wire fences. The most common type of perimeter fence for holding sheep, at least for the bottom half, is woven wire. (The woven wire is usually combined with either barbed or electric wires at the top and bottom). This type of fencing keeps the sheep in and helps to keep predators out. It comes in different weights and styles, including a high-tensile version that isn't damaged by ice like the common galvanized wire. High-tensile wire should last for 30 years or more. High-tensile woven wire has the added benefit of being lighter, which is nice on the old bones when you're building a lot of fence. The up-front cost pays for itself over the long haul. Woven wire needs to be stretched taught to perform correctly.

Smooth-wire electric fences. Smooth-wire electric fences are gaining popularity for all types of livestock operations. These fences are relatively easy to construct and do the job well. For sheep, a smooth-wire electric perimeter fence should probably be at least five strands tall — and the sheep must be trained to respect the electric wire before they're first let out.

Because wool acts as an insulator, it's easiest to train your sheep to electric fencing right after they are sheared, but if you have no other choice than to train fully fleeced sheep, wet them down so that their wool no longer acts as an insulator. Once the sheep receive a good shock, they will avoid the fence. Training is best done initially in a small pen that is well secured by panels or regular sheep fence surrounding the electric wire.

High-tensile wire can be stretched very tightly without breaking, and the stretching makes an attractive fence. Low-tensile wire contains less carbon, which makes the wire softer and more prone to breakage. The low-tensile wires should actually be left with a little slack in them, so if animals run into the fence, it has some give — almost like an elastic band. The slack gives the low-tensile wire fencing less eye appeal than the high-tensile variety, but it's quicker and easier to install. The smooth wire for electric fencing that you find

Benefits of Aluminum Wire Fencing

The one type of low-tensile wire we use regularly on smooth wire electric fences is aluminum wire. Although this type of wire is a little more expensive, it's unbelievably light. Another advantage is that it "whistles" with the slightest breeze and "glows" in the dark, so animals (both yours and wild animals, like deer) always seem to know where the fence is. If they aren't able to see the smooth electric wires, deer can pop them off the insulators by running into them, but with an aluminum fence you'll have little problem — the deer will jump the wire if they know where it is.

in regular farm-supply stores, hardware stores, and outlet/discount lumber facilities is generally low-tensile galvanized.

Temporary fences. *Polywire* and step-in plastic or fiberglass posts, or *polynet*, can be used to construct temporary fencing. Polywire is an electroplastic twine that consists of strands of wire twisted with strands of polyethylene fibers. Actually, electroplastic fencing is available in a wire style, a cord style, and a tape style, and there are a variety of types on the market that vary by weight, size, color, and number of strands. Polywire (in all its varieties) doesn't last forever, but the better-quality brands last much longer than the cheaper brands found in discount stores. In our experience, the cheap kinds become virtually useless in 3 to 5 years, whereas the better brands last about 10 years.

Polynet is a great temporary fence alternative for sheep, especially if you want to move them into an area that isn't well fenced on the perimeter. We always "mowed the lawn" in Minnesota — where lawns need mowing at least once a week in the summer — with some of our ewes and lambs in a polynet enclosure. These fences have built-in posts.

Soft steel cable is another alternative that's especially attractive for semipermanent installations, such as lanes and permanent paddock subdivisions. This type of cable costs about the same as polywire or polynet, and it conducts electricity better. However, it kinks and breaks easier if you take it down and put it up often.

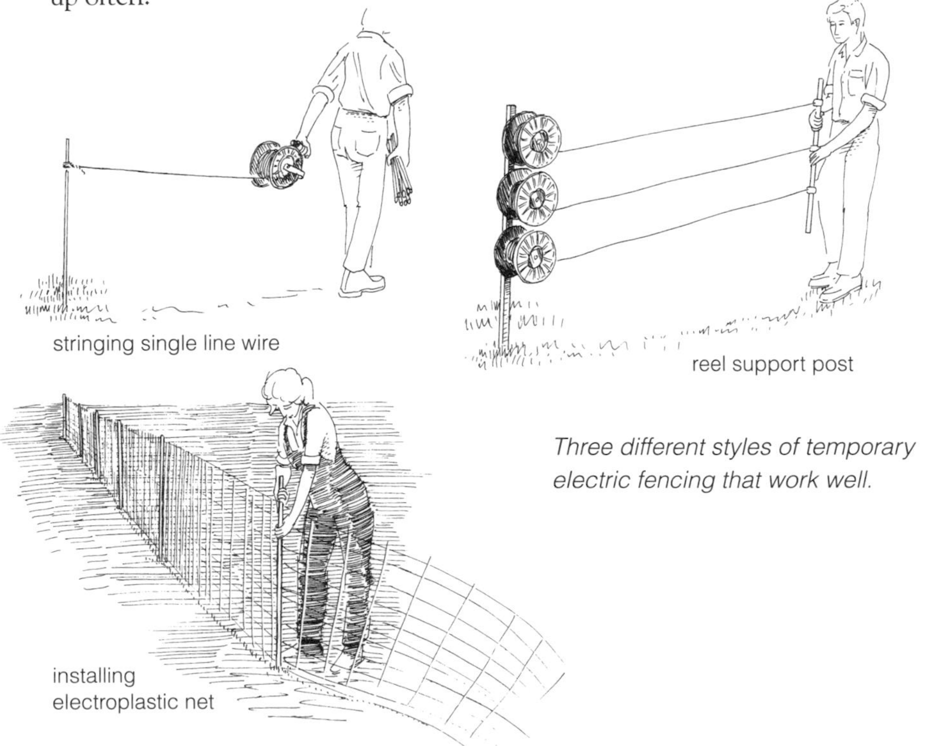

Three different styles of temporary electric fencing that work well.

Posts

Fences are only as good as the posts that hold them up, so it's worth getting the right kinds of posts for the job. Permanent fences can be constructed by using wooden posts, metal T posts, or a combination of the two.

Unless you happen to be blessed with an abundance of Black Locust or Orange Osage trees on your land, it's probably best to buy your posts. These two species of trees make almost-indestructible fence posts that are known to last for decades without any treatment whatsoever. No other North American trees can compete. If you do want to try cutting your own posts from other species, you'll have to cut, de-bark, dry, and treat with chemicals if you want them to last for a long, long time.

When you purchase posts, look for ones that have been pressure treated. The chemical that seems to offer the most protection with the fewest negative side effects is chromated copper arsenate (CCA). No one knows exactly how long a CCA post lasts, but some of these posts have been in the ground since the 1930s and are still doing their job — so your posts will probably outlast you. A note on safety: Although CCA is one of the least toxic preservatives used to treat posts, it's still a good idea to wear gloves when working with treated posts because some people are allergic to the chemical. Please note that if you want to be certifiably organic, you cannot use chemically treated posts.

Metal T posts. Metal T posts can be driven directly into the ground with a fence-post driver. Just don't get your hand between the driver and the post. I can attest to the fact that this is a very painful place for your thumb — and in my case it earned me a cast for 6 weeks. In spite of that caveat, T posts are really great. When building a smooth-wire electric fence, attach the wire to the T post with a plastic insulator.

Fiberglass and plastic posts. Posts that are made of fiberglass or plastic come in a step-in style (our strong preference) or a pound-in style. These posts are great for temporary fences. Fiberglass posts are readily available at farm-supply stores, but sunlight breaks down the fiberglass particles, so after a time you pick up painful fiberglass splinters. High-quality plastic posts are available from several of the catalog suppliers listed in the resources.

Corners and Ends

With the exception of really small enclosures, all permanent fences need to be constructed with well-braced and sturdy corners and ends. A good corner or end post is made with a wooden post that is at least 6 inches (15 cm) in diameter that's braced by another wooden post that is at least 4 inches

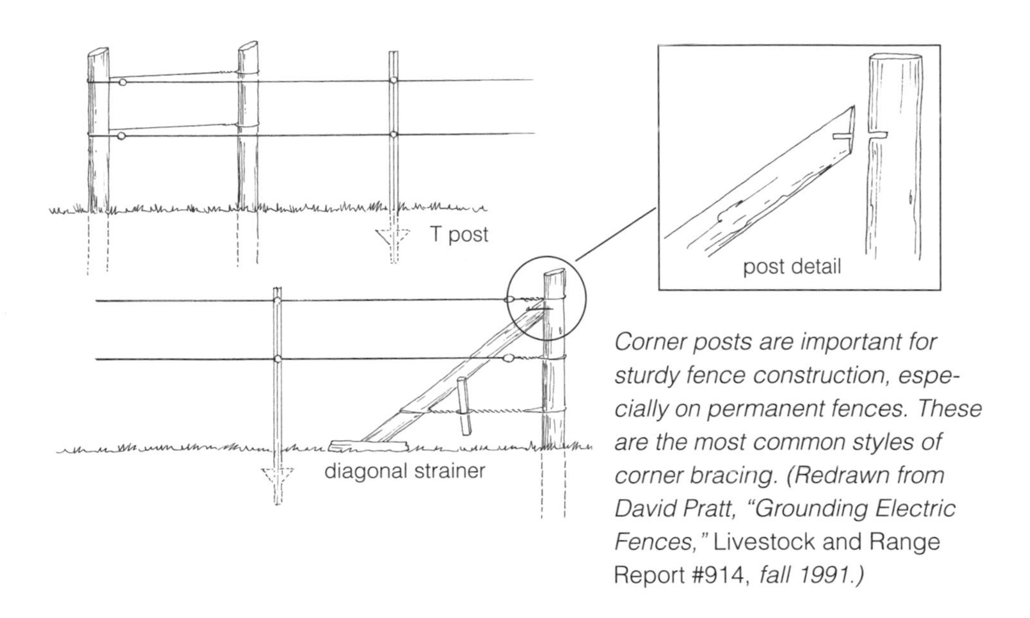

Corner posts are important for sturdy fence construction, especially on permanent fences. These are the most common styles of corner bracing. (Redrawn from David Pratt, "Grounding Electric Fences," Livestock and Range Report #914, *fall 1991.)*

(10 cm) in diameter. For small enclosures, you can get away with unbraced 4- to 6-inch posts, as long as they're deeply buried. Small enclosures can also be constructed with just T posts, and T posts make good corners and ends for temporary fences.

Rules of Thumb for Fence Corners and Ends

Wooden posts that are used for corners and ends should be buried to at least ⅓ of their height, and a little deeper doesn't hurt anything. So an 8-foot post (2.4 m) should have at least 2.5 feet (76 cm) below the surface, and 3 feet (90 cm) is even better. If the post needs to support a heavy wooden or metal gate, it needs to be at least 4 feet (1.2 m) deep to counterbalance the weight of the gate. The soil that you replace around the post needs to be well tamped for compaction. If you live in a really sandy area where adequate compaction is hard to get, there are a variety of earth anchors that can be used to help secure the posts. Some folks place wooden posts in concrete, but that's a bad choice: They spend unnecessary labor and expense, and wooden posts in concrete often rot at the ground level sooner than posts that are simply buried in the soil.

Facilities

There are things that you've got to have to call yourself a shepherd: you, your sheep, some land, and some fences. Everything else — buildings, handling systems, farm equipment, and all the other odds and ends you think you might need to raise sheep — *can be done without!* That's right — you don't have to have a single building, you can get by without any handling structures, and you don't need a whole bunch of fancy equipment. Don't get me wrong, some facilities can make life easier for you and the sheep, and others become absolute necessities if you choose an intensive management approach, like winter lambing. But if your heart's set on sheep, you can have them without having to spend a small fortune on fancy facilities.

So deciding what's really necessary and important on your operation is a matter of choice. The choices are based on your goals. When deciding what you need, keep in mind the following questions:

- What's your style of farming? (Are you trying to make a living as a commercial shepherd, or do you want to keep a dozen sheep for fun and mowing services?)
- How's your financial health? (Do you have an outside job or a big trust fund, or are you relying on your sheep to make a profit?)
- How much time can you spend caring for your sheep? (Is your outside job 10 hours per week or 50? Do you have other obligations that will keep you away from the flock at certain times?)

Natural Behavior and the Design of Handling Facilities

Considering a sheep's natural behavior will aid you in designing good handling facilities. The following list of principles is supplied by the American Sheep Industry Association's *Sheep Production Handbook* (Englewood, CO: ASIA, 1996, p. 212):

- Sheep move toward other sheep and follow one another.
- Sheep prefer to move uphill and toward open spaces.
- Sheep move away from buildings.
- Sheep move better around slight corners or curves where they cannot see what lies ahead.
- Sheep move away from things which frighten them.
- Sheep have legs and move themselves around.
- Sheep do all these things instinctively.

Buildings

On a sheep farm, barns generally meet two needs: storage for feed and supplies, and a place for winter lambing. Therefore, whether you need any buildings at all depends primarily on the time of year you'll be lambing.

For small flocks that lamb in late spring or early summer on pasture, no buildings are necessary. Grain and minerals for a small flock can be stored in large plastic or metal trash cans, which keep moisture and pests (for example, bugs and rodents) out. Remember that if feed is stored in cans, the lids must be fastened *very* securely. If the sheep gain unfettered access to feed, overconsumption can be fatal. Keep the cans securely lidded and out of the sheep's reach. Hay for a small flock can be stored under a tarp. Some folks who do pasture lambing have portable temporary structures or they use tepees in the pasture.

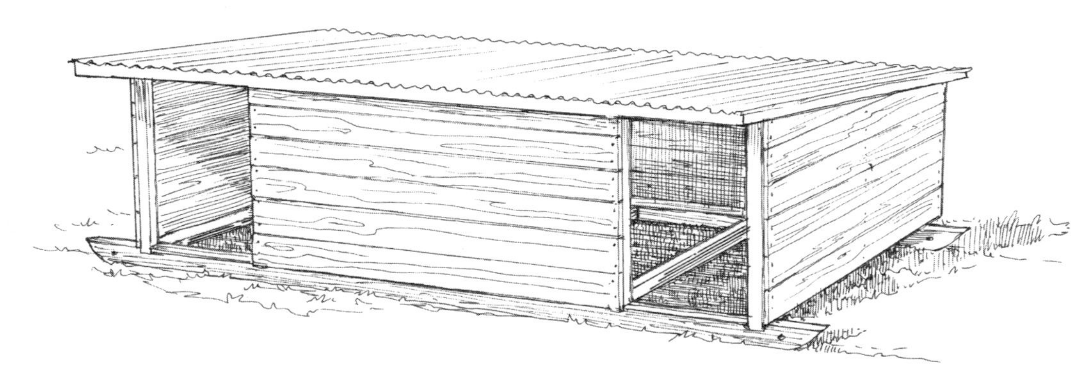

Portable structures provide flexibility and inexpensive shelter for sheep and lambs on pasture. Because sheep only use the shelter in bad weather, up to six ewes and their lambs can share one shelter.

For large flocks that lamb on pasture, a small sheep and lambing shed comes in handy as a place to store feed and supplies and as a place to take care of sick or hurt animals. This type of structure provides flexibility for the shepherd. A design for a small lambing shed is available from the USDA plan service. This design works well for small- to medium-sized flocks that will be lambing during inclement weather.

Old farm buildings can often be remodeled to meet the shepherd's needs, or inexpensive, alternative types of buildings are gaining acceptance. For example, shepherds are beginning to use "hoop" houses (which are like a greenhouse made with plastic sheeting) or straw bale structures instead of a conventional building. (Whether you're thinking of constructing a new building or adapting an old one, check the resources for more information.)

Sheep and lambing shed for 30–36 ewes

This small sheep shed can be convenient for small-flock owners and even works well for larger flocks that are lambing on pasture. It has a separate room for feed and supply storage, as well as lambing pens, a creep-feeder area, and an open area for sheep to feed at a feed rack. (This shed is based on USDA plan #5919, which can be ordered from your county Extension agent or from the Superintendent of Documents, Government Printing Office, Washington, D.C.)

Lambs artificially reared in a greenhouse barn at Old Chatham Shepherding Company (Chatham, NY). Note the nipple buckets, which make feeding milk replacer to large numbers of lambs relatively easy. (Photograph by Elizabeth K. Luke)

Jugs

Ewes lambing for the first time may be nervous or confused because of their lack of experience or underdeveloped maternal instincts. They should be alone in the jug with their lambs for at least 3 days until they become accustomed to the nursing lambs. Mature ewes may need to be alone with their lambs for only 1 day. After you're confident that the ewe has bonded with her lambs, she can be sent to a pen with other ewes and their new lambs. The size of the groups of ewes depends on how long it's been since they lambed.

Some shepherds have their ewes lamb in the jug, while others use it immediately after the ewes have lambed. Ewes prefer a larger area for the actual lambing, where they can walk around freely before labor. The one advantage of lambing in a jug as opposed to lambing on pasture is that it provides a confined space if help is needed for a difficult birth. In addition, good light is available for watching the ewe's progress. Because of the trend toward larger sheep, recommendations for jug size have been increasing. The larger pen is definitely best if you want to have the ewe confined in the jug for lambing.

The jug allows the ewe and lamb to bond without distraction, keeps the lamb from getting separated from its mother (especially in the case of twins or triplets where the ewe cannot count beyond "one"), and protects the lamb from being trampled by other sheep or becoming wet and chilled. Ordinarily, the new family is penned together for up to 3 days so that they can be easily observed and treated if complications arise. Do *not* allow dogs or strangers to approach the jug area, especially with nervous ewes. Frightened or nervous ewes can quickly turn a serene, protective environment into a "lamb blender," with fatal results.

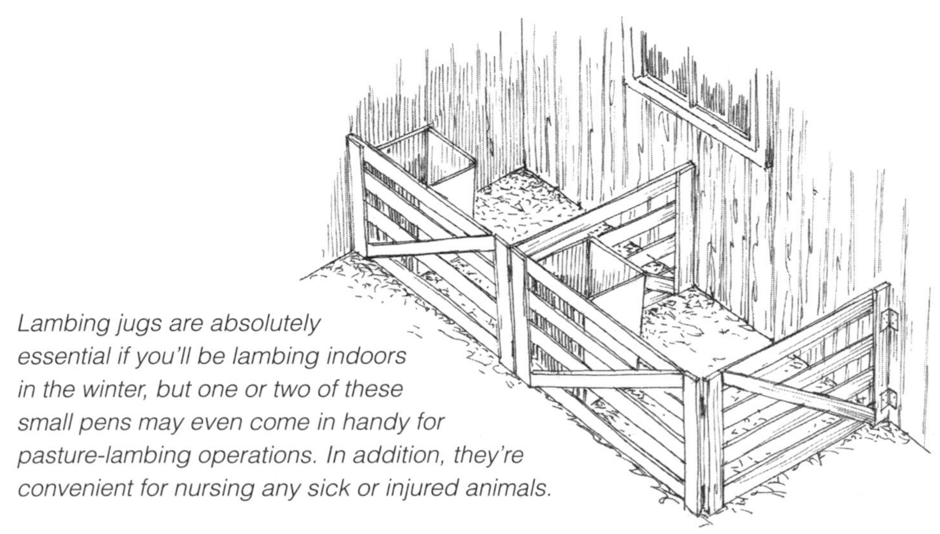

Lambing jugs are absolutely essential if you'll be lambing indoors in the winter, but one or two of these small pens may even come in handy for pasture-lambing operations. In addition, they're convenient for nursing any sick or injured animals.

If the ewe lambs outside, it is not difficult to get her to the jugs nearby. Carry the lamb slowly, close to the ground so she can see it and follow. Since lambs do not "fly," the ewe will instinctively look for the lamb on the ground. If the lamb is raised more than a foot or so off the ground, the ewe may "lose" it and run back to where she dropped it. When this happens, you will need to go back and begin again. If the lamb calls out to the ewe along the way, she will normally follow readily. There are commercial "lamb cradles" and "lamb packers" available, which allow you to carry the newborn lamb inches off the ground as if it were a suitcase. Using lamb packers has the advantage of being easier on your back, with less distraction to the ewe from your humped-over appearance. From the ewe's viewpoint, it appears that the lamb has suddenly begun to follow you and she will instinctively follow it.

Jug Requirements

When lambing is going to take place in a barn, you need to have *jugs* (or lambing pens or "claiming pens") ready for the newborn lamb and its mother. The pens should be prepared with:

- Clean bedding
- A small hay feeder
- A container of water that cannot be spilled; a plastic 5-gallon (19 L) pail is ideal — but remember, to avoid drowning, the bucket must be tall enough and bedding must not be built up around it.

As a general rule, you need approximately one jug for every ten ewes in the flock. If you own a small flock, you should be prepared to have at least three jugs, and each should provide at least 24 square feet of floor space for small breeds and 40 square feet of floor space for large breeds (2.2–3.7 square meters, respectively).

Consider the lambing-barn environment. A healthy barn is clean, dry, and free of drafts but not warm. Drafty or warm barns can cause pneumonia in young lambs and sometimes in ewes. A closed barn without proper ventilation allows ammonia from fecal decay and urine to build up, which can irritate eyes and lungs, predisposing an animal to pneumonia and respiratory disease.

There are two approaches to maintaining a healthy barn environment. In the first method, the barn is cleaned out each day and a small amount of lime and fresh bedding are placed on the floors. The second approach is called the deep-bedding method, and it's the one we prefer when any animals are kept in a barn. It not only provides a good environment for the critters, it cuts down on daily chores.

Maintaining a Healthy Barn Environment: The Deep-Bedding Method

Build up an 8-inch (20 cm) thick layer of bedding (straw, wood shavings, sawdust, shredded paper or newsprint, or dried leaves all work). Once every day or so, clean the dirtiest spots off the surface of the bed, and then add just enough fresh bedding to the surface to create a clean, dry environment. Once every year or two, you'll need to clean out all the bedding down to the ground and start again. The bedding you remove is already partly composted, and if you pile it up outdoors and let it compost for 1 more year, you've got black gold for your garden. Sweden has some of the toughest animal-protection laws in the world, and Swedish farmers are required to use deep bedding for animals kept indoors.

Handling Facilities

Handling facilities are a wonderful resource for shepherds with small flocks and an absolute necessity for large flocks. They allow you to gather, sort, perform medical procedures, and shear your flock with a minimum of aggravation and with less chance of injury to you or your sheep. These facilities don't have to be extravagant to be effective, but if you do invest in them, you'll be happy you did.

Well-designed handling facilities consist of a gathering pen, a forcing pen, chutes, and sorting pens. All pens should be designed so that there are no sharp corners or right angles, and they should have at least one gate that is wide enough to drive a vehicle or tractor into. Long, rectangular pens with curved ends work better than circular pens, though either design will do.

Handling facilities can either be purchased as prefabricated panels that are connected to each other or permanently constructed on site. The prefabricated models have the advantage of being portable, relatively lightweight, and flexible to meet changing needs (you can always buy more panels as your flock size increases), but they are rather expensive. The type that is constructed on site doesn't provide the same degree of flexibility, but it can be constructed for less cash outlay, especially if you use recycled building materials.

The gathering pen should be large enough to accommodate all the sheep you'll ever have at one time, with lots of room to spare. We keep water tanks and salt blocks in gathering pens, and feed treats there, so the animals are accustomed to going right in. When we need to catch them, we simply place some treats out, wait until everybody's in and chomping happily, and then close the gate behind them. In a home-constructed system, the gathering pen can be made out of woven wire, seven strands of smooth wire, or rails. A good size for gathering pens is 5 to 6 square feet (about 0.5 m^2) per mature sheep and 3 to 4 square feet (about 0.3 m^2) per feeder lamb. If the gathering pen is 12-feet (3.7 m) wide or narrower, you can reach sheep on either side of you with a crook when you're standing in the center. Typically, the outer walls of all pens are 3.5 to 4 feet (1–1.3 m) tall.

The forcing pen is used to confine smaller groups, and when necessary, to force them into chutes. For small flocks, forcing pens can also be used for collecting sheep for shearing or for placing them in sheep chairs for hoof trimming. The forcing pen should have solid sides, and for on-site construction can be made out of plywood, metal, or boards. The forcing pen should be heavy-duty, because the pressure of the sheep against the sidewalls can break down a poorly constructed wall.

Chutes are used for medical treatment and sorting. The chute should also have solid walls, and it should be narrow enough to ensure that the sheep enter single file. Gates can control ingress and egress from the chutes. For small

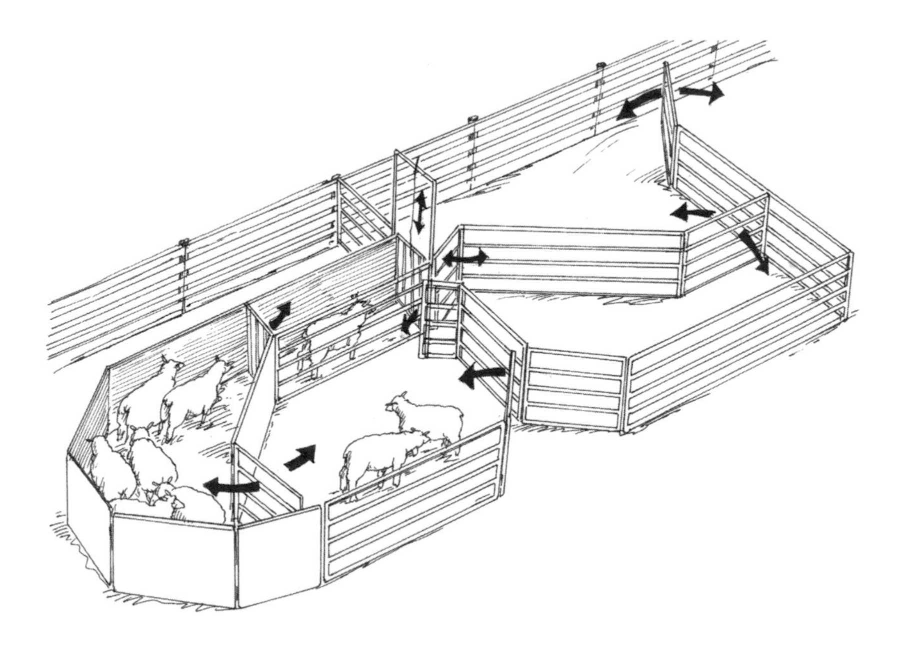

Good handling facilities can make your life a lot easier and hassle free. This design has several different holding pens for sorting animals, and gates that move in either direction. By placing water tanks and salt, or treats, in holding pens on a regular basis, the sheep will be motivated to move in and out of the pens and are then fairly easy to catch in the pens when you need to work with them.

flocks, a 15- to 20-foot- (4.5–6.1 m) long chute is adequate; owners of large flocks (more than 150 head) can benefit from increasing the chute length to 15 feet (4.5 m) per hundred animals. Like the forcing pen, chutes should be solid sided, but a 4-inch (10 cm) gap at the bottom allows air to circulate through the chute. The sidewalls of chutes can be 3 feet (0.9 m) high for most sheep, though for especially tall breeds, increasing the height of the walls slightly may be advisable.

Sorting gates and pens are designed to ease the job of dividing the sheep into groups. For example, running the flock through the system when it's time to wean the lambs allows quick separation of ewes and lambs. The sorting gates should be lightweight and easy to use but strong enough to stop oncoming animals. Although gates can be made of wood, wooden gates are heavier and slower than steel or aluminum gates. The sorting pens can be constructed like the gathering pen, but internal fences that separate sorting pens can often be shorter — say 3 feet (0.9 m) tall, which allows you to cross between the two pens by hopping the fence.

Scales

For larger flocks, incorporating a scale into the handling facilities may be advantageous. Scales allow you to track production, assess feeding programs, and assure honesty in transactions. A good scale built for use in a handling system is expensive, running in the thousands of dollars.

For small flocks, lambs can be weighed by making a sling out of plastic or burlap and using a hanging scale. Though hanging scales are economical, they are really only practical for weighing younger lambs. Some hanging scales are capable of handling the weight of larger animals, but for the shepherd, getting a full-grown ewe into a sling and then onto the scale is about as easy a job as Atlas had, lifting the weight of the world. It may be possible, but it's probably not something you really want to do.

Shrinkage

Sheep weigh a little less upon arrival to a sale barn or butcher than when they were first shipped. This is caused by shrinkage, and the extent to which shrinkage occurs depends on how far they've been transported, the weather, general stress, whether feed and water were readily available, and other similar issues. Shrinkage may be as high as 10 percent, but in most cases it averages about 3 percent.

Estimating Weights

Don't have a scale to weigh adult sheep? Try this simple method that gives a good estimate of weight:

1. Measure all the way around the sheep's body with a tape measure just behind the front legs. Measurement = C. (If the sheep is in full fleece, part the wool so that the measurement is accurate.)
2. Measure the length of the body from the point of the shoulder to the point of the rump. Measurement = AB.
3. Multiply (C x C x AB) and divide by 300.

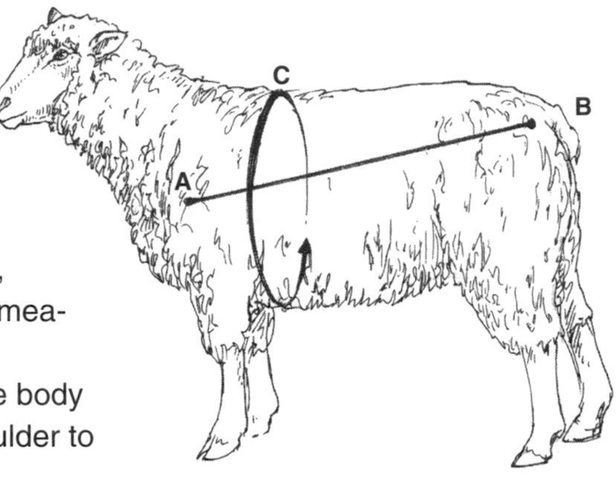

If you don't have a scale, you can estimate an animal's weight using the technique described here.

Restraining Devices

Restraining devices make it easier on you to handle sheep for medical purposes and foot trimming. For a very small flock — say less than a dozen animals — you can probably get by without any restraining devices, but as the number of animals increases, the value of these tools increases exponentially. These devices run the gamut from high-dollar turning cradles that can be built into a handling facility to inexpensive gambrel restrainers.

Turning cradles and freestanding tilt tables are well suited to larger operations. In fact, with prices that run up to $1,000, they're impractical for smaller flocks. A really nifty invention that hasn't been around as long as turning cradles or tilt tables, the sheep chair serves the same purpose for smaller flocks. Sheep chairs take advantage of the fact that sheep can't right themselves when they're on their rump (the same position that's used for shearing). The chair hangs over the edge of a panel and adjusts to comfortably fit all sizes of sheep. Best of all, at less than $100, it is a viable option for shepherds who have smaller flocks.

A turning cradle, or tilt able, is commonly used for foot trimming and veterinary procedures, but is somewhat pricey for a small flock.

A more economical approach to restraint for small-flock owners than the turning cradle, the sheep chair immobilizes the sheep when "sitting," which makes giving shots and trimming feet surprisingly easy.

A gambrel restraint is a plastic device that was invented in New Zealand. At less than $20, this is the least-expensive restraining device and is a good device for owners of really small flocks. The gambrel restraint controls the animal by immobilizing its front legs. The bad news is that this restraining method can be tedious if you're doing something to a bunch of sheep at the same time, for example, pregnancy checking the whole flock.

Feeding Facilities and Equipment

The only special feeding equipment that you really need for small flocks is a few rubber feed pans for feeding grain and treats and a slightly larger rubber pan for water. These pans are readily available from farm-supply stores and catalogs, are nearly indestructible, and are really inexpensive. If the water freezes in the pan during the winter, just turn it over and stomp on it.

As flock size increases, you can invest in a variety of specialized feed equipment that makes it easier for you to feed and more equitable for the less assertive sheep in your flock. There are many designs for feed troughs and self-feeders. *The Sheep Housing and Equipment Handbook* of the Midwest Plan Service is a good source for plans (see Resources).

Creep-feeders provide the opportunity for lambs to enter and eat all they want, but ewes cannot enter because of the size of the openings. The creep

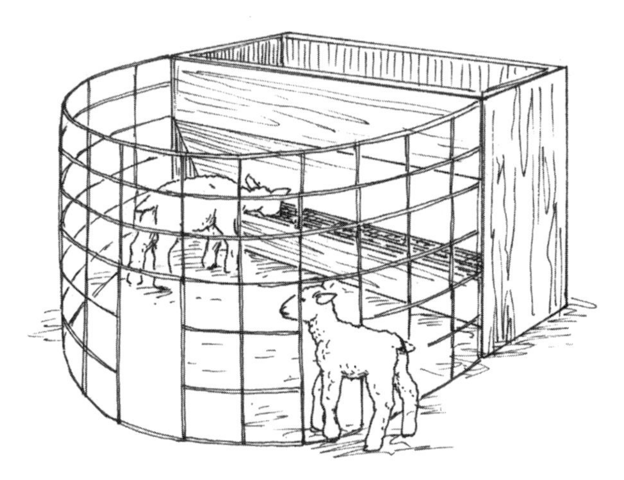

Creep feeders allow lambs to enter and get extra feed, while keeping larger animals out. This inexpensive design comes from MidWest Plan Service (see Resources) and is fairly inexpensive to make.

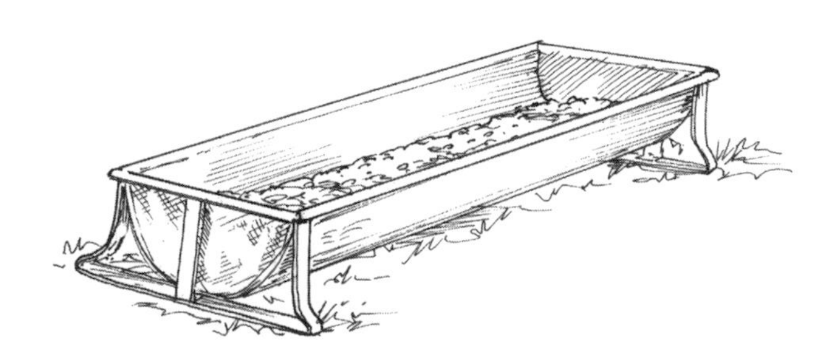

Using a feed trough for grain saves feed and cuts down on parasite problems. Feed troughs can be made at home or bought in a store and, happily, don't cost a fortune.

should be sheltered, with good, fresh water provided daily, and it should be well bedded with clean hay or straw. The heavy stems of alfalfa that are left uneaten in the ewes' hayrack make good creep bedding. If the creep is in the barn, it should be well lighted, because that is the lambs' preference and they will eat better. Hanging a reflector lamp 4 or 5 feet (1.2–1.5m) above it will attract the lambs. They can start using the creep when they are about 2 weeks old.

Farm Equipment

A shepherd doesn't have to have any tractors, planters, harvesters, or other heavy farm equipment. Tractors and implements are expensive, require lots of maintenance, and depreciate. Given time, they become worthless piles of rusted metal, parked along a fence line or off in the woods. Instead of purchasing farm equipment, lease machinery when you really need it or contract with a neighbor to do the work for you.

FOUR

HERDING DOGS

Sheep have a relatively strong desire to group together in response to a threat, real or perceived, and that trait makes them well suited to being worked with dogs. How useful a dog can be to a shepherd depends on many things, especially:

- The quality and training of the dog
- The suitability of the farm to a working dog, for example, how paddocks, fencing, and handling facilities are arranged
- The shepherd's understanding of the dog's ability
- Most important, the shepherd's willingness to work with a dog

A well-trained stock dog can be an enormous help to you as a shepherd. It can greatly reduce the amount of equipment required for sheep handling. The dog can drive stock from one pasture to another, load one sheep (or hundreds) into corrals or stock trailers, or work with you as you operate a squeeze chute for pregnancy checking, shearing, or worming. A good dog can single out one animal without moving the entire flock to a sorting facility. During lambing, it can help bring in expectant mothers. A herding dog can also help you count your livestock by filtering them along a fence and help you at feeding time by keeping them away from the feeders while you're spreading grain or hay. You will truly know the value of a herding dog when it comes to escaped sheep — it regathers them easily, whereas without a dog, you could have serious trouble capturing the wayward sheep. In short, a good stock dog can be more help than several human helpers, as the effect of the dog on the stock and its physical abilities differ greatly from those of a person and are generally more useful than a second pair of hands.

While some herding dogs may act to a certain extent to protect their sheep, they should not be expected to function as livestock guardian dogs. (See chapter 5 for more about guardian dogs.) In fact, a herding dog shouldn't be left unsupervised with the sheep. Because it's been bred and trained to move and control stock, if left unsupervised, it is unlikely that the dog will allow the flock to remain "uncontrolled" for long. This is especially true of young herding dogs, which need very close supervision. Luckily, as most herding dogs mature, they learn when their intervention is required and when they are "off duty."

Traits of Herding Dogs

A well-trained, well-bred dog allows one person to manage a very large flock with minimal physical effort and aggravation, but any dog that has the respect of the sheep may be of some use in the management of a flock. For example, an untrained dog, regardless of breed, that barks and chases after the sheep can still be of some assistance in pushing them into chutes or squeeze gates but may need to be restrained on a leash.

With some training, almost any dog that has a desire to "go after the sheep" can be taught to be of assistance in moving sheep around. However, a nonherding breed of dog thus encouraged to go after the sheep must be carefully watched. Such a dog, unrestrained by hundreds of years of selective breeding, may be motivated solely by a predatory instinct and could become a threat to the sheep. Indeed, any dog that's encouraged to work the sheep has been given permission by the owner to interact with them and should be supervised or restrained when not working. In many breeds, there is very little separation between the instincts to chase and play, hunt the livestock, and herd the livestock.

The amount of assistance the dog provides the shepherd is determined by the natural ability of the dog to work livestock and how well it has been trained. When a dog is being acquired to work livestock, the best policy is to buy one from bloodlines of the breed most often used in your area for herding.

Characteristics of Herding Dogs

Herding dogs have been bred for hundreds of years to accentuate certain qualities of assistance to the shepherd. There are accounts of dogs used to herd sheep at least as far back as the sixteenth century. These qualities and instincts are generally lumped together and referred to as "herding instinct" or "working ability." This inbred instinct should include the dog's having some idea of what to do with the stock. The dog should instinctively want to keep the sheep in a group and to head off any sheep that tries to escape from the group. The dog should also have a desire to work with the shepherd, to accept commands while herding the sheep, and be willing to curb its desires in order to respond to the wishes of the handler.

Breeds

Many of the traditional herding breeds are now most often bred for the showring or as pets. This has resulted in a loss of herding qualities in some of these breeds. If a shepherd already has a dog of an old "herding" breed and it shows an interest in working the sheep, the dog may be taught to provide some assistance to the shepherd.

Herding breeds are often divided into three very general types: gathering dogs, tending dogs, and driving dogs. Any good herding dog, regardless of the

SHEPHERD STORY

Betty Levin raises sheep in Lincoln, Massachusetts — once a farming area but now a bedroom community for Boston. Betty raises Border Leicester sheep for their wonderful fleeces, working with a flock that she's been breeding for more than 25 years. Betty has used wool quality and ease of lambing as the primary criteria for selection in her breeding program.

Keeping sixty sheep on a limited amount of pasture and working alone (at a youthful 70 plus) leaves little doubt in Betty's mind that she couldn't possibly keep her flock without the assistance of her dogs. Her sheep graze small pastures along roads and in vacant fields around neighbor's homes, fenced only with temporary electric fencing. This provides maximum use of a small amount of available grazing, but requires frequent movement of the flock. Because of her managed-grazing program, Betty frequently sorts her sheep into different size groupings, to fit the available paddocks. She does all of this work alone aided only by her dogs.

Once the ewes are heavy in lamb and carrying their valuable fleeces, they must be handled carefully to protect her entire year's investment. The ewes are brought into the barn daily to be fed and examined for signs of approaching lambing. Any ewes that are close to lambing are sorted into small groups and held close to the barn for frequent checking, while the remaining sheep are moved up the hill for the remainder of the day. These jobs require quiet dogs that can work close to the sheep in the barn and the small pens without upsetting the pregnant ewes. This proximity to the flock also protects Betty from being trampled by overeager eaters and aggressive new mothers.

As much as Betty relies on her dogs to aid her in her daily chores, she also finds great joy in raising and training talented, young Border Collies. After 30 years of raising and working with herding dogs, Betty finds great satisfaction in seeing a good dog doing the work for which it's been bred for generations.

type, should be able to do all of the herding-related work on a farm. That is, gathering dogs should be able to drive, and driving and tending breeds should be able to gather.

Gathering Breeds

Dogs like the Border Collie, which were originally bred to work in wide spaces, are considered gathering dogs. Such dogs were bred to gather semiwild sheep off large, open pastures. While the Border Collie is the preeminent gathering breed, there are also some excellent Kelpies bred in Australia and the United States that are still working. In sum, gathering-dog breeds are:

- Border Collies
- Kelpies
- Australian Shepherds
- Collies
- Bearded Collies

Border Collies

In North America, the only breed that is still bred primarily for its herding instinct and working ability is the Border Collie. Because of this breeding, Border Collies are often the only choice for a herding dog. This selection for working traits, as opposed to being bred for the showring or pet market, has given these dogs other attributes beyond the simple herding instinct. For example, a natural, inherited ability to gather sheep out of a field or an innate sense of the distance to work off the sheep to keep them quiet are still traits often demonstrated by the Border Collie.

A Border Collie at work (Courtesy American Sheep Industry Association 2000)

Tending Breeds

Dogs that are considered tending breeds are those that were developed in Europe to help in the grazing of sheep in areas around crops. Dogs of this type customarily took their sheep out to graze each day and then patrolled along the grazing area to keep the sheep restricted to the unfenced space that they were supposed to graze. The following dogs are considered tending dogs:

- Belgian Malinois
- Belgian Sheepdogs
- Belgian Tervurens
- Bouvier des Flandres
- Briards
- German Shepherds
- Beauceron Pyrenean Shepherds
- Pulis

Driving Breeds

The breeds that are now known as driving breeds were originally developed to help drovers move sheep to market along open lanes or for use in stockyards. The New Zealand Huntaway is a driving breed that's used today in New Zealand to drive very large flocks of sheep by barking and moving back and forth behind the sheep. Other common breeds of this type are:

- Rottweilers
- Welsh Corgis (Cardigan or Pembroke)
- Old English Sheepdogs
- Australian Cattle Dogs

Selecting a Dog

There are two basic approaches for obtaining a herding dog: Start with a puppy, or buy a mature, trained dog. The puppy route is, or course, less expensive up front, but the puppy won't be ready to work for quite some time, will require intensive training, and may never work out well. By purchasing a mature, trained dog, you can be sure that the dog knows its stuff, but be prepared to pay significantly higher prices for a trained dog than for a puppy

Starting with a Puppy

When selecting a puppy as a future working dog, it's impossible to assess its natural herding ability. Thus, choosing a puppy is really a process of selecting the parents that are most likely to produce a puppy that suits your special needs.

In selecting the parents, your primary consideration should be given to their working ability. Both parents should be seen working the type of stock that the puppy will be expected to handle. At the very least, both parents should be able to:

- Gather a group of sheep a few hundred yards away from the shepherd and fetch them to the shepherd in a quiet, controlled manner
- Hold the sheep in a group for the shepherd
- Be able to single out an individual sheep and control it without the use of force or excessive aggression
- Move the sheep without difficulty or the use of force
- Demonstrate their ability to move the sheep in the way that would be appropriate for ewes during their last weeks of pregnancy — gently but firmly

The prospective buyer should interview the dog breeder to learn about his or her breeding program. Learn what traits are being bred for and what type of puppy the breeder hopes to produce. While breeders cannot guarantee the herding qualities of the pups they produce, they can at least discuss why they have bred two particular dogs and what they hope the puppies will turn out like. There are plenty of top working dogs around. Do not settle for a puppy whose parents don't demonstrate the qualities that you require.

It is important to feel comfortable with the breeder. The breeder should be willing to provide the names of prior customers and inform the buyer how previous puppies have performed.

Both parents of the puppy should have had their hips x-rayed to determine if they have canine hip dysplasia and ideally should have a rating of their hips from the Orthopedic Foundation of America. Dogs with any sign of hip dysplasia should not be bred. Both parents should also have had their eyes examined by a veterinary ophthalmologist and have been certified as clear of progressive retina atrophy (PRA) and collie eye anomaly (CEA). Both of these conditions are hereditary eye disorders that can lead to some sight loss and even blindness. Only a veterinary ophthalmologist can determine that a dog is free of these disorders.

Always keep these health concerns in mind when selecting a puppy. Regardless of a dog's talents, it is useless if it is physically unable to perform the job for which it was bred.

Disease Control in Puppies

All puppies should have been routinely treated for hookworm and roundworms from the time they were 3 weeks of age. All pups should have received their first inoculation for a variety of highly contagious diseases, including canine distemper, parvovirus, leptospirosis, coronavirus, hepatitis, and various respiratory diseases. The inoculation given for these diseases before the pup leaves the breeder is just the first of a series of inoculations needed at 3- to 4-week intervals until the puppy is 16 weeks old and full immunity is reached. At 16 weeks, the puppy requires a rabies inoculation.

Training

Initially, puppies should be taught good manners: to come when called, to walk on a leash, and to lie down on command. Formal, herding training can't start until the dog "begins to work," or "turns on to sheep." Most well-bred herding dogs begin to show a desire to work sheep between 8 weeks and 12 months, but it's important not to leave the puppy unsupervised with the sheep. The puppy could be hurt by the sheep, or the sheep could be hurt by the puppy; it could also learn bad habits by working unsupervised. The best age to start formal training is usually between 10 and 12 months. This is because the first instinct exhibited is the desire to get to the head of the flock and turn back the escaping sheep, but this can be hard to accomplish until the puppy is fast enough to catch the sheep. Dogs that begin training before they can outrun the sheep quickly learn to chase them rather than to try to turn them back to the shepherd — a very bad habit indeed.

If the shepherd plans to do anything with the dog other than just let it chase the sheep, then the dog is going to need to develop instincts for tactics beyond chasing. The quality of the other instincts and the cleverness of the trainer ultimately determine the quality of the dog.

Once the young dog begins to work the sheep, he should be encouraged to keep them in a group. This is also a good chance for the shepherd to help the dog "break" the sheep. The sheep need to learn to respect the dog and learn to

Basic Training

When the dog has learned to hold the group of sheep together, it is possible to begin to back away from the sheep with the dog on the opposite side. The dog should be encouraged to walk up to the sheep and move them to the handler. This is called fetching or wearing. Much of the beginning training of gathering dogs is based this technique. The young dog should learn to move the sheep with purpose, but quietly and not too fast. The dog should also learn to stop and walk up on command. Ideally, early training of a young dog is best begun in an area with the following characteristics:

- Small and round
- About 100 to 300 feet (30.5–91.5 m) in diameter with no corners
- Fenced well enough to prevent the sheep from feeling that they can escape from the dog

Once the dog has learned to go around the sheep and keep them in a group it is time to move to a larger area.

move away from it. Young pups that don't know how the sheep are supposed to respond may require your help to move the sheep along — at least until it gets the idea of what's required. This is especially true if you're also using "unbroken" sheep that don't know how to respond to the dog. If possible, training a pup is easier to do with a flock that's used to being worked by a dog. But if you're locked into a new pup and unbroken sheep, the best solution may be to ask a friend or acquaintance with a mature working dog to come over and help break the sheep for the new dog. If help is not available, take ten or fifteen young sheep, put them in a small pen, and use them for the beginning training of the pup.

Once these early, basic lessons are learned, it's time to move on to directional commands, the outrun, and driving. All of the dog's training should be done with dog and the sheep close to the shepherd. Once lessons are learned, they can be perfected at a greater and greater distances from the shepherd.

It is important not to expect too much from the young dog at the beginning of training. Allow him to gain confidence in his ability before expecting him to move difficult sheep. Patience at the beginning of training will be more than repaid at the end with a powerful dog that can walk up to any sheep and be confident that the sheep will move away.

A well-bred herding dog should learn the early lessons quickly, and by the time the dog is 12 or 13 months old, it should already be a useful helper on the farm. Remember though, the dog is young — be ready to offer assistance as needed until all of the jobs are learned and the dog has matured.

There are many expert trainers of stock dogs that offer lessons and clinics to assist beginners in training their dogs. It is possible to attend such lessons almost anywhere in the United States as either a student or an observer. A few such lessons are usually a big help for shepherds with their first dogs.

Buying a Trained Dog

There are always trained dogs available for sale. The level of training varies from "started" to "fully finished."

A started dog will generally gather sheep at about 200 yards (182.8 m) and bring them to the handler, requiring few commands to do so. The dog will stop on command and walk up to the sheep on command.

A dog that is fully trained to the "open level" of trials (competitions) should be capable of placing "in the money" at an open trial of fifty to sixty dogs. Such a dog is able to gather sheep at any distance (in some cases up to half a mile [0.8 km] or more away). The dog should be able to drive sheep in a controlled manner several hundred yards away from the handler. A fully trained dog should also be able to shed and control a single sheep.

Traditionally Used, Basic Commands

Like most specialized skills, training and handling herding dogs comes with its own unique vocabulary. Some of the common commands used in working with herding dogs include:

- **"Come-by."** Travel around the sheep in a clockwise direction.
- **"Way-to-me."** Travel around the sheep in a counterclockwise direction.
- **"Down."** Stop moving, either stand or lay down.
- **"Walk-up."** Move toward the sheep.
- **"Look-back."** Leave the sheep and go look for another group of sheep elsewhere.
- **"That'll do."** Stop working and return to the handler.

The prices of started and trained dogs vary according to the quality of the dog, the level of training, and the part of the country. Dog trials are excellent places to inquire about available trained dogs and to see the actual work of the dogs that are for sale.

Dog Trials

During the late nineteenth century, sheepdog trials became a popular activity among European shepherds and farmers. The trials in the British Isles developed quite a bit differently from those in Europe. Most dog trials in the United States are based on the British model.

While many excellent stock dogs never compete in sheepdog trials, such trials remain an excellent place to see a variety of dogs at work and to learn more about the breeds from the handlers and breeders that frequent the competitions. A good trial will show off a variety of dogs with different styles of working and different levels of training. Most regions of North America have Border Collie organizations that host competitions. In addition to Border Collie trials, there are now dog trials for other herding breeds sponsored by the American Kennel Club or the American Herding Breed Association. (See the resources for contact information.)

British Sheepdog Trials

A series of dogs are run one after another over a set course of obstacles, using three to five sheep (or cattle) from the same flock (or herd) to test each dog's abilities against the others. The shepherd commands the dog to do the work while remaining in a distant place. The shepherd is only allowed to assist the dog for the final penning, or *shedding*, as it is called in trials, of the sheep. Continental trials usually use a larger flock of sheep and allow the shepherd and dog to move the sheep together, using natural features of the terrain to test each dog's ability to control the movement of the sheep.

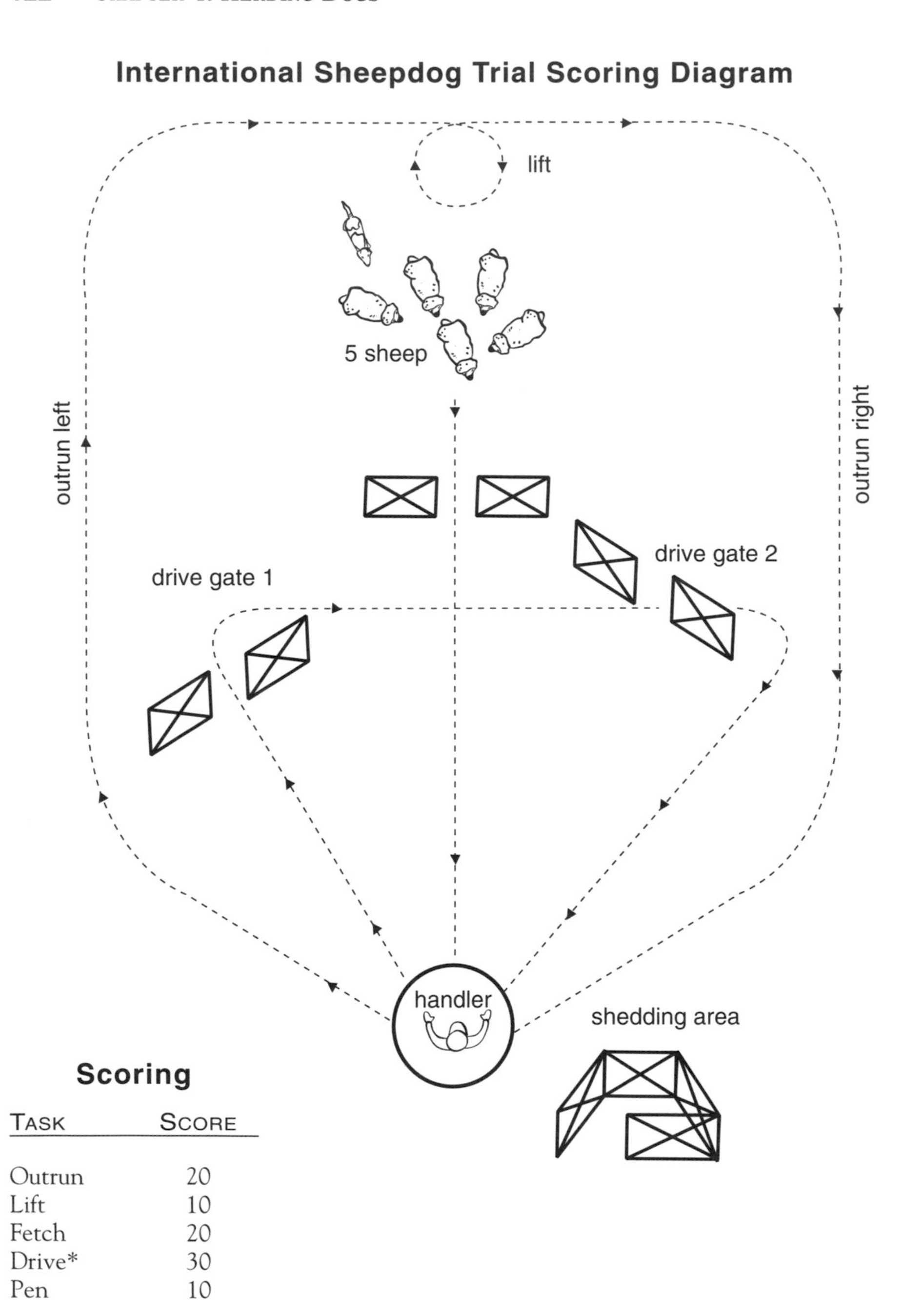

Scoring

TASK	SCORE
Outrun	20
Lift	10
Fetch	20
Drive*	30
Pen	10
Shed	10

**15 points are awarded for each gate.*

Terminology

Folks who work with herding dogs share their own lexicon. Here are some of the terms you may hear when attending trials or visiting with trainers:

- **Balance.** The place the dog needs to be in relationship to the flock in order to control their movement.
- **Distance.** How far off the sheep the dog is working; this varies greatly from dog to dog and flock to flock; it is a part of balance.
- **Driving.** Moving the sheep in any direction except toward the handler.
- **Eye.** Herding dogs can be divided into two general types: dogs with "eye" and dogs without "eye"; eye is how a dog "stares down" the stock — most Border Collies and some Kelpies have eye and work their stock from a distance by stalking.
- **Fetch.** Bringing the sheep to the handler.
- **Gather.** Going out and finding the sheep, gathering them into a flock, and fetching them to the handler; three related terms are *outrun, lift,* and *fetch.*
- **Lift.** The dog's first contact with the sheep on a gather, when the sheep determine how they will react to the dog.
- **Outrun.** The dog leaving the shepherd and running out around a flock of sheep; the outrun should be wide enough to find all of the sheep the dog needs to retrieve without chasing any away, but not so wide that time and energy are wasted.
- **Power.** The amount of influence the dog has over the sheep without physical contact; some dogs are more intimidating to sheep than others.
- **Style.** How the dog moves; stylish dogs seem to be stalking — they keep their head down low and their eyes focused but maintain the speed needed to keep on the sheep if they try to break.
- **Unbroken sheep.** Sheep that have never been worked with a dog; such sheep may run away in panic or attempt to stand and fight the dog.
- **Wear.** Fetching sheep to the handler while the handler walks away — a sort of parade of handler-sheep-dog; this is a useful exercise for training young dogs and for moving sheep around the farm, as the sheep readily learn to follow the shepherd in response to the presence of the dog.

FIVE

PREDATORS AND PROTECTION

Predators are a potential problem for all shepherds. Although coyotes kill more sheep than dogs do, predation by dogs actually impacts more shepherds. Bears and wildcats can also create nightmares for a shepherd, and occasionally birds of prey (eagles and hawks) and carrion birds (vultures and ravens) are the culprits.

According to the American Sheep Industry Association, predators in the United States killed about 368,000 sheep and lambs during 1994, representing a loss to their owners of close to $18 million. It is estimated that coyotes were responsible for about two-thirds of those losses.

Managing for Predators

Not all "predators" actually kill sheep, and predators are important members of the food chain, creating a balance that nature depends on. Predators keep populations of wild herbivores, such as deer and elk, from overpopulating their ecosystems, and they feed on lots of small rodents and rabbits. They'll also eat insects and carrion, which is often abundant along highways. When they do kill, they aren't trying to ruin your day, cut into your profit, or break your heart; they're simply following their survival instincts.

Predator species tend to be opportunistic animals, seeking the easiest target to meet their needs. In other words, they usually go for young, old, weak, or sick animals first, though some have been known to attack mature, healthy animals that are in their prime. All predators become more aggressive as their

SHEPHERD STORY

For Ian Balsillie and Karen Bean, the idea of keeping sheep and goats was appealing when they purchased 20 acres of logged land in Washington State, but they knew there was no way they could do it without some help. "Our 20 acres is in the foothills, on the west side of the Cascade Mountains," Ian told me, "and the land around us still supports a full range of predators. The day we picked up our first sheep, we also picked up our first guardian dog."

"We've never lost a sheep or a goat to a predator, with the dogs on duty, though we did lose our second dog. She was kind of a nutcase, very aggressive, and chased a cougar for too long. She met her demise."

Ian stressed that working with guardian dogs can be challenging — especially at first. "When they're young, they're sort of like teenagers; they can be obstinate. They try to tunnel out of fences or wander off. But usually by about 18 months old they understand their job."

Ian also says that breeding and training can be rewarding, but they're also really hard work. "Breeding and training guardian dogs isn't easy money. We mainly bred in the first place so we could keep back our own additions, but you can't keep a whole litter of pups."

The number of dogs required to protect the flock varies with circumstances, but Ian and Karen have found that in their case, where predator numbers are high, three dogs are ideal. With three dogs, they can split up; one or two dogs can chase off the predator, while the other stays with the flock to guard its flank. But again, Ian mentions that a good guardian doesn't chase the predator too far.

"By keeping the dogs, we can coexist with wildlife. Last summer we had a bear living right at the edge of our property. There was an agreement between the bear and the dogs: You stay away from our sheep, and we'll leave you alone. It worked for everybody."

hunger increases, like during a drought, and may attempt to take anything they can get their paws on.

As a shepherd, you can learn to manage your flock so that a predator will decide that eating at your house is a lot harder than chasing mice and rabbits. The following box will give you some ideas on minimizing predation. Also keep in mind that your flock will suffer from less predation if it is strong and healthy, so good feed and adequate health care pay in more ways than one.

Discouraging Predators

Predators can be discouraged by:

- Keeping guardian animals, like dogs, donkeys, and llamas.
- Using lighted night corrals with high, predator-tight fences.
- Putting bells on some of your sheep. You can hear the bells if the sheep are being chased. High-frequency bells have also been tried with some success, especially for warding off dog attacks, as the sound is unpleasant to the animal. The high-frequency bells haven't worked as well for coyotes or bears, though in research centers, no sheep that was wearing a bell has ever been killed.
- Having sheep in an open field in sight of your house.
- Using coyote snares along fence lines. These will catch both dogs and coyotes. Check the legality of snares, in your area, before using.
- Having a gun. Even a pellet gun can drive off an attacking dog. Although a dog running through a flock of sheep is not an easy target, most predators spook at the sound of a gun, shot into the air.
- Using "live traps" (cages) for trapping dogs, which allows harmless animals to be set free. These traps are of little value with coyotes, which are too wily to be caught. State wildlife officers may supply live traps for bears or wildcats that are repeat offenders.
- Using propane exploders (which produce loud explosions), radios, and other noise-making devices provides temporary relief, though predators generally lose their fear of these unless their placement, volume, and timing are changed often.
- Using a combination strobe light and siren. This device has been developed and tested by the USDA and seems to significantly reduce predation.
- Removing the carcasses of animals that have died from an area where living animals are kept reduces scavenging, which can evolve into predation.
- Scheduling lambing later in the season if you lamb outdoors on pasture. In early spring, most predators are hungry after a long winter and are feeding their own demanding offspring, yet other feed may still be scarce. By late spring and early summer, other prey (deer or elk, rabbits and rodents, and so on) are more abundant, so the predators aren't as frantic in their efforts to feed.

Identifying Predation and Predators

Sometimes predators get a bum rap: If the corpse of a dead sheep has obvious bite marks, it's natural to think that a predator was the perpetrator. But remember, sheep die from a number of causes, and unless you actually see a predator attacking a live animal, the sheep may have died of natural causes and then been fed on by scavengers.

How to Tell If Your Sheep Have Been Victimized

When you find a dead sheep and suspect predator damage, assess the scene. Signs of a struggle, such as drag marks, torn wool left on brush, or spots of blood in various places all point to predation. If there are no signs of a struggle, examining the carcass may help. Animals that have been fed on by scavengers after dying do not bleed under the skin at the bite marks. This type of bleeding, known as subcutaneous hemorrhage, only happens when the heart was beating while the bites were inflicted. When subcutaneous hemorrhage is present, the next step is to try and confirm the kind of predator. Close examination of tooth spacing and size, feeding habits, and pattern of killing can help correctly identify the type of predator responsible for the kill.

Coyotes

Wile E. Coyote may have looked the fool in all of his encounters with the Road Runner, but he's not a good example of the species. Coyotes are intelligent, curious, and adaptable. In fact, they're expanding their range into urban and suburban areas of the United States and finding satisfactory homes there.

Coyotes usually attack at the throat, but sometimes they'll grab at a haunch, bite the top of the neck, or attack in the soft flesh under the belly. They generally select lambs over adults, unless hunger has made them desperate. They tend to first eat the organs and then the flank or behind the ribs.

Smaller lambs and those born to young, old, or crippled ewes are more commonly victims than those of middle-aged and healthy ewes. Researchers have found that coyotes are more likely to take the smallest lamb from a set of twins or triplets than to take a larger, single lamb. If a coyote is preying on a flock, one approach to dealing with it is to place a livestock protection collar on all susceptible, small lambs. The collar contains a poison that the coyote ingests when it attacks the lamb's throat. The advantage of this collar is that it only targets the killer and does not injure other coyotes or critters that aren't killing the sheep. However, these devices are illegal in some states and must be used by a trained and certified applicator.

Dogs

Dogs are a special class of predator for shepherds, and according to the American Pet Products Manufacturers Association, there are plenty of these predators out there. The U.S. "owned," or pet, dog population is estimated at more than 62 million, which means that there are more than six dogs for every sheep (and that number doesn't include the countless feral and abandoned dogs).

These predators can be more dangerous than coyotes, as one or two dogs can maim and destroy dozens of sheep in one night. One dog attack on a flock can make the difference between a profitable and an unprofitable year, and many people have been driven out of the sheep business because of dogs.

Fido and Spot don't have to be wild, vicious, or even brave to chase sheep. When dogs chase sheep, they're following their natural impulse to chase whatever runs. Unfortunately, sheep run at the slightest disturbance. The offending dogs are not as much at fault as owners who don't keep their pets at home.

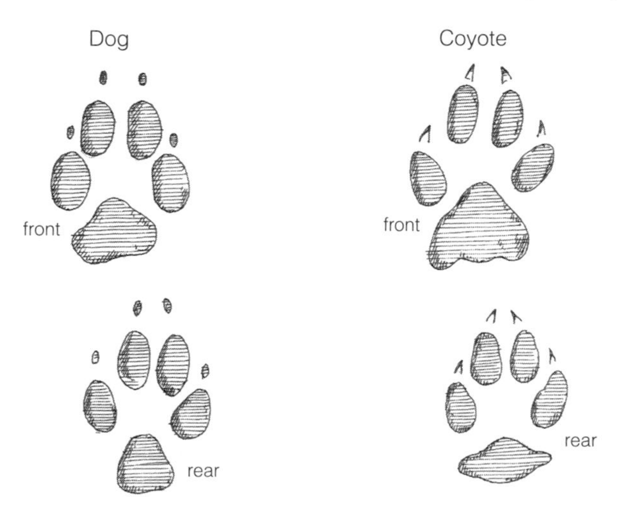

If you have problems with predators, it helps to be able to identify their tracks. The coyote track is about 3 inches (7.5 cm) long for the front foot, and 2½ inches (6.3 cm) for the rear. Dog tracks can vary from slightly larger than a coyote track to quite a bit smaller, depending on the size of the dog, so attempting to differentiate a dog from a coyote based on track size alone doesn't work. Notice that the pads vary. If you live where wolves present predator problems, look for a doglike track that is about 5 inches (12.7cm) long in the front.

Dogs attack rather indiscriminately — they grab at any part of the sheep they can. Sheep often survive dog attacks, but may be badly injured. Even dogs that are too small to physically kill or maim a sheep can still cause heart failure in older ewes and abortion in pregnant ewes. Broken legs also often result when dogs chase sheep. Dogs that actually kill and eat sheep display similar feeding habits to coyotes, going first for the organs and then the flank.

Other Predators

Wolves and foxes are less problematic than coyotes for most shepherds, though both are similar to coyotes in their style of predation and feeding. The big bad wolf is able to — and will — easily take down adult sheep. Wolves tend to hunt in packs of two to four animals.

Because of their small size, foxes only take fairly small lambs. (We've had lots of foxes around and never lost any sheep or lambs to them, but they annihilated our domestic ducks.)

Bears and wildcats are common in more remote areas, especially in the West. These animals can kill perfectly healthy, strong adult sheep as easily as a young or an old sheep. They often kill more than one animal in a single attack and then feed on their favorite parts of each kill.

Bears usually use their massive paws to strike a sheep down. First they eat the udders of lactating ewes, and then they eat the organs. They may stop at this point and move on to the next sheep, or they may feed on the neck and shoulders as well. Occasionally, they'll consume the entire sheep in one feeding.

Cats generally attack by biting the top of the head or neck. They have a habit of burying partially eaten carcasses under grass and brush to feed on again later. They first feed on the organs and then the flank.

Eagles and other birds of prey occasionally kill lambs. They attack by dropping out of the air at high speed and closing their talons into the head. Eagles will strip a carcass and leave little except the skin and the backbone.

Laws

Many wild predators are protected or controlled by federal and/or state laws and regulations. If you have, or suspect you have, a problem with wild predators, call the Wildlife Services office of the USDA or your state's wildlife office to learn about specific remedies and laws in your area.

Your county Extension agent should be able to tell you the county's dog laws or, better yet, give you a copy of the county or state laws. You may find that they are strict and well spelled out but lack enforcement.

State Laws and Dog Control

Sheep owners should know the law and work for better dog-control legislation if necessary. Some states permit the elimination of any trespassing dog that is molesting livestock. Others require that the owner have the dog destroyed or be charged with a misdemeanor. Most states allow the livestock owner to recoup payment from the dog's owner (if known) for both damage and deaths to livestock. If a dog is chasing or killing sheep, promptly contact your local sheriff or animal-control officer. They can assist you in determining the dog's owner, impound the dog, and press charges against the owner on your behalf.

SHEPHERD STORY

Getting into agriculture was like jumping off a cliff," Becky Weed, a geologist by training, told me. "Neither I, nor my husband [Dave Tyler] grew up on farms, so I don't know exactly where this obsession with sheep came from, but we've been farming pretty much full time since 1993, though Dave still works out about half time as an engineer, and I do a little consulting."

In the 1980s, Becky and Dave started a small sheep flock on a Montana farm, but within a short time, they lost 20 percent of their flock to coyotes. "At first, we called the Animal Damage Control agent, and he trapped and shot a couple of coyotes. I knew this couldn't be a long-term solution. We probably couldn't trap and shoot all the coyotes that came through, even if we wanted to, but I didn't really want to do it anyway. I began reading about guardian animals, and decided that was probably a better approach."

The first guardian animals Becky and Dave got were a pair of burros. The burros were cheap, and they worked well, though sometimes one of the burros, Jack, would get annoyed with lambs and kick at them. Becky never saw him actually land a connecting blow, but it worried her a little. Then, due to changes in Dave's job, they had to sell all the critters and move. In 1993, they moved back to a farm and started a new flock. This time, they decided to try a guardian llama.

"We got Cyrus as a weanling, and basically just tossed him in with the flock. He loved the sheep instantly, and the sheep loved him. When a ewe lambed, Cyrus would go right up and start licking off the lamb, and the ewe's never seemed bothered by his attention. Now, it's like the sheep consider him to be the sheep-God — they follow him everywhere.

Guardian Animals

Sheep have been bred for thousands of years to be docile, a trait that makes them easy victims. However, there are other species that become quite aggressive when predators invade their territory, and shepherds have harnessed this trait to defend their sheep for almost as long as they have been keeping sheep.

During the early and mid-1900s, shepherds switched from using guardian animals to using guns, poison, and traps. They adopted the philosophy that any predator was a bad predator and that elimination of all predators was their goal. Now, the use of guardian animals as protectors of sheep has garnered renewed interest. This partially results from desperation, as poisons and traps are now outlawed in so many states and many species of predators are protected by laws like the Endangered Species Act. But shepherds are also becoming more interested in strategies that protect their sheep without indiscriminately killing all predators.

"Cyrus is really amazing with the sheep. He's always attentive, always looking for an anomaly. One day I was bringing the flock up to move them to a new paddock. He was leading, but suddenly stopped and wouldn't move anymore. Finally, I noticed that one lamb was lagging way behind. It turned out to have an injured hip. As soon as I went and got it, he began leading the flock again to the new paddock."

Although Cyrus is completely effective against dogs and coyotes, Becky notes that llamas aren't as effective against bears and mountain lions. "We had a bear last summer and did lose a couple lambs to it. We've also lost some lambs to bald eagles. If we lived in an area where bears were more common, I think dogs would probably work better."

Today, Becky and Dave market wool and lamb from their 240-ewe flock of Corriedale–Columbia crossed sheep through a cooperative of "predator-friendly" ranchers. "We knew if we wanted to make it with sheep, we had to carve out an alternative marketing niche. The predator-friendly approach brought ranchers and environmentalists together. The group stressed issues like open space and habitat protection, economic viability for family farmers and ranchers, and practicing agriculture that protected these values."

At first, the group had some blankets made to market their wool, but people were interested in smaller items and wearables. Now the group has its wool processed and then a group of local women knit sweaters, scarves, gloves, hats, and pillows from the wool. These items are sold from the group's Web site *(www.lambandwool.com)* and through some small stores. Several environmental groups also market these predator-friendly products in their catalogs.

Attacks usually occur at night or very early in the morning when you're normally asleep. A guardian animal is on duty 24 hours a day and is most alert and protective during the hours of greatest danger.

Few guardian animals actually kill predators, but their presence and behavior can reduce or prevent attacks. They may chase a trespassing dog or coyote but should not chase it far. Chasing for a prolonged distance (or time) is considered faulty behavior because the guardian should stay near the flock, between the sheep and danger. The best guardians balance aggressiveness with attentiveness to the sheep.

Whichever type of guardian you're considering, remember the following:

- The guardian needs to bond with the sheep, and the bonding process can take time.
- Guardians should be introduced slowly, across a fence; it's usually easiest to make the introduction in a small area rather than in a large pasture.
- One guardian is generally sufficient on a farm; and on open range, at least two are required. On open range or in large pastures, bigger animals (like donkeys and llamas) are more effective than dogs — though dogs can significantly cut down on losses, even in large areas.
- Each animal is an individual and will react differently in different situations. Some *individuals* don't make good guardians!

Animals That Can Be Used as Guardians

Dogs, ponies, llamas, donkeys, mules, and cows can all be used as guardians. I've even heard of people using geese to guard sheep. Though I don't think they'd be effective against any wild predators, they'd probably do the trick with domestic dogs.

Of all the species used as guardians, donkeys are usually the least expensive (Courtesy American Sheep Industry Association 2000)

Guardian Dogs

Guardian dogs are most effective on small farms or ranches with good perimeter fences (preferably electric or electric added to woven wire). It could be suggested that a dog is doubly effective if it is protecting its own well-defined territory as well as its flock. Users of guardian dogs say that they cannot place a dollar value on the peace of mind they have from knowing that the dog is with their sheep. It can mean greater utilization of pasture, too, since the sheep don't have to be herded into a night corral.

A dog needs to save very few lambs to justify its yearly upkeep. The cost per year for a dog decreases the longer it lives, as the purchase price is amortized over a longer period. Let all the neighbors know of the dog's presence when you acquire it so it will not be shot as an intruder.

To select a dog, you need to find a reputable breeder who regularly supplies dogs for guarding sheep or goats. A dog breeder who also raises sheep is probably a good choice.

Starting with a Puppy

Maremma pup meets lamb at Hampshire College's Livestock Dog Project (Amherst, MA).

Purchasing a puppy costs less up front than does purchasing a proven guardian dog, but for a new shepherd, it's probably worth the extra expense to purchase a proven dog from a seasoned shepherd who raises and trains guardian dogs with the flock. These shepherds generally guarantee the dog's abilities as a guardian, and not all pups work out as good guardians.

If you decide to go the puppy route, the ideal age to remove a pup from the litter is when it is about 8 weeks old, although some claim that dogs placed with sheep before they are 2 months old do better than those reared with sheep when they are older than 2 months.

While there is no real way to test the puppy before you raise it, you can observe it in relation to its littermates. Also, observe its parents. They should not be overly shy or aggressive and should be free from hip dysplasia, a hereditary joint problem common to large breeds. Most breeders guarantee that a pup

Benefits and Problems with Guardian Animals

Although guardian animals can be a great help to shepherds, keeping them may have some drawbacks as well. The benefits of using a guardian animal include:

- Reduced predation
- Reduced labor and fencing costs
- Increased utilization of pastures
- Environmentally benign predator control

Some of the potential problems include:

- Playfulness, which can be deadly to sheep
- Lack of guarding ability—some guardians don't guard; they're disinterested, or they roam from the flock
- Aggressiveness with people
- Interference—some guardians interfere with working or moving the flock
- Destructive behavior—some guardians destroy property (chewing, digging holes, and so on)

will remain free from dysplasia until at least 18 months of age. Pups should have had their shots by 8 weeks of age, confirmed by a veterinarian certificate.

Guardian Dog Training

A guardian dog can be trained primarily through being raised with sheep. The process involves supervision to prevent bad habits from developing and to establish limits of acceptable behavior.

The Bonding Process

The sheep–dog bonding process requires training of both the sheep and the dog. Sheep will initially accept a puppy more easily if they become acquainted in fairly close quarters; however, they may take a long time to accept the dog if it is turned into a large pasture with them. The normal procedure is that when the pup is 6 to 9 months of age, put it in a safe enclosure in the sheep area with some sheep as young as 4 months old.

Remember that a guardian dog isn't a pet — it must bond with sheep, not humans. You don't have to be mean or abusive to a guardian dog, but his place is outside with the flock at all times, not in the house. Discouraging bonding

Using a Dangle Stick

The dangle stick is a thin board or stick 18 to 30 inches (45.7–76.2 cm) long. It hangs from a swivel hook and chain on the dog's collar. When the dog stands upright, the stick should hang 3 or 4 inches (7.6–10.2 cm) above the ground.

This device allows the dog to eat, drink, and display submissive and investigative behavior toward sheep, but when he tries to run (that is, chase), the stick gets tangled around his legs. The device provides immediate discipline and prevents playful chasing. Use it on a playful pup for 3 to 4 weeks and then remove it in stages. First remove the stick but leave the dangle chain, then take away the chain when the playful behavior has stopped.

with humans will keep the dog with the sheep and reinforce its protective instinct. Once the sheep and the guard dog have formed a strong relationship, the sheep will seek out the dog, running to it if there is any disturbance.

Some experienced guardian dog trainers say that you should not handle, pet, or pamper a guardian dog in any way. Others say that not only is it okay to handle and pet the dog, but it's also important. You want a guardian that can be caught, put on a leash, and walked on the leash when necessary. If you never interact with the dog in a positive fashion, you won't be able to catch it for vaccinations or medical procedures, to hold it back while the sheep are being sheared, or to handle it in other cases when you need to. The dog should bond to the sheep but be friendly and comfortable with you. You can have this type of dog by offering praise and a pat when it's good.

Other Training Considerations

It is best not to try to train two pups at once, because they will be too inclined to play and may molest the flock in the process. One pup alone will also bond more readily to the sheep. Pairing a young dog with an older, more experienced dog works better if you wish to use two dogs. In this scenario, the pup is "trained" by the older dog.

Because of the potentially high mortality rate and the lengthy training needed, those who rely on guardian dogs as their primary means of predator control should consider having a ready replacement available.

Most folks who use guardian dogs recommend that working dogs be neutered to avoid the problems encountered when the guard dogs or neighboring dogs are in estrus. Neutering is normally done at about 4 months of age.

No amount of proper training or early exposure to sheep can guarantee that a dog will become a good guardian. The instinctive ability, strong in the traditional guardian breeds, plays a great part in success. The main attributes needed are:

- Attentiveness — bonding to sheep and staying with them
- Appropriate aggressiveness — growling, barking, and fighting if necessary
- Defensiveness — staying between the sheep and danger
- Trustworthiness — must not harm the sheep
- Reliability — wary of unfamiliar humans, but slow to attack

Guardian Dog Breeds

The various breeds of guardian dogs share similar behaviors. These dogs are ordinarily placid and spend much of the daylight hours dozing. Despite their calm temperament, all of the breeds are fierce when provoked and are wary of intruders, both animal and human. Good-natured breeds are best for small farms, while more aggressive breeds are needed for large ranches and open range.

In USDA trials, success rates of guardian dogs did not differ significantly among breeds or between sexes. Here's a quick rundown on the main breeds available in North America.

Guard Dog Tip

To obtain the most effective guardian, it is best to avoid too much training because that may interfere with its instinct and independent intelligence. Essential commands include "come," "stay," and "no." Basic training involves walking the dog on a leash when necessary and habituating the dog to handling.

Akbash

Male Akbash (Anatolian Shepherd) with sheep on Toni Tooker's ranch in Somerset, CO.

Some folks consider the Akbash to be a white-headed variety of the Anatolian Shepherd, but others think of it as a distinct breed. Either way, it has proven itself to be a very effective guardian. In fact, in a 1995 survey of Colorado shepherds who use guardian dogs, the Akbash was rated as the overall most effective guardian against all predators, including bears and lions.

Anatolian Shepherd

Originating in Turkey where it is known as "Coban Kopegi" (or, shepherd dog), Anatolian Shepherds look just like the Akbash, but with a black head, and are among the more aggressive of the guardian breeds, even to human strangers. However, they can be trained by socialization to be friendly toward visitors if this is desirable. Strangers must be introduced to the dog, with the owner making sure the dog accepts them. These dogs are extremely possessive of family, property, and livestock.

Briards

Probably the least-common breed serving as guardians in North America is the Briard, though their use in this capacity dates back to early in the history of the United States when Thomas Jefferson imported them for the job. These dogs look sort of like "Benji" the movie dog, only bigger. Briards originated in France, and the French traditionally used these dogs for both protection and herding services. Briards are slightly smaller than other guardian breeds. They might not be as effective in areas where large predators — bears or cougars, for instance — are a problem, but they should work well where domestic dogs are the primary perpetrators. They may also be of some help where coyotes are a problem.

Great Pyrenees

The Great Pyrenees is a native of the Pyrenees Mountains between Spain and France and is said to have a common ancestry with the Saint Bernard. They are mostly pure white, have a rough coat, and are a most impressive size, weighing from 100 to 125 pounds (45.4–56.7 kg).

This is the gentlest of the guard dogs, probably because they have been bred in the United States as pets for many generations. It is also the breed most commonly used as guardians in North America. In USDA trials at Dubois, Idaho, the Great Pyrenees was the only breed that did not at any time bite a human. Although they are separated from their traditional guardian ancestry, Great Pyrenees have generally proven to be reliable when raised and bonded to sheep at an early age.

Great Pyrenees with sheep at Ambrosia Farm, Chepacket, RI (Courtesy American Sheep Industry Association 2000)

Komondor

Another common working guardian breed is the Komondor (plural is Komondorok). It is considered to be of Hungarian breeding, but may have been brought there by Turkish Kun families, who migrated with their sheep and dogs in the thirteenth century. Its name means "corded coat" — it has a tremendous coat of hair that hangs in locks similar to that of an Angora goat. The coat may require some maintenance to take out burrs.

Adult Komondor at USDA Sheep Experiment Station, Dubois, ID.

These dogs have a very serious disposition and are devoted guards, wary of strangers, and independent thinkers. While used in Hungary to guard herds, they were also used to protect property and factories. Tests done by the USDA have found these dogs to be more successful with pastured sheep than with sheep on open range. This breed has been bred for 1,000 years to be independent, but that independence must be carefully channeled by a firm and loving master for the dog to be effective.

Kuvasz

A Kuvasz (plural Kuvaszok) has a rough, white coat and dark lips, eyes, skin, and nails. The males weigh 100 to 130 pounds (45.4–59 kg), the females 90 to

Kuvasz with lamb at Lala Kingsbury's Spinning Wheel Sheep Farm, Frankfort, ME.

110 pounds (40.8–45.4 kg). This breed is probably a native of Hungary and was used there for many years. Many Kuvaszok were killed there during World War II, sadly depleting the original stock.

The Kuvasz is independent and not easily obedience trained — "no" must be strictly enforced. It is very protective of its own property. Once they learn the boundaries, these dogs protect them fiercely. The females seem more alert, while the males are more apt to kill predators. They are able and agile runners and catch or corner a predator easily. While capable of functioning without supervision (after proper training), this breed seems to have an emotional need for a certain amount of human company.

Maremma (Maremmano Abruzzese)

Maremma guard dogs at Woodsedge Wools, Stockton, NJ. Adult, left; 5-month-old pup, right.

The Maremmas have a sleepy-eyed, relaxed look and a rough coat that is usually white. These guardians have been used in the mountains of Italy to guard sheep for centuries. Usually two or three per flock are used to protect all sides from wolves. These dogs protect very well as a team. In Italy, their ears are usually docked as pups to prevent a wolf from getting a grip on the head.

These dogs are independent but obey single commands they were taught as a puppy. They interpret commands in terms of context and duty — loyalty to the flock always prevails. Maremmas are one of the most successful breeds used in the Livestock Guard Dog Program and are known to be among the calmest of the guardian breeds during the daytime, but with instinctive nocturnal alertness.

Shar Planinetz (Sarplaninac)

A Shar puppy exhibiting proper "sheep attentiveness." (Pat Devore's New Horizons Sheep Farm, Peace Valley, MO)

The first Shar known to be imported into the United States in 1975 was carried down the mountains of Macedonia in a basket on the back of a donkey. This breed has been used traditionally by shepherds of Macedonia and is reported to have been a court guardian of kings. Its name comes from the mountain range of Macedonia in southeastern Europe, in the area of Yugoslavia. While similar to the Maremma or Pyrenees, it is a bit smaller. Its coloring is usually tan to dark brown and is often black. This dog has a quiet, gentle temperament, and many have been trained and distributed by the Livestock Dog Project at Hampshire College in Amherest, Massachusetts.

Tibetan Mastiff

Four-year-old Tibetan Mastiff, very patient with his sheep and his puppies on Michael Morgan's farm in Tonasket, WA.

Although it is one of the oldest breeds in existence today, the Tibetan Mastiff is rare in North America. Its lack of genetic problems is evidence of centuries of natural selection and survival of the fittest. It is black with tan markings, including tan spots over the eyes. It has a distinct double coat with a ruff around the neck and shoulders and carries its full tail well over its back. In its native land, these dogs travel with caravans of the Tibetan sheepherders and traders, defending the herds and the tents of their masters from such predators as wolves, snow leopards, and robbers. They are loyal to master and flock, with antipathy to strangers. The bitches go into estrus only approximately every 10 to 12 months (the mark of a primitive breed) and lack dog odor.

SIX

FEEDS AND FEEDING

The old saying goes, "You are what you eat," and that's true for your sheep, too. Good nutrition results in:

- Higher levels of fertility and multiple births
- Greater milk production and nursing ability
- More wool production and better wool quality
- Fewer troubled pregnancies and less health problems in general
- Quicker lamb growth

Flocks that are malnourished, however, suffer from every imaginable problem, including higher predation, disease, abortion, and premature lambing. Undersized lambs that haven't reached full size before birth have less chance of survival and lose more body heat after birth than big, healthy lambs.

Raising sheep is an efficient way to convert grass into food and clothing for humans, but pasture alone is seldom adequate to feed sheep 12 months of the year. Thus, some supplements (grain, hay, minerals) are necessary. Feeding time is a good time to check on your sheep, feel the udders of ewes close to lambing, and note eating habits, which greatly reflect their state of health. Count the sheep, particularly if you have any wooded pasture where one could get snarled up or be down on its back and need help.

Digestion

Although a few "foods," like sugar water, can be absorbed directly from the stomach into the bloodstream, most foods are unusable until they are broken into molecules, which are made up of groups of atoms. This process is called digestion, and it has mechanical, chemical, and biological components.

Like cows, goats, and deer, sheep are members of the class of animals known as ruminants. Ruminants have a unique, four-stomach digestive system.

The First Phases of Digestion

The top of the mouth of ruminants is a hard palate — they don't have any teeth. The bottom teeth tear and grind feed against the top palate, providing the mechanical component of the digestive process. Initially, the food is only lightly chewed and combined with saliva to form a small ball, or bolus, of feed. The sheep swallows this bolus and it enters the rumen, or first stomach.

SHEPHERD STORY

Shepherds are discovering that sometimes the best forage is somebody else's forage. For the small-scale shepherd, this might be mowing lawns or ditches for a neighbor by using sheep and portable electric fencing. But some folks are taking this approach to a new level.

Dick Henry has had a dual-career tract: He's been a commercial sheep farmer and he's managed environmental organizations. Both interests have now joined together like a magnet to steel.

"As the Director of the Audubon Society of New Hampshire, I was working with Public Service of New Hampshire (PSNH) on air-pollution issues. I knew PSNH had problems with maintaining right-of-ways under their power lines; deciduous trees would grow under lines and short them out, so they were spending a lot of time and energy spraying with herbicides or mechanically mowing the right-of-ways," Dick explained to me.

"Research from Britain and Australia showed that using mob grazing techniques with sheep could produce the same result that PSNH was looking for: a low-growing vegetation cover. I ran the idea of doing this past the PSNH staff, and they were willing to try it. If we could make it work, it would do what they needed done, in an environmentally friendly fashion."

Mob grazing uses a large flock of animals that concentrates in a small area for 24 hours and is then moved off. Timing of moves is controlled so that the flock comes back to the same place when the plants that need to be controlled are at a vulnerable point — so timing differs slightly from that of managed grazing on a pasture.

Even with the support of a major company, Dick points out that this isn't an easy process to get going. It's capital intensive and requires trained shepherds and herding dogs, guardian dogs, portable fencing, and equipment to work the animals out in the field. It took Dick almost 2 years of planning and preparation to get Bellwether Solutions, LLC, his company that provides vegetation control with sheep, operating. Part of the planning involved figuring out just what the heck he could do with 1,500 sheep, shepherds, dogs, and equipment in the winter when they couldn't be grazing New Hampshire right-of-ways!

The Rumen

The rumen is like a biological factory, where microorganisms work to break down feed through a fermentation process. We're inclined to think of "bugs" (that is, bacteria and other microorganisms) as bad, but those that normally reside in the rumen are not only beneficial, they are absolutely essential to the animal's survival. These bugs are called *flora* and have evolved through a mutually beneficial, or *symbiotic*, relationship with the sheep over many aeons.

"Although this isn't a magic bullet, a number of nonnative invasive species of plants — like leafy spurge, spotted knapweed, and kudzu — are plants that sheep eat and that can respond to correctly timed, mob grazing. Since kudzu is taking over the Southeast (best estimates are that it currently covers over 7 million acres and is spreading at a rate of 120,000 acres per year), grazing kudzu-covered ground provides us with our winter program."

Dick stresses that his approach differs significantly from commodity-based agriculture because he's not selling a commodity, like lambs or wool, but is selling a service. "With commodities, if you have a good year, then chances are everybody else has had a good year, too, and the price goes down. When you have a bad year, so have all the other farmers, so the price goes up, but nobody has product to sell. We concentrate on the service. In fact, we started out using wethers so we wouldn't have to deal with lambing. Now we are running some ewes as well, so we can build our flock from within, but we're still not interested in selling a commodity."

By selling a service rather than a commodity, Bellwether Solutions has signed contracts for several years in advance and has been able to have a steady cash flow throughout the year — two things most commercial farmers would love to have.

Dick points out that there are more opportunities out there for service-oriented agriculture. And there's the potential to greatly expand the type of sheep- and/or goat-grazing programs that Dick and a few other practitioners have begun around the country, but there are some serious constraints on this new "industry."

"There is a real lack of trained shepherds in North America, and for this industry to grow, we're going to need to develop those skills and build a talent base. I'm always looking for shepherds, and I'm willing to train, but this is serious work. I've had people write letters saying they'd like to work for me 'this summer, because I want to finish my book.' Well that's not what this is about. This is a 7-day-a-week, 24-hour-a-day job; but for some people it's the perfect job — a nice way to make a living outdoors, surrounded by animals."

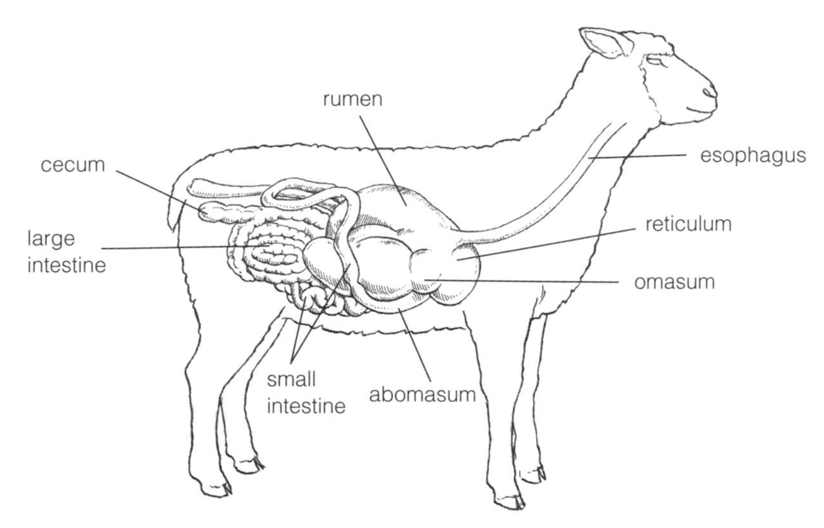

Digestion in sheep is a complex process that takes place in a four-stomach system. The stomachs are the rumen, reticulum, omasum, and abomasum.

The rumen of a mature sheep has a capacity between 5 and 10 gallons (18.9–37.8 L), and each gallon (3.8 L) has about 200 trillion bacteria, 4 billion protozoa, and millions of yeast and fungi. Keeping this work crew active and healthy so they can do their jobs well is an important function for shepherds, even if they don't usually think in these terms. A key to rumen health is making dietary changes slowly, so that the flora have a chance to acclimatize to new feed regimens.

Some heavy items, like whole grain or stones, may bypass the rumen and go directly into the second stomach, or reticulum. Grain that bypasses the rumen won't be as thoroughly digested, which is why it's a good idea to feed cracked grain or whole grains mixed with hay. That way, most of the feed will get some time in the rumen.

The fermentation that occurs in the rumen produces a significant amount of gas, which the animals must pass by belching. Occasionally, an excessive accumulation of gas and/or foamy material builds up in the rumen, causing bloat.

The Other Stomachs

From the rumen, a slurry of well-fermented feed passes onto the reticulum, then to the third and fourth stomachs, or the *omasum* and the *abomasum*. The abomasum is sometimes called the true stomach, because it functions in a way that's most similar to that of single-stomached creatures, including humans.

The Dangers of Bloat

Bloat is very serious in ruminants — it can be fatal in only a few hours. Too much of almost any feed can cause bloat, but it most commonly results from lush pasture. Legume pastures, such as very leafy alfalfa and clover, are far more dangerous than grass pastures, which is another good reason to develop mixed pastures of grasses and legumes.

When changed from sparse to lush pasture, sheep may gorge themselves *unless* they have been given a feeding of dry hay prior to turning out on the new pasture. Sheep seldom bloat when they are getting dry hay with their pasture. The coarse feed is thought to stimulate the belching mechanism while keeping the green feed from making a compact mass. Some sheep seem more prone to bloat than others, possibly due to a faulty belching mechanism. (See chapter 7 for more information on bloat and how to treat it.)

Digestion in these latter stomachs is primarily a chemical process during which enzymes break down the feed as it passes through.

The four-stomach system allows ruminants to eat — and digest — feeds that other critters can't take advantage of. Specifically, it enables them to digest cellulose, which accounts for 50 percent of the organic carbon on earth (but we can't digest it). Being able to digest cellulose allows sheep to receive maximum benefit from all of the amino acids that are present in their feed.

Final Phases of Digestion

The cecum and intestines are the final places the feed passes through. These organs provide one last chance for some breakdown to occur through both chemical and biological processes.

The small intestine is like Grand Central Station — it's the place where most transfers occur. Available nutrients, which are now in their elemental and molecular forms, change trains, switching from the digestive "track" to the bloodstream, or commuter trains. This commuter system (veins, arteries, and capillaries) disperses the nutrients to points throughout the body where they disembark as needed, "feeding" the organs, muscles, and tissues. The other function of the intestinal system is really important: It controls disposal of the waste products that the body can't use.

Cud

One unique characteristic of ruminant digestion is that the animals have to "chew cud." Cud is a bolus of partially digested material that's regurgitated into the mouth from the rumen. Unlike the first chewing, which is quick, cud is chewed very thoroughly and then reswallowed. This process serves two purposes: It allows additional mechanical grinding, and provides a continuous source of large amounts of saliva for the rumen, which helps maintain a healthy environment for the flora. Cud chewing takes place off and on during the day; all together, it takes up about 6 hours on a daily basis.

Digestion in Lambs

At birth, lambs lack a fully developed rumen. In fact, at first the rumen is relatively small (25% of total stomach capacity), and the abomasum is relatively large (60%). By the time the lamb is 4 months of age, the rumen is almost fully developed (75% of stomach capacity) and the abomasum is about 10 percent of stomach capacity. This means that lambs can't digest cellulose early on.

Lambs shouldn't be weaned until their rumen is fairly well developed; the earliest point at which this occurs is around 45 days. Lambs that are weaned too early are likely to have stunted growth, and some die.

Right after birth, the lamb is capable of absorbing antibodies from the mother's milk directly into its bloodstream through the large intestine. Typically, antibodies (more about these natural disease fighters in chapter 7) would be broken down in the rumen, but the lamb's rumen isn't working yet. Also, antibodies are normally too large to be absorbed through the intestines. To adjust for this, nature provides a brief window of opportunity when the antibodies can pass through the intestinal wall and into the bloodstream as a way to jump-start the baby's immune system. Nature also made special accommodations by pumping up the first milk, or colostrum, with extralarge doses of the antibodies that float around in the ewe's system.

Colostrum

Lambs must receive colostrum as soon as possible after birth, because the intestinal lining begins shutting down from the moment of birth until it can no longer allow the passage of antibodies. Picture a sieve, which is at first large enough to pass marbles but that is continually closing down until not even one grain of sand can pass through. This closing process takes anywhere from 16 to 48 hours.

Although lambs can survive without the nutrition provided by the colostrum, it is very difficult for them to survive without the disease-protecting antibodies that it contains. When your ewes have been vaccinated prior to lambing season, they'll pass along immunity from the vaccine through the colostrum. If it's not possible for the lamb to receive colostrum from its mother (for such reasons as death, disease, or rejection), then it needs colostrum from another ewe. (Collect and freeze some colostrum for such emergencies.) Cow or goat colostrum may be substituted for ewe colostrum in a pinch and is often available for the asking from local farmers, especially if there are some dairy farms in the area.

There are commercial preparations that are useful when no other colostrum is available. These milk-whey-antibody products transfer a certain amount of immunity to the newborn when mixed with a milk replacer (or diluted canned milk or cow milk) for the first day. These products are available through a veterinarian or mail-order catalogs (see Resources).

The advantages of ewe (or even cow) colostrum from your own farm is that its antibodies are "farm specific" and can protect the lamb against any organisms that the ewe (or cow) was exposed to on your farm.

Emergency Newborn Lamb Milk Formula

While there is no satisfactory substitute for colostrum, in the worst-case scenario a lamb can be fed the following mixture for the first 2 days, rather than just starting out with milk replacer. This recipe packs a little extra punch that benefits weak babies.

26 ounces (769 mL) milk (prepare by mixing ½ evaporated, condensed milk with ½ water)
1 tablespoon (14.8 mL) castor oil (or cod liver oil)
1 tablespoon (14.8 mL) glucose or sugar
1 beaten egg yolk

1. Mix well, and give about 2 ounces (59.1 mL) at a time the first day, allowing 2 to 3 hours between feedings. Use a lamb bottle, or in a pinch a baby bottle will work (enlarge the nipple hole a little by making a small X opening with a knife). When the lamb is older, lamb nipples, which are larger, should be used.
2. On the second day, increase the feedings of the formula to 3 ounces (88.7 mL) at a time (or 4 ounces [118.3 mL] for a large, hungry lamb)

2 to 3 hours apart. On the third day, the formula can be made without the egg yolk and sugar, and the oil can be reduced to 1 teaspoon (4.9 mL) per 26 ounces (769 mL) of milk.

3. After the third day, you can gradually change to lamb milk replacer. Do not use milk replacer that is formulated for calves; it is too low in fat and protein. Your local feed store, or veterinarian, can special order lamb milk replacer if they don't have it in stock. Goat's milk is also a good lamb food, so if you have it available, you wouldn't need powdered milk replacer.

Raising a Colostrum-Deprived Lamb

It's possible but difficult to raise a colostrum-deprived lamb. Colostrum provides the lamb with antibodies against common environmental pathogens, which protect it during the first few weeks of life until its own immune system can begin producing them. The lamb is too young to receive vaccinations, but you may be able to administer antisera to provide temporary protection against common lamb diseases. Your veterinarian can help you evaluate the possible benefits of antisera.

Is the Lamb Getting Milk?

One surefire way to tell if the lamb is actually getting milk while it's nursing is to watch its tail: For lambs that are getting milk, those little tails swing back and forth like a flag in a good breeze. If the tail is not in motion, you may have a problem. Constant crying may be another indication that something is amiss, but not always — some lambs will starve to death without making a sound. For the first week or so, make sure that the ewe has milk (by hand-milking some out of each nipple) and that the lamb is actually getting some. A lamb that's getting milk will have a puffed-out belly, whereas one that's not getting any has a sunken belly and its skin piles up in folds. Some ewes may come into milk only to dry up after a day or two, so never assume that a ewe will continue to milk after the first day. Be vigilant while the lambs are little.

The ewe's milk should be sufficient if she is well fed. However, if a young ewe does not have sufficient milk, supplement it with a couple of 2-ounce (59.1 mL) bottle-feedings for the first 2 days, preferably with milk taken from another ewe or with newborn milk formula. Insufficient milk letdown can sometimes be resolved by injections of oxytocin, available from your veterinarian. If the quantity still is not sufficient for the lamb, supplement it with a couple of 4-ounce (118.3 mL) feedings of lamb milk replacer during the first week, then increase to about 8-ounce (236.6 mL) feedings when the lamb is 2 weeks old. Poorly fed old ewes also may have scant milk supply. If you are feeding an orphan or a lamb whose mother has no milk, see the box on page 272.

Bottle-Feeding

Bottle-fed lambs require extra care. Never overfeed. It's better to underfeed than to have a sick lamb, which happens easily with bottle-fed babies. Overfeeding can cause a type of scours, or diarrhea, that can be deadly. Bottle-fed lambs are also more prone to bacterial and viral infections.

During bottle-feeding, cleanliness is critical to lamb survival. Keep bottles, nipples, and milk containers clean. Keep milk refrigerated, *warming* it at feeding time. Use care if warming in a microwave oven, because it can easily become too hot, and a lamb with a scalded mouth won't do well.

Nutrients

Nutrients perform three basic functions for an animal: They support structural development, which includes strengthening bones, muscles, tendons, wool, and skin; provide energy; and regulate body functions. Although each type of nutrient (proteins, vitamins and minerals, water, carbohydrates, and fats) helps with more than one of these basic functions, each has a role at which it excels.

Energy

Energy keeps the body warm and enables the animal to do work (like growing or reproducing). It's also considered the most common limiting nutrient for sheep, meaning that it's least often found in sufficient quantities in their diet. Fats are the prime source of energy, with carbohydrates also providing a significant amount. Grains like corn and oats, sweet treats like apples and molasses, and beans and nuts can all supplement energy needs. Pasture that's growing well and kept in a vegetative state provides lots of energy, but as it becomes rank and overly mature, the energy level falls off.

Terminology

These are a few terms that are used specifically when discussing feeds:

- **Feedstuff.** Any food intended for livestock consumption.
- **Ration.** The combination of foods in a specific diet for a specific animal or class of animals at any given time. Includes everything the animal is receiving.
- **Forage or roughage.** The hay or pasture portion of the ration.
- **Concentrate.** The grain or grains being fed as part of the ration.
- **Supplements.** The vitamins, minerals, or protein added to the ration to provide additional nutrition.
- **Energy.** The part of the ration that is made up of sugars, fats, fatty acids, and starches that are used by the body for muscle and nerve activity, growth, weight gain, and milk secretion.
- **Fiber.** The part of the ration that comes from cellulose and hemicellulose in plant matter; is broken down by ruminants and horses to create additional sugars and fatty acids.
- **Protein.** The portion of the ration that contains amino acids; required by the body for cell formation, development, and maintenance, especially for muscle and blood cells.
- **Balanced ration.** A ration that provides all necessary nutrients in the proper proportions (including energy, fiber, protein, vitamins, and minerals) for the animal's needs on the basis of its age and level of work.
- **Dry matter.** The mass of the ration or feedstuff if the water is "baked-off." For example, a sample of mixed meadow hay might contain 85 percent dry matter, so your 60-pound (27.2 kg) bale of hay would actually weigh 51 pounds (23.1 kg) on a dry-matter basis (0.85 x 60 pounds [27.2 kg]).
- **Total digestible nutrients.** The part of the ration that the animal can actually take advantage of. Feed reports, feed tags, and feed charts report the total digestible nutrients (TDNs) of the feedstuff. If the TDN on a sample of hay was tested as 60 percent on a dry-matter basis, the bale would contain 30.6 pounds (13.9 kg) of digestible nutrients (0.6 x 51 pounds [23.1 kg]).

Body Composition of Lambs at Various Ages

	WEIGHT		WATER	FAT	PROTEIN	ASH
STATUS	POUNDS	KG	%	%	%	%
Newborn	9	4.1	72.8	2.0	20.2	5.0
Feeder	65	29.5	63.9	17.0	15.7	3.4
Fat lamb	100	45.4	53.2	29.0	15.0	2.8
Very fat lamb	125	56.8	39.0	44.0	14.4	2.6

Proteins

Unlike carbohydrates, which may contain as few as twenty atoms, proteins are made up of thousands of atoms. But constructing something out of so many parts can be made simpler by constructing prefabricated substructures. The structures that are used to make up protein molecules are called amino acids, and there are about twenty essential amino acids. Proteins are constructed by altering the combinations of these acids. Ruminants can easily obtain all the necessary amino acids from plants. Most sheep have a higher requirement of protein than other species of livestock, because protein is a major constituent in the development of wool.

Protein is higher in legumes than in grasses and can be supplemented by feeding legume hay or cubes, field peas and soybeans, sunflower seeds (which are also really high in energy), and brewer's yeast or grain. Some commercial protein supplements contain animal proteins, such as bone and blood meal. Although these often provide a less expensive source of protein than plant supplements, they may contribute to disease transmission. (Scientists suspect that feeding these types of animal proteins contributed to mad cow disease in Europe.)

Vitamins and Minerals

Vitamins and minerals could be considered the "regulators" of a sheep's diet. These regulators can be likened to switches in a house; they turn things on and off when needed, adjust the temperature to keep things comfortable, and help process information. Vitamins, minerals, some forms of protein, and enzymes are all critical to regulation. Vitamins are usually adequately supplied in good, green feeds, like pasture and hay. Minerals can be supplemented with a trace mineral mix or block — just make sure the one you choose is specifically for sheep, because those prepared for cattle and horses may contain too much copper.

Vitamins and minerals are only required in very small quantities, but shortages (or excesses) of a critical vitamin or mineral can have grave impacts on your flock's health. Deficiencies of vitamins cause diseases like rickets and anemia. Overdoses are toxic. Deficiencies and excesses are usually the result of soil mineral imbalances, which vary from farm to farm and from region to region.

Water

The first and most important nutrient (though people don't always think of it as such) is *water!* Animals can live for up to 10 days without food but may not survive for 2 days without water. Water, in adequate quality and quantity, must always be available. Water serves the following functions:

- Keeps sheep cool in hot weather (but even in the dead of winter, it's necessary)
- Aids in transporting nutrients throughout the body
- Carries waste out of the body
- Is required for the chemical reactions that take place throughout the body
- Keeps cells hydrated and healthy

Water should be kept clean. The ideal temperature for water is about 50°F (64°C). (In northern climates, sheep may meet quite a bit of their water need by eating snow, but they should still be given an opportunity to drink water at least once a day.)

Moisture in feeds also affects the sheep's drinking habits; very moist feeds reduce water intake, and dry feeds raise it. Hot summer pasture has very little moisture in it.

To cope with heat, sheep lose moisture through their skin, which adds to the need for ample water. Providing shade helps keep down moisture loss, but the sheep still need clean, fresh water. Also, access to shade should be limited and regulated, like controlling movements through the pasture, because shaded areas that are overused contribute to parasite problems.

Ewes with nursing lambs need extra water to make milk. On average, mature sheep drink between 1 and 2.5 gallons (3.8–9.5 L) of water per day. Late-gestation and lactating ewes are toward the top of the scale.

Carbohydrates

Carbohydrates make up about three-fourths of the dry matter in plants, so they're one of the most significant nutrients in a sheep's diet. Sugars, starches, and fiber are all classes of carbohydrates, and their proportions in an individual plant vary depending on the plant's age, the environmental factors, and the type of plant. For example, sugars make up a higher proportion in young plants, and fiber makes up a higher proportion in older plants.

Carbohydrates are named after their atoms of carbon (atomic symbol C), which are attached to molecules of water (H_2O). A simple sugar molecule might be made up of six carbon atoms attached to six water molecules ($C_6H_{12}O_6$), and it's a readily useable nutrient, providing an instant burst of energy. Starches are composed of groups of sugars, strung together like Christmas lights. Since both sugars and starches are easily digested, they provide high feed value.

Fiber is mostly made up of lignins and cellulose. Lignins are virtually indigestible, but the cellulose is readily digested by the bacterial fermentation that takes place in the rumen. Sheep, like other ruminants, can make use of this feed supply that's abundant in grass, hay, and other forages.

Fats

Like the carbohydrates, fats (and fatty substances) are made up of carbon, hydrogen, and oxygen, but the proportions are very different. For example, olein, a kind of fat that's commonly found in plants, has a chemical formula of $C_{57}H_{104}O_6$, meaning that there are 57 carbon atoms, 104 hydrogen atoms, and only 6 oxygen atoms.

Fat is an essential nutrient — especially for young and growing animals (including people). It provides almost twice as much energy as carbohydrates do, and it helps an animal control its body temperature.

Feeding Practices

What and when you feed is dependent on the animal's stage of life. The same quality and quantity of feed supplement is not needed at all times. Lambs and young animals need more and higher-quality feed (relative to their size) than mature animals, lactating ewes need more than dry ewes, and during the breeding season rams need more than they do during the off season.

Feeding Behavior

Sheep prefer to eat during daylight hours. They begin about dawn, and when given free choice, as on a pasture, they eat off and on throughout the day until dusk. They eat at night only if they have no choice, and then won't eat as much.

Because sheep must have time during the day to rest and chew cud, the feed must have sufficient nutrients to meet all of the sheep's needs during the day. If pregnant ewes are fed poor-quality hay in winter, when there are fewer hours of daylight, their nutritional needs will not be met and they will require supplementation with grain.

Sheep are gregarious and eat more and better when they are in a group than when they are alone. Thus, if you're just starting out, buy two or three lambs instead of one. They like to eat as a group, but won't mix with other flocks very well once a bond has developed in the existing flock. Place two flocks together, and they'll each stake out a "home territory," separate from each other, even if they're in the same pasture.

Sheep avoid grazing near their own feces but don't seem to mind grazing around the feces of other species. They like higher ground better than lower ground and do well grazing with other species, such as cattle, goats, horses, and even pastured poultry. In fact, there is a tremendous benefit from grazing several species together:

- Since they don't share the same internal parasites, multispecies grazing reduces parasites in all species.
- There are fewer predation problems when sheep are grazed with cows, horses, or llamas.
- Since different species eat slightly different amounts and varieties of the available feed, the land can carry more animals by weight of multiple species than it can of a single species; in university research, it's been found that meat production per species is increased by up to 125 percent (that is, lambs and calves grow quicker) when several species are grazed together.

Feed Changes

A sheep's stomach can adjust to a great variety of feed, provided that changes are made *gradually*. A sudden change of ration, such as sudden access to excess food, can cause death. Rumen flora can adapt to the diet, but they cannot adapt quickly.

Any kind of abrupt change of diet will distrub the rumen, and those types of disturbances not only cause acute problems, like bloat, but also can cause chronic problems. Sudden changes interfere with the synthesis of A and B vitamins. Vitamin A in particular acts as an anti-infection vitamin. Insufficient vitamin B results in lack of appetite, emaciation, and weakness.

Guidelines for Feed Changes

A good rule of thumb is to make any changes to a feed program no quicker than 10 percent per day. If you're purchasing new sheep, try to find out what kind of feed they were eating before you get them home so you can slowly switch them to your own program.

Feeding Program

Assuming you have good-quality hay, the following general guidelines are appropriate. If your hay is of poor quality, you may need to feed a little more grain than the amounts specified below. If the reproductive cycle is timed to the peak of high-quality grass production in the spring, you'll need little if any supplemental grain for lactation. If you plan to lamb in the winter, you'll definitely want to follow the prebreeding guidelines.

1. Seventeen days before turning the ram in with the ewes, give up to ½ pound (0.23 kg) of grain per ewe, starting gradually for the first few days. This is called *flushing,* and it increases ovulation.
2. The ram should start receiving up to ½ pound (0.23 kg) of grain also, but you can wait until about a week before he's turned in to begin supplementation. After breeding season, begin reducing his feed until he's back to just hay or pasture.
3. Continue giving the ewes the ½ pound (0.23 kg) for about 4 weeks after mating, then taper off gradually. This may prevent resorption of the fertilized ova.
4. Feeding light grain, say ¹⁄₁₀ pound (0.05 kg) per ewe per day, is okay until the last 5 weeks of pregnancy.
5. During the last 5 weeks of pregnancy, the ewe should be on an increasing plane of nutrition to prevent pregnancy disease, gradually working up to ½ pound (0.23 kg) or more per ewe.
6. For the 6 to 8 weeks of lactation, ewes with single lambs should have approximately 1 pound (0. 5 kg) of grain per day, while a ewe with twins should have 1½ to 2 pounds (0.7 to 1 kg), plus hay for each. Again, taper off as the lambs eat more grain and hay (in their creep-feeder).
7. Start diminishing the quantity 10 days before weaning until the ewes receive no supplemental grain, leaving feed in the creep-feeder for the lambs.

See pages 350–353 for feed charts that can help you balance a ration to optimize performance. More information about grain is given under Types of Feed.

Optimizing Performance

Getting the most from your feed is really a two-part affair:

1. Make sure your flock is eating what it needs when it needs it.
2. Try to do so economically.

You may opt out of the second part. For example, if you have a really small flock and still have an outside job, you may want the convenience of purchasing prepared feeds, even if they cost a little more.

Regular Feeding (Time and Amount)

Measure the quantity of grain given each day, by using the same container, or number of containers, for each feeding. Sheep do not thrive as well when the size of their portions fluctuates. If they are fed in the evening, it should be at least an hour before dark.

When given regular feedings at an expected time, sheep are less likely to bolt their feed and choke. Too much variation in feeding time is hard on their stomachs and the rest of their systems.

Although feeding at an expected time is important for all sheep, for pregnant ewes it also makes a difference what *time* you feed. In some university trials, regular feeding of ewes at around 10:00 A.M. helped to reduce the incidence of night-to-early-dawn lambing. Other recent tests suggest late-afternoon feeding, shifting to even later times as lambing approaches. Both feeding schedules concentrated lambing primarily into daylight hours.

Types of Feed

Although pasture should be a primary feed source, grains, hay, and a variety of vegetables can also be put to good use. Remember though, no sudden changes in diet — a sudden change can paralyze the sheep's digestive system and cause death from acidosis, impacted rumen, enterotoxemia, or bloat. With four stomachs, acute indigestion is not a minor illness for a sheep.

Grains

Whole grains are better for sheep than crushed grain. For instance, rolled oats have so much dust that they can cause excessive sneezing, leading to prolapse in heavily pregnant ewes and breathing problems in lambs. Unprocessed corn and wheat still contain the valuable germ, or embryo, that's located within the grain kernel and is typically removed during milling. This germ is rich in vitamin E — a vitamin that helps ewes protect their lambs from white muscle disease.

Whole grains, fed with hay, promote a healthy rumen, where the feed gives better conversion to growth. On the other hand, a diet consisting mainly of pelleted or finely ground feed causes the rumen to become inflamed. The inflammation traps debris and dust, causing a vicious cycle of further inflammation.

Grain Mixtures

There are lots of possible combinations of grains to use as supplemental feed for sheep, but the following recipes are generally suitable.

- Mix 50 lb (23 kg) of shelled corn, 20 lb (9 kg) of oats, 20 lb (9 kg) wheat bran, and 10 lb (4.5 kg) of linseed meal.
- Mix 75 lb (34 kg) shelled corn, 10 lb (4.5 kg) of cottonseed hulls, 14 lb (7 kg) of soybean meal, and 1 lb (0.5 kg) of molasses.

Hay

Hay should be stored in the darkest part of the barn to preserve the vitamin A content, which is depleted by exposure to sunlight. Careful storage is necessary to avoid weather damage and nutrient loss. Exposure of hay to rain cannot only leach out its minerals, but can also result in moldy hay, which is a cause of abortion in ewes. If you have no barn, hay can be stored under tarps.

The lower the hay quality, the more you have to feed. Lots of heavy stems in the hay mean that the sheep will eat less. A certain amount of hay is always discarded, some pulled out onto the ground and wasted (pile this in your garden twice a year) and some uneaten stems (take the clean ones out of the feed rack for clean bedding for lamb pens). If you buy two different kinds, or grades, of hay, save the best for the pregnant ewes. Late in pregnancy, hay must be of high quality, as the growth of the lamb will crowd the ewe's stomach and leave little room for bulky low-nutritive feed.

The Benefits of Alfalfa Hay

One reason alfalfa hay is such a superior feed for sheep is because it contains nine vitamins, especially vitamin A, which is so lacking in winter pasture grass. The greener the hay, the higher the vitamin content. Alfalfa hay is also high in calcium, magnesium, phosphorus, iron, and potassium. Protein content is from 12 percent to 20 percent depending on what stage it is cut (highest protein yield occurs when it's cut young, or in the bud) and on its subsequent storage. Alfalfa got its name from an Arabic word meaning "best fodder," which is most appropriate.

Extras

Windfall apples, gathered and set aside out of the rain, can be a welcome addition to the winter diet, but in limited quantities. Sheep love apples — they even prefer the overripe and spoiled ones — and a few apples a day adds needed vitamins. An excess of apple seeds, however, especially the green seeds, can be toxic.

Fresh pomace from apple cider making is good feed for sheep, in small quantities, if you have not sprayed your apples. Fermented pulp is not harmful if fed sparingly, but decomposed pulp is toxic.

Molasses is another treat for sheep and is a good source of minerals. The sugars enter the bloodstream quickly, so it is of value to ewes late in pregnancy to prevent toxemia — but not in excess.

Discarded produce from the grocery store is another treat. Lettuce, cabbage, broccoli, celery, and various fruits, past their prime for human consumption, are often available at the local store. Fed sparingly, or regularly in measured quantities, they are a good addition to the diet.

Plant a small area of carrots, rutabagas, turnips, or beets for a succulent treat in fall and winter. These provide good roughage and variety in the diet. Avoid potatoes, as once sprouted they can be toxic to sheep and cause birth defects.

Salt and Minerals

Salt is another year-round necessity for good health. When sheep have been deprived of salt for any length of time and then get access to it, they may overindulge and suffer salt poisoning. The symptoms of this disorder are trembling and leg weakness, nervous symptoms, and great thirst. It can be treated by allowing access to plenty of fresh water, but better yet it can be prevented by keeping salt available at all times. Mineralized salt, especially that containing selenium, is recommended. Just make sure it is a mineral salt for sheep, *not* cattle, as minerals for cattle can contain toxic levels of copper. Regular access to salt is said to be useful, along with roughage, to prevent bloat, which is one of the most serious digestive upsets.

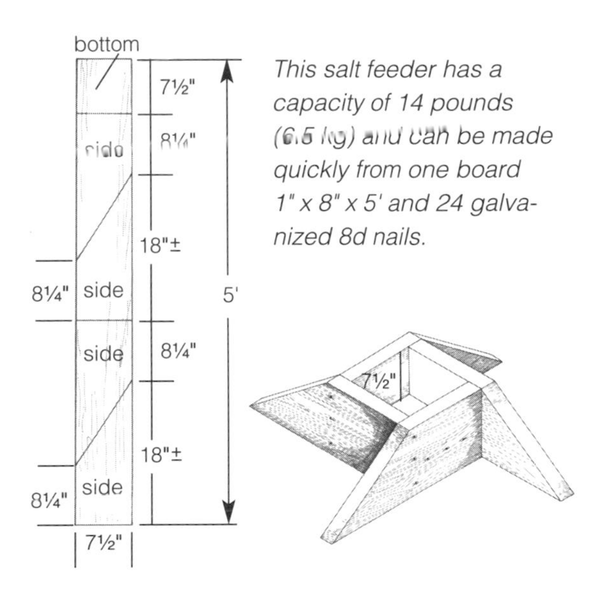

This salt feeder has a capacity of 14 pounds (6.5 kg) and can be made quickly from one board 1" x 8" x 5' and 24 galvanized 8d nails.

Poisonous Plants

All areas of North America are host to some poisonous plants. The good news is that animals usually don't chow down on these plants, unless they're very hungry. But sometimes an individual animal will start eating a poisonous plant even if there's plenty of good feed available.

Check with your county Extension agent to learn which plants in your area are poisonous (make sure you specify that you're interested in sheep, as some plants that are poisonous for cattle or horses don't affect sheep at all). With any new pasture, walk around it and note any unusual or unfamiliar plants. If there are plants you cannot identify, you can send several fresh whole plants (stem, leaves, flowers, seeds) to the state agricultural college, wrapping them in several layers of newspaper. They should be able to identify them and advise about their toxicity.

After you know what toxic plants are around, develop a plan for dealing with them. Some plants are only toxic during certain stages, so you can keep animals out of the area during that period. Others are mildly toxic and require ingesting huge amounts, so you'll just need to be aware and keep on eye on things. Some are highly toxic and are best eradicated before pasturing sheep in the area where they're growing. In learning about these plants, also learn how to handle them — some can be poisonous to humans just from handling.

When plant poisoning is suspected, call your veterinarian promptly. Knowing what's causing the poisoning can increase the chances of effective treatment. Keep the sick animal sheltered from heat and cold and allow it to eat only its normal, safe feed.

Again, in most instances, animals do not happily eat toxic plants. However, if they are overly hungry and better food is not available, they may eat anything at hand. If they have been deprived of sufficient water for an extended period, this can cause them to reduce their feed intake. Then, when they suddenly get enough water, their appetites increase greatly and they may devour almost anything. Have water and feed available at regular times in needed quantities.

Overgrazing of pastures, which means shortage of grass, can cause sheep to eat plants that they would otherwise avoid. It is better to keep fewer sheep well fed and healthy than to keep more than your pasture and pocketbook can sustain.

Poisonous Plants

PLANT	EFFECT
Red maple leaves	Kidney damage
Acorns	Kidney damage
Rhododendron	Toxicity
Mountain laurel	Toxicity
Azalea	Toxicity
Yew needles	Extreme toxicity
Apple seeds	Poisonous, especially the green seeds
Nightshade	Poisonous to animals and humans (goats like it)
Skunk cabbage	Birth defects
Wild tobacco stalks	Birth defects
Potato sprouts	Birth defects
Death camas	Poisonous, especially in the spring
Horsebrush	Poisonous, especially in the spring
Lupine	Poisonous, especially in summer and fall
Milkweed	Poisonous, especially in summer
Chokeberry	Poisonous, especially in spring
Loco	Poisonous
Halogeton	Poisonous
Water hemlock	Poisonous, but not palatable
Larkspur	Poisonous during some seasons — dangerous (sheep like it)
Tansy ragwort	Poisonous

Toxic Substances

Other kinds of poisoning occasionally happen. For these, prevention is definitely the best cure. Store all chemicals and cleaning supplies where animals can't get into them, and always, always, always properly dispose of waste containers! Look out for the following substances, which are often found around the farm or ranch:

- **Waste motor oil,** disposed of carelessly.
- **Old crankcase oil** (high lead content).
- **Old radiator coolant or antifreeze** (sweet and attractive to sheep).
- **Orchard spray** dripped onto the grass.
- **Weed spray** (some have a salty taste).
- **Most sheep insecticidal dips and sprays.**
- **Old pesticide or herbicide containers, filled with rainwater.**
- **Old car batteries** (sheep like the salty taste of lead oxide).
- **Salt** — sheep require salt for health and when deprived, and then allowed free access, they may ingest large quantities, causing salt poisoning.
- **Commercial fertilizer.** Be careful not to spill any fertilizer where sheep can eat it, and store the bags away from the sheep. They may nibble on empty bags. Several rainfalls are needed after fertilizing a field, and it still may not be safe unless the pasture grass is supplemented with grain and hay. Symptoms of fertilizer toxicity are weakness, rapid open-mouthed breathing, and convulsions. For a home remedy, use vinegar, 1 cup (236.6 mL) per sheep, as a drench. A veterinarian's treatment will be more effective if it can be started soon enough.
- **Cow supplements containing copper.** Some cattle mineral-protein blocks contain levels of copper that are lethal for sheep. Some mixed rations intended for cattle may also have copper and should not be used for sheep.
- **"Empty" lead buckets filled with rainwater.** (Rainwater is soft and readily dissolves lead; lead-laced rainwater can kill a sheep that drinks it.)

SEVEN

General Health Considerations

This chapter and the next talk about health concerns surrounding sheep. This chapter covers health concerns that are universal to all sheep, regardless of sex or age, and concludes with a discussion of drug treatment options. Chapter 8 covers issues that are specific to rams, ewes, and lambs. (See chapters 9 and 10 for issues relating to reproduction and lambing.)

Some of the topics in these chapters are unpleasant, but all are subjects about which a shepherd really needs to be informed. Most important, however, is that these chapters emphasize management strategies that help to keep your sheep healthy in the first place. As C. E. Spaulding, DVM, says in his fine book, *A Veterinary Guide for Animal Owners*, "The most important 'drug' you can give your animals is good husbandry."

Successful treatment of illness requires detection as early as possible, before the sheep is "down." With the development of new medications, it's no longer true that "a down sheep is a dead sheep," but the chance for recovery is much better if illness is diagnosed and treated quickly, and prevention is always better than treatment.

My first and best advice to a new shepherd is: *Find a good veterinarian, with whom you feel comfortable, and seek his or her help often when you're first starting out.* As your competence level increases, your need for the veterinarian's assistance will decrease, but be sure to budget for frequent visits during the initial phases when you bring your animals home for the first time. (Luckily, vets work far cheaper than people doctors — although most are plenty smart enough to have gone that route instead. As a rule, they really love animals, and that's what directed them to their calling.)

Most rural areas still have large animal veterinarians who come right to your farm. These professionals are generally willing to answer even your "dumbest" questions when you're a rank novice, so don't be afraid to ask for their input, but do expect to pay them for their time.

Most veterinarians prefer to help you establish a program that keeps your flock healthy from the get-go, rather than responding to 2 A.M. emergency calls where everything around them is dying or dead. (Though most veterinarians will show up for an emergency at all hours of the day or night on any day of the year.)

If you happen to come across a veterinarian who doesn't want to answer questions or recommend management strategies, try another one. Your veterinarian should be able to:

- Help you decide on a vaccination program
- Understand poisonous plants or soil mineral imbalances in your area, and
- Offer input on other kinds of questions

My second piece of advice is to get a couple of other books on health care (see Resources for some suggestions). This is a big topic, and no one book can do it justice. The more you educate yourself, the better off both you and your critters will be.

Healthy Strategies

There are many causes of health problems. Some are lack of exercise, unsanitary housing, moldy or spoiled feed, toxic plants or other poisonous substances, improper diet (insufficient water or feed or overeating), parasites, injuries, infection from assisted lambing, germs from other sick sheep, abrupt change of feed, and stress (from weather, shipping, predators, and so on). So, if health is the goal, what are the strategies that help you meet the goal?

1. **Practice good sanitation.** As Norman Gates, DVM, says in *Sheep Disease Management*, "Without exception, the environmental factor most often associated with sheep diseases is 'poor sanitation.'" Clean bedding in barns; fresh, clean water and feed; and properly cleaning items such as needles and lamb bottles all go a long way toward keeping your flock healthy. Carefully store and dispose of regular cleaning products, herbicides, pesticides, pharmaceuticals, batteries, and anything else that may be poisonous. Empty packaging from these products should also be carefully discarded.

2. **Provide appropriate and adequate nutrition for all animals.** Animals that receive high levels of nutrients have much stronger immune systems than those that do not.
3. **Create a healthy environment. Provide adequate space for animals.** If they'll be kept in buildings at all, make sure there's good ventilation, but no drafts. When they're kept outdoors, make sure they can get out of the wind. They're flock animals, so provide them with "company" whenever possible. Avoid situations that can cause more stress than necessary. Manage pastures so animals aren't forced to eat too close to the soil or too close to their own manure.
4. **Always be observant.** Healthy animals are bright and attentive, have good appetites, and generally appear to be enjoying life. Their manure should be solid and pelleted. Their eyes and ears should be responsive to their environment. They should keep up with the rest of the flock.
5. **Purchase healthy animals.** When acquiring new stock, consider health to be one of the most important criteria. Any animals that don't appear to be completely healthy should be left with the seller. Review the seller's health records — oops, he has none? Go somewhere else. If you're purchasing an entire flock from one seller, consider making the purchase contingent on a veterinary examination. The money you spend for the examination up front could save you thousands of dollars down the line. Never purchase breeding stock at a sale barn or through a "jockey." (Jockeys act as middlemen, purchasing animals here and there, and then reselling them. A few are reputable and guarantee what they sell, but most don't.)
6. **Select healthy animals.** Health should also be one of the main criteria you use in making breeding selections. A ewe with a prolapsed uterus, a ram that suffers from foot rot, or a group of lambs that has scours are all good candidates for culling.
7. **Quarantine all new animals.** A quarantine period (at least 14 days) for new animals gives you the chance to look for signs of illness without danger of the new sheep spreading something to the rest of the flock. Even if they were purchased directly from someone you know well and you believe them to be in perfect health, they should still be quarantined. The person you bought them from may have an illness starting within the flock and not even know it. (This is the voice of experience talking: We purchased calves one time from a farmer we knew well and respected completely, but a disease was just starting in his herd. The farmer and I discovered it about the same time, and it

cost both of us money and some calves, not to mention causing lots of aggravation. Guess what — the disease was brought into his herd by some calves he purchased from someone he knew well, who had bought sick calves from someone he knew well. It was a vicious cycle, which any of us could have broken by following the quarantine rule, but we knew and respected the people we were buying from!)

8. **Treat all new animals.** The day you bring a new critter onto the farm, deworm it then and again exactly 14 days later. Also on the first day, trim the hooves and spray foot rot spray (10% zinc sulfate in water is very effective and doesn't stain wool). Vaccinate new sheep as appropriate. If you're bringing home your very first sheep, ask the seller to help you deworm and take care of the feet before the sheep are loaded in the trailer. This gives you the opportunity to learn how to do these procedures from an experienced hand.
9. **Isolate animals that appear ill.** If one sheep seems to be coming down with something, try to isolate it until you're sure. Again, you may prevent a disease from running through the entire flock by isolating the first one or two cases you see. And you'll be able to make good treatment decisions when the sheep is isolated in a controlled environment.
10. **Vaccinate.** Where you live and your goals for the flock will have bearing on your vaccination procedures. For example, if you plan on showing animals, you'll need a different regimen than if you plan on running a *closed flock* and never taking them anywhere. (A *closed flock* is one in which no new animals are introduced from outside the flock.) Your veterinarian can help you decide on an appropriate vaccination program that suits your goals and geographical location.
11. **Maintain a closed flock if possible.** Once you have healthy animals, one of the easiest ways to keep them healthy is to avoid unnecessarily introducing outside animals to the flock. Every time you bring in a new animal, you increase the likelihood of disease.
12. **Apply appropriate control techniques.** When problems crop up, *use the right medicine, at the right time, and in the right quantity!*
13. **Have your veterinarian perform a postmortem examination.** When an animal dies under any sudden or suspicious circumstances, the veterinarian should do a postmortem examination (or necropsy). This is especially important if the animal was mature and apparently healthy. The examination must be done quickly, as animal carcasses deteriorate quickly after death.

Recognizing Sick Sheep

You must be able to recognize normal behavior of your sheep, even for each individual animal, to know when one is acting abnormally. Have some quick and easy way of catching them when needed, like a corral where they can be fed and then enclosed. Signs of abnormality are loss of appetite, not coming to eat as usual, and standing apart from the group when at rest. Be concerned if a sheep is lying down most of the time when the others are not. Any weakness or staggering, unusually labored or fast breathing, change in bowel movements,

SHEPHERD STORY

Controlling internal parasites in sheep can be a real challenge, particularly in the humid Southeast, but Linda Phillips and partner Susan Gladin have turned this challenge into a small business.

When Linda first got into raising sheep, she struggled against internal parasites. "I used a commercial product to control the parasites, but I would still lose sheep to worms. I'd worm the flock, and 3 or 4 weeks later, lose a sheep. Granted, I didn't really know what I was doing yet — like, I didn't know to worm them once and then worm them again exactly 14 days later — but it was still horrible and expensive."

Linda began an extensive search for better ways to control the parasites. She researched all the information she could find, both within conventional and alternative systems. She studied the life cycles of parasites, and began having manure samples tested for fecal egg counts. In discussions with staff at the University of North Carolina, she learned that 1,000 eggs per gram of stool is the point at which worming is recommended.

By managing grazing she could reduce worm loads. "Sheep will eat right in the area where they just passed manure, so they pick the worms right back up. By moving them often, they are eating off cleaner ground." But she also learned that a grazing program by itself doesn't solve the problem. "The thing with parasite control, particularly alternative control, is that you have to have a real program."

Linda's search next led her to Susan, who was using a blend of herbs to control parasites. Susan had a basic recipe that included wormwood and some other herbs. The biggest problem with the recipe was that it wasn't very palatable, so the animals didn't want to eat it.

"Susan and I did more research into Chinese medicine and preparation of traditional herbal remedies. We played with the recipe until we came up with the one we're now using, which is more palatable and requires fairly small doses that can be mixed with grain or minerals. The animals eat it without any trouble."

change in "personality," wool slipping, hanging the head over the water source, or a temperature more than 104°F (40°C) all indicate possible problems.

The normal temperature of a sheep (except in very hot weather) ranges from 100.9 to 103.0°F (38.3–39.4°C). A veterinary rectal thermometer has a ring or a hole at the outer end, so you can tie a string for easy removal.

If you need to collect urine for a sample, such as for use with the pregnancy toxemia (ketosis) strips or glucose strips for enterotoxemia, try a plastic cup fastened to the end of a shepherd's crook handle. Impatient? Try holding the sheep's nostrils closed for a moment. This stress sometimes triggers urination.

Next, Linda and Susan began looking for a way to make on-farm, inexpensive test kits for checking fecal samples. "Some veterinarians charge up to $50 to do one test. We knew that an inexpensive method would improve the control." They developed their own kits.

The herbal preparation and the test kits were doing the trick, and through word of mouth, people were coming around to purchase them. So, in 1996, Linda and Susan began marketing their products through their own company, Farmstead Health Supply.

In 1997, the University of North Carolina ran a study comparing one of the top-selling, commercial worming preparations with their product. The sheep that the University used in the study were carrying egg loads right around 1,000 when the study began. After the sheep were treated with the commercial preparation, the egg load dropped down to 180 eggs per gram but climbed back up during the test period to over 1,000. The sheep treated with the herbal preparation only dropped down to about 600 eggs per gram, but unlike the commercially treated animals, their egg count remained constant

Linda says, "Some people put down alternatives. They say that they won't work, and that herbal medicine is just quackery. But, we have relied on our products, almost exclusively, for over 3 years, and all my sheep are still out there in the pasture breathing. If our products didn't work, I wouldn't have any sheep left."

Despite her reliance on herbal preparations, Linda indicates there are times when the knockdown power of the commercial preparations still have a role. "When we moved to our new place, I wanted the animals to come to the land with as low an egg count as possible, so I wormed them with a commercial product twice, 14 days apart."

Linda has the following recommendation for any animal health program: "Really watch your animals. If an animal looks like it's getting behind, test for parasites. With sheep and goats, if they're getting a heavy parasite load, the skin under their eyelids begins to look really pale, instead of pink."

Alternative Health Practices

The topic of alternative health practices is controversial in some circles. Many people scoff at practices such as homeopathy, acupuncture/acupressure, and use of herbal remedies. Personally, I don't agree — we have had good luck with alternative practices, and they have been in use in other parts of the world for considerably longer than our "modern" medicine. I concur that these methods aren't a replacement for current medical practices, but I believe that many alternatives deserve serious consideration.

Our earliest forays into alternative medicine began with homeopathy. Homeopathic preparations are made from a wide variety of natural substances. One of the first preparations we began using in our quest to minimize the use of antibiotics was homeopathic sulfur, and our success rate with it was good enough to convince us that alternative medicine wasn't simply quackery. Over the next few years, with lots of further study, we began using a wide variety of preparations for certain problems that crop up from time to time. Learning more about alternative practices takes time and research, but may be worth it for you.

Natural Defenses

Natural defense begins with physical barriers. Skin, wool, and hooves actually do a pretty good job of keeping most organisms away.

The second defense is simply washing away invaders. Bleeding, saliva, tears, and urine all help to flush invading organisms out and away from the body. Enzymes, which are chemicals that naturally occur in bodily fluids, also

Alternative Nutrition and Mineral Supplementation

An important part of alternative health practices relates to nutrition and minerals. Some farmers and researchers in New Zealand are reporting that by feeding fish meal to lambs (in other words, boosting their nutrition) they are seeing a reduction in worm problems. Doug Gunnick, who raises lambs organically in Minnesota, reports with a mixture of garlic and cayenne pepper added to his lambs' feed he has no trouble with worms and uses no chemical wormers. Other folks feed diatomaceous earth and report that it helps to reduce worm counts. (We use it ourselves.) Diatomaceous earth is fossilized shells of *diatoms,* which are prehistoric one-celled organisms.

help fight invaders. A wounded animal licks its wounds — not only does the licking remove pathogens, but the enzymes in the saliva help to disinfect the area.

Through self-interest (that is, by protecting their own territory),the normal flora help to keep yeast and fungi from getting out of control. This is why a long run of antibiotic use often results in yeast infections — it compromises the normal bacteria as well as the pathogenic bacteria.

Over time, constant exposure to organisms in the environment (like bacteria, viruses, and worms) results in the development of a degree of resistance, or immunity, to the organisms, especially in mature sheep. When a bad "bug" has breached the previous defenses, then the immune system kicks into gear.

White blood cells are always ready to mount a quick attack when an invader shows up — these cells are the first responders of the immune system. They fight the invader by mounting an inflammatory response, which is why redness, swelling, and heat around a wound happen quickly after injury. During an illness, the heat of an inflammatory response presents as fever, but not all illnesses are accompanied by a steady fever and fevers sometimes cycle between normal and elevated temperatures. Cycling is an interesting phenomenon: The body's temperature rises in an effort to kill multiplying bugs. The increased temperature reduces the number of bugs, and the body's temperature goes back down. If any bugs remain, however, they multiply again until the body responds by cranking up the heat.

In some minor situations, like a skin abrasion, the white blood cell response is enough to take care of things, but when an attacking virus or bacteria has gotten a strong start, the white blood cells might not be able to squelch it. While the inflammatory response is going on, the body starts building *antibodies*. If the white blood cells are the cavalry, then the antibodies are the army. It takes antibodies a little longer to mobilize, but when they hit the scene, they come in force. Antibodies can form to attack parasites, allergens, pathogens, and even cancer cells.

Antibodies are highly specific to the various invaders they fight and can take up to 2 weeks to form. But by using vaccinations, we can preprogram the antibodies. Through preprogramming, antibodies instantly recognize an invader and mount a full-blown attack almost immediately. Thus, the response time is cut from 2 weeks to mere hours, which is almost always fast enough to quell an invading organism's assault.

Strong, healthy animals have strong, healthy immune systems. Their bodies take care of almost all potential problems, but occasionally, even antibodies aren't enough to overcome an invading organism. At this point, medical intervention is needed.

Causes of Illness in Sheep

Chemical, biological, and physical "agents" cause illness. Chemicals include toxins such as pesticides, cleaning products, and batteries. Biological agents are probably the most common cause of illness and include parasites, bacteria, viruses, fungi, and yeasts. Physical agents are environmental factors, like drafts, mud, handling stress, and improper diet. Physical factors rarely cause illness or death by themselves (though hypothermia can readily kill), but they most definitely worsen situations in which a chemical or biological agent is involved by causing stress and weakening the animal's immune system.

Chemical Agents

Chemicals cause poisoning. Some chemicals are biological in nature — for instance, poisonous plants — but it's the chemicals in the plant that cause illness. Poisoning can be avoided by diligently paying attention to what the sheep can access. Remember to properly handle and dispose of anything that's chemical in nature. Try to be conscious of the environment where your sheep spend time; for example, if the sheep spend time in an old barn with lead paint, you could have trouble. And be sure to learn what poisonous plants grow in your area.

Biological Agents

When a biological organism causes disease, it's called a *pathogen*. Most often, pathogens are introduced organisms, though sometimes, even normal flora can become pathogenic. Most of the diseases we think of, like pneumonia and scours, are caused by bacteria or viruses, though parasites cause more general health problems in sheep.

Bacteria are single-celled organisms. Some are crucial to good health, but others are pure trouble. Bacteria can be treated with antibiotics, though some bacteria are resistant to certain antibiotics, meaning that the antibiotic doesn't work against that particular bug.

Unlike bacteria, viruses do not respond to antibiotics at all, period, end of story. (Those antibiotics that the doctor gave you for a sinus infection won't help with your daughter's cold. In fact, taking antibiotics for a virus may increase bacterial resistance.) Scientists and medical folks now believe that a lot of the antibiotic resistance we're seeing stems directly from improper use in animal agriculture, such as routine feeding of antibiotics and not following withdrawal times.

Yeasts and fungi don't often cause problems for shepherds, though they can cause some skin problems, respiratory infections, and *mastitis* (that is, an infection in the ewe's udder). Yeast and fungal infections often follow extensive use of antibiotics.

Parasites

For shepherds, internal and external parasites are the most common biological bad guys. In comparison with other animals, sheep are more resistant to bacterial and viral diseases but are more susceptible to parasites. Although parasites can affect all ages of sheep, they are deadliest for lambs and young ewes.

Parasites aren't a single class of organisms, but run the gamut from protozoa (single-celled members of the animal kingdom) to far more complex organisms, like worms and insects. In simple terms, a parasite obtains food and/or shelter from another organism. Most parasites are relatively benign (for example, normal flora meet this definition). Some, like biting flies, are mainly a nuisance. But others, including intestinal worms, may cause serious (and even fatal) illness. Sheep that become weakened and run-down by a parasite infestation can be killed by the parasites themselves or fall victim to a secondary bacterial or viral disease.

Parasites are capable of attacking most parts of the body. Luckily, some of the worst parasites aren't found in North America, though in the southeastern United States, internal parasites are considered a limiting factor in sheep production. Strong, healthy animals that are managed on clean pastures are less likely to have severe parasitic infections, and some breeds — like the Gulf Coast Natives — are known for a high tolerance to parasites.

Internal Parasites

A heavy load of internal parasites causes a vicious cycle of undernourishment of the sheep that in turn makes them more vulnerable to parasites. Deaths from parasites occur most often in lambs, yearlings, extremely old sheep, and poorly fed sheep. Internal parasites reduce productivity and cause anemia, wool break, progressive weakness, and sometimes death.

Worms are the predominant internal parasite in sheep, and there are eleven species that cause problems for shepherds in North America. They typically inhabit the abomasum (true stomach), small and large intestines, heart, and lungs.

Young lambs have far less tolerance to a worm infestation. A worm load that would have no impact on an adult can quickly kill a lamb, and the dying lamb may not look thin. To reduce infestation in lambs that are on pasture with the ewes, practice what is called *forward creep-grazing*; that is, let the lambs graze each clean pasture ahead of the ewes through a creep-type fence opening that the ewes cannot get through.

Identifying Worm Infestations

Signs of a heavy worm load in sheep include:

- Anemia
- Scours (diarrhea)
- Coughing
- Weight loss
- Potbelly
- Wool break
- Bottle jaw

Some of these signs are self-explanatory, but the others may be new to you.

Anemia. Anemia is usually the first symptom of roundworm infection, though it's not always easily discernable. Anemia shows up as a very pale, grayish color of the inner, lower eyelids and gums, which in healthy animals are a fairly bright pink. This disorder is the direct result of intestinal worms drinking the sheep's blood — up to a pint (0.5 L) a day in heavy infestations. The sheep becomes listless, has pale mucous membranes, and loses body condition. It may waste away and die if it is not dewormed.

Scours. The small, brownish stomach worm, *Ostertagia*, is the main culprit in cases of scours. This worm is so perfectly camouflaged against the walls of the sheep's intestine that it may be difficult to spot even in a postmortem examination.

Roundworms: Common and Problematic

Roundworms that inhabit the digestive system are the most common internal parasites in North America. *Haemonchus contortus,* the large stomach worm, is common and problematic for sheep in high rainfall areas, and *Ostertagia circumfecta,* the brown stomach worm, is more common in drier areas. Roundworms feed on blood and bodily fluids from the stomach lining, causing anemia and serum loss. Although each worm only takes a few drops of blood per day, in a heavy infestation thousands are present and the blood loss can quickly become overwhelming.

Potbelly. Potbelly is a visible phenomenon in which the animal appears really skinny but has a great big belly. If members of the flock that aren't pregnant appear to be, then potbelly is what you're seeing.

Wool break. In wool break, the fleece begins falling out at the roots or breaking off just above the roots, making the sheep look quite motley. A few breeds of sheep do shed their wool in the spring, and some breeds in areas of heavy rainfall lose wool along the backbone — but if you don't happen to have one of those breeds, all fleeces should look solid and healthy.

Bottle jaw. Bottle jaw is visible swelling under the jaw, and it's one of the last symptoms to manifest itself — so if you see it, consider it a final warning that the sheep have worms severe enough to cause deaths.

Less Common Internal Parasites

The stomach worms are the most common troublemakers for sheep, but a number of less common parasites can also cause problems.

Lungworms. Prevalent in low-lying or wet pasture, lungworms live in air passages and cause coughing, rapid breathing, and sometimes discharge from the nose. The coughing can precipitate prolapse during pregnancy. The small lungworm (hair lungworm) can cause pneumonia and bronchitis.

To prevent lungworm infestation, keep the sheep away from ponds and wet areas where snails can be found. Several species of slugs and snails act as intermediate hosts for the lungworms — that is, the parasite spends part of its life cycle in another creature. This is something to consider when buying sheep from a farm having low-lying pastures.

Tapeworms. The feeding head of the tapeworm injures the intestine and is thought to facilitate absorption of the toxin involved in enterotoxemia. (If you have vaccinated against enterotoxemia, then this isn't a problem.) Tapeworms are not usually the primary worm infestation in a sheep, but since tapeworm segments passed in the feces are large enough to be seen in the sheep droppings, their presence is alarming. A moderate level of tapeworm infection causes little damage to adult animals but can seriously retard the growth of lambs.

Liver flukes. Liver flukes require a snail or a slug to act as an intermediate host. These hosts are found on wet, marshy land. Ponds, ditches, and swampland all provide a breeding place for the snails, so fence sheep out of these areas if possible. Snail-destroying chemicals are available but must be used with caution: Most of these chemicals contain copper sulfate, which would poison the pasture for the sheep. In addition, they cannot be used in areas that drain into water inhabited by fish or water that humans or livestock use for drinking.

As with other parasites, liver flukes sometimes cause potbelly and bottle jaw during the earlier stages, followed by loss of body condition, diarrhea, weakness, and death. Liver fluke infection can be diagnosed accurately in the liver of a slaughtered sheep and sometimes can be diagnosed by microscopic examination of feces.

Coccidiosis. Coccidia are microscopic protozoan parasites that are present in most flocks but rarely cause problems.

It once was believed that each species of animal had its own type of coccidia and that cross-infestation did not happen. Later experiments have proven that some types of coccidia (there are a number of them even within those that infest sheep) are transmissible to different animal species, which act as intermediate hosts. When coccidia-infested muscle tissue (or even intestinal tissue) was fed raw to dogs, they became infected and passed sporocysts. However, cooking or freezing apparently renders these parasites noninfectious, so meat fed to dogs or cats that come into contact with livestock should be previously cooked or frozen.

The symptoms of coccidiosis are:

- Diarrhea
- Diarrhea with straining
- Chronic dark-green or bloody diarrhea
- Loss of appetite
- Death on some occasions

Causes and Prevention of Coccidiosis

Outbreaks of coccidiosis happen mainly in 1- to 3-month-old feedlot lambs being raised in crowded conditions other than in the pasture arrangement of a farm flock. Any rapid change of feed ration may predispose the lambs to coccidiosis, which usually appears within 3 weeks of the time that they are brought into the feedlot. Other factors are chilling, shipping fatigue, and interruption of feeding during shipping. Small amounts of coccidial oocysts may be found in most mature sheep, but these sheep develop immunity, so they seldom show symptoms of infestation. However, they can contaminate their surroundings so that weak lambs may become infected. To prevent this, lambs should be fed during shipping and should not have their ration changed too abruptly from grass to concentrated feed. Overcrowding and contamination of feed and water must be prevented, for this is the main source of infection.

The lambs that recover usually become immune. Routine fecal samples that show evidence of the parasite allow you to start treatment at an early stage.

Life Cycle of Worms

The life cycle of most worms is similar. Millions of eggs from these worms are passed with the feces. (For example, each *Haemonchus* worm is capable of producing 500 eggs per day!) Under favorable conditions (that is, warm, humid weather), the eggs hatch into infective larvae in 5 to 10 days. These larvae migrate onto the moist sections of the grass and are ingested by the sheep. Once swallowed, they invade the tissues of the digestive tract where they undergo a maturing stage and emerge as adult worms in about 21 days.

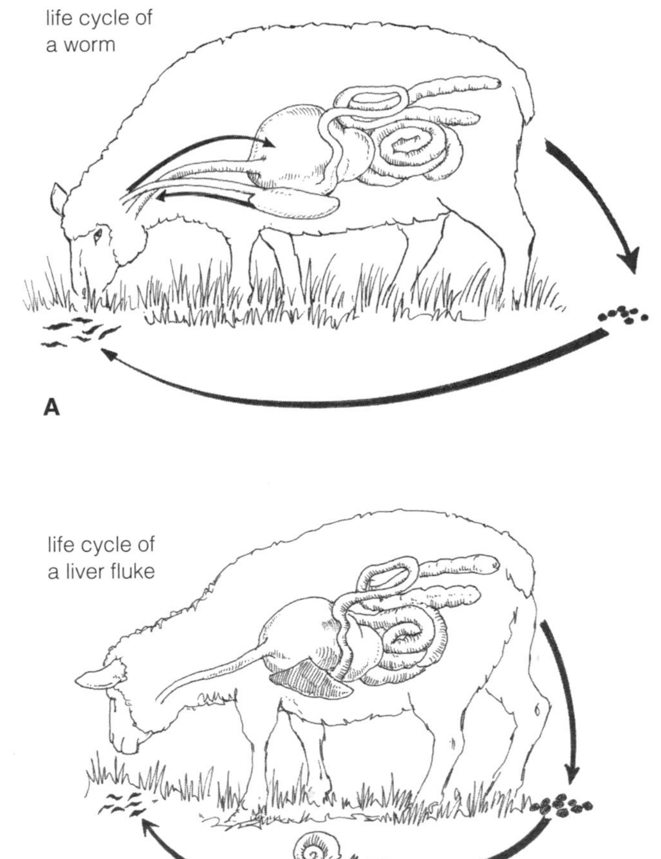

A. *Most internal parasites have similar life cycles. Sheep ingest worm larvae while eating, and pass eggs in their manure. The eggs hatch in the soil, and the larvae migrate on the grass.* **B.** *Flukes require an* intermediate host, *in this case a snail that picks up the eggs and discharges the larvae.*

Eggs or larvae or both usually die during cold, freezing temperatures or hot dry weather; in many parts of the United States, these weather conditions tend to "sterilize" the pasture. However, nature has devised a diabolical survival mechanism for these worms that allows them to survive adverse conditions by hibernating as immature worms in the tissues and emerging weeks or months later when conditions for survival are more favorable. Hibernating larvae are discussed in greater detail later, as they figure strongly in the design of an effective worming program.

Internal Parasite Control

Sanitation is critical for internal parasite control. Feed shouldn't be placed on contaminated ground or bedding. Water tanks and creep-feeders should be kept clean and free of contamination by fecal matter.

On pasture, especially set-stocked pastures, a main factor contributing to heavy parasite loads is population density. A few sheep in a given area deposit fewer eggs than lots of sheep in the same area. Thus, fewer infective larvae will develop. Sheep that are set stocked generally have lower-quality feed, which weakens them. They're also forced to crop the grass that is too close to the ground, which results in ingesting more larvae, increasing the worm load, and aiding parasites in completing their life cycle. When you move sheep from one paddock to another with managed grazing (see chapter 3), you gain some measure of control by allowing time for the worm larvae to die. Sheep managed this way are also generally better nourished, which makes them stronger.

The eggs and larvae of many worm species can survive as long as 3 months in cool, damp weather but may die within a few weeks during dry, hot weather. According to an old Scottish adage, you should "Never let the church bell strike thrice on the same pasture." Thus, take these weather-related influences into account when planning your sheep's movement. In moderate, humid climates, turn the sheep out in the spring onto a paddock that was used in the heat of summer instead of the one they were on during the fall. If you live in the far north, you can turn your sheep out in the spring on the last paddock they grazed in the fall.

Malnourished sheep cannot tolerate as much worm burden as well-nourished sheep can. Lack of proper nutrients, insufficient protein, and unbalanced nutritional elements, including lack of vitamins and minerals (such as selenium), can leave them more vulnerable to worm damage.

Worming

There are many drugs available to help control worms. (For more about specific worming drugs, see the Drugs for Sheep section at the end of this chapter.)

Until recently, the common approach for shepherds was to set up a worming schedule and follow it religiously by worming every 3, 4, 6, or 8 weeks. The problem with this method is that it is costly and has led to drug resistance in the parasite population.

Today, more shepherds and the veterinarians who work with them are coming to terms with flexible worming programs that are tailored to the needs of the individual farms. For example, on farms where worms are only an occasional hazard — such as when pastures are understocked, sheep and cattle are grazed alternately, and pastures are rotated — treatments may only be needed when parasitism becomes evident through fecal tests. For areas in which worms are especially bad, such as the Southeast, regular use of herbal products (see the Shepherd Story, pages 166–167) and regular testing of fecal samples provide a good strategy. If the fecal samples have particularly high worm counts, then supplemental use of a pharmaceutical wormer (or *anthelmintic*) is in order.

Keep in mind that frequent worming increases selection pressure on the worm population, and resistance may become a problem sooner for you than for your neighbor who worms less often. "How often do I need to worm?" has the same answer as the question, "How often do I need to wash my car?" When necessary!

Lambs that are run on pasture with ewes are more prone to problems than mature animals, so you may need to worm them separately and more frequently than the other animals. If you use pharmaceutical wormers for lambs, do so at about 2½ to 3 months of age, and be sure to note withdrawal days on the label before slaughter.

Worm resistance. Resistance is a process of selection rather than one of development. Contrary to popular opinion, worms, insects, and bacteria cannot alter their genetic makeup and become resistant to a particular drug or insecticide simply because they have been exposed to it — for example, in the manner that your skin develops calluses and thus resistance to blisters by exposure to a shovel handle. Instead, individuals with higher-than-average resistance to something are more likely to survive and pass that trait on to their offspring.

Resistance increases with the frequency of treatment because we keep killing the susceptible worms and leaving the resistant ones to regenerate the population. Resistance can also develop from improper use of a drug, so always follow instructions to the letter. This is part of the reason why you should never give too much or too little of any kind of medication. By not following the recommendations, you may allow a *marginally* resistant worm to survive and propagate offspring with greater natural resistance when it might have been susceptible to the full dose.

How Resistance Develops: An Example

If we used a fictitious wormer called XYZ, it would kill all the worms susceptible to XYZ. The few individuals that were naturally tolerant of XYZ would not die, and their offspring would carry the resistance factor. If we continued to use XYZ to worm the sheep, the percentage of XYZ-resistant offspring in the population would increase with each generation until the majority of the population was XYZ resistant. It would be like having a drug that prevented white sheep from conceiving, but black sheep were naturally resistant to it. If we fed that drug to a flock with both black sheep and white sheep, sooner or later most of the flock would be black — not because the black sheep developed resistance by their exposure to the drug, but because they were selected from the population.

However, every type of parasite does not develop resistance to a particular wormer any more than every species of bacteria develops resistance to a particular antibiotic. Some species do, and some species don't. Do not assume that just because you use a wormer that the parasites are necessarily developing resistance to it. In fact, the old recommendation to change wormers often to avoid developing resistance is wrong — evidence suggests that this practice may be the fastest way to cause the worm population to become resistant to everything. However, it is not recommended that you use the same wormer until you see resistance beginning to develop. When you do change wormers, make sure to switch the class of wormer because resistance usually develops along chemical class lines, not brand names. Read the fine print for the generic name or chemical class. If you can't figure it out, ask your veterinarian or county Extension agent for help.

The more effective a wormer is on all the different species (broad-spectrum) the less chance of selection for resistant strains. With a highly effective drug, the worm numbers become so depleted that they lack the genetic variability required for selection for resistance in a short time.

To determine whether the worm is becoming resistant to the drug you are using, you must do an egg count. If egg counts are done just before and then 1 week after administration of the correct dose of a wormer and the percentage decrease in the number of eggs is less than 80 percent, the presence of anthelmintic-resistant parasites must be strongly suspected.

To avoid introducing resistant strains when bringing new sheep into your flock, treat all incoming sheep, preferably before you bring them to your farm.

Depending on your locale, you may benefit from treating with two different wormers. Ask your veterinarian. After treatment, the animals should be penned in a dry, grass-free area for 24 hours to avoid contamination of pasture with viable nematode eggs that did not pass out of the sheep when the worms are killed.

Targeted worming. In most sheep-rearing areas, only approximately 5 percent of the worm population survives on the pasture during the winter months. Thus, the remainder of the spring's worm population is in the sheep in the form of *hypobiotic* (arrested) larvae that are encysted in the tissues. Once the sheep gain access to the pasture during favorable weather conditions, the ratio reverses, with 95 percent of the worm population in the grass and 5 percent in the sheep. It is possible to worm the sheep, but impossible to worm the pasture. Logic then dictates that the most opportune time to deal a severe blow to the new season's worm population is before the sheep begin grazing the pasture, so that they cannot transfer or "seed" the pasture with a new worm population. The sheep should be wormed 2 or 3 days before turning out on pasture so that the eggs are left in the feces in the barn or lot where the larvae cannot survive.

Many people practice a double-drenching with an oral worm medication in the summer in dry climates, which they say reduces the worm burden for the season. They worm the ewes during dry weather, and following up with a worming 6 weeks later; this normally reduces the worm burden below harmful levels. Hot, dry weather significantly reduces the larvae population in the pasture, thereby reducing the infection rate in the ewe. It is also helpful to move sheep to a clean pasture 24 to 48 hours after each worming to help keep the pasture clean.

External Parasites

Most external parasites are more of a nuisance than a major health threat for sheep, but there are some exceptions. However, the following external parasites can all have substantial economic effects on a shepherd's operation and can occasionally be deadly.

- Sheep ticks
- Lice
- Maggots
- Nose bots
- Scab mites

Sheep Ticks

The sheep tick is not a true tick, but a wingless parasitic fly that is known as a *ked*. These ticks pass their whole life cycle on the body of the sheep. They lay little brown eggs that are white inside. The eggs hatch into almost-mature keds in about 19 days.

Sheep ticks suck blood and roam all over the sheep, puncturing the skin to obtain food. These puncture wounds cause development of firm, dark nodules, which damages the sheepskin and reduces its value. (These defects are called "cockle" by leather dealers and at one time were thought to be caused by nutritional problems.)

The ticks produce irritation and itching, causing the sheep to rub, scratch, and injure their wool. Some sheep bite at themselves to relieve the suffering and sometimes become habitual wool chewers. Wool chewers may get impacted rumens from eating the wool.

Ticks reduce weight gain by causing anemia and by impairing the quality and yield of the meat. Tick feces diminish the value of the wool by causing stains that do not readily scour out. Wool stained by ticks is sometimes referred to as "dingy."

Ticks can be easily eradicated with systematic treatment. Mature ticks lay only one egg per week, for a total of a dozen or so in their lifetime. The eggshells become attached to the wool about ½ to 1 inch (1.3–2.5 cm) from the skin. Therefore, most of them are removed in shearing, making them easy to eliminate. Newly hatched ticks die within an hour unless they can obtain blood from a sheep. A mature tick cannot survive more than 2 to 4 days away from the sheep.

To be effective, the whole flock must be treated for ticks at one time, oth-

Tick Treatment: A Historical Perspective

In the nineteenth century, adult sheep were seldom treated for ticks. Since the shearing was done later in the spring than is common now, the heat of the sun and the scratching of the sheep drove most of the ticks onto the nicely wooled lambs. Herders waited a few weeks after shearing, then dipped the lambs in a liquid tobacco dip, sometimes with soap added. The vat used was a narrow box, with a slatted, grooved shelf at one side. The lamb was lifted out and laid on the shelf. Then the workmen squeezed the fleece, letting the dip run back into the box. By reusing the dip, 5 or 6 pounds (2.3–2.7 kg) of cheap plug tobacco could treat 100 lambs and was quite effective on the ticks, although the mature sheep still had enough ticks left to get a good start on the next infestation.

erwise the untreated sheep pass the ticks back to the treated sheep. Examine a new lamb or sheep before turning it in with your own, and treat it even if you find a single tick.

Don't make the mistake of leaving any of your sheep with ticks. Every sheep must be treated in one session.

Lice

Next to sheep ticks, lice are probably the second most troublesome external parasites. One species of biting lice and several species of sucking lice affect sheep.The eggs are attached to the individual wool fibers and hatch into the nymph stage 1 to 2 weeks later. After several molts, which require another 2 to 3 weeks, the nymphs emerge as adults. One interesting fact about lice is that their most active time is often during the winter months, making them the only pest that is active all winter.

Through feeding, lice cause intense irritation and itching, which results in restlessness, constant scratching and rubbing against walls and fences, interrupted feeding, weight loss, and severe damage to the wool. The cardinal sign of lice in the flock is hundreds of telltale tags of wool hanging from things like fences and trees where the sheep have been rubbing.

Lice are susceptible to the commonly used insecticides, but two treatments are often needed to kill newly emerged nymphs (as the egg is a protected stage). Once removed from the sheep, they will not return unless you introduce lice-infested sheep into the flock. If in doubt, treat any new animals before placing them in your flock to prevent reinfestation.

Maggots

Maggots are the larvae of several types of flies known as blowflies. These flies are about twice the size of a big housefly, and most have distinctly shiny bodies that are colored blue, green, or silver. They appear in the spring and reproduce from that time through the hot weather. Flies lay their eggs at the edges of wounds or in manure-soiled wool, and when they begin to attack, it's called flystrike. The eggs hatch in 6 to 12 hours, and the maggots feed on the live flesh at the edge of the wound. They enlarge the wound and, if not detected, can eventually kill the animal. Maggots often infest dog bites, so if your sheep have been chased by dogs, check them for unnoticed wounds and flystrike.

Maggots are not necessarily a big deal — the most important aspect of dealing with them is overcoming your own revulsion (they are rather disgusting!). You can get rid of them quite easily if the sheep is not too heavily infested. The real danger is not knowing they're there. To avoid needlessly losing animals, catch and examine them if anything looks at all suspicious.

Watch for flystrike, which is usually indicated by large numbers of flies continually harassing an individual animal. Observe the sheep for moist areas in the fleece, and monitor injuries for infestation. Notice if animals scratch excessively on fences. Look for white specks in the wool that resemble individual curds of cottage cheese — these are the egg sacks that the flies deposit.

When you locate an infestation, clip all the wool around it and spray with a strong hydrogen peroxide solution. If the maggots have gone "deep," some other product, such as an avermectin product may be necessary (see Drugs for Sheep on page 194). Depending on the nature of the wound, you may need to spray the sheep with a fly repellent until the wound heals. Even when the sheep is sprayed, it should be kept under close observation for a few days and treated again if needed. If the sheep has not been sheared, you might want to shear it after treating the area and removing all the maggots to make it easier to spot other infestations.

The following measures will lessen the chances of trouble with maggots:

- Keep rear ends of ewes regularly tagged, especially any time that droppings become "loose" from lush pasture or stomach worms. Urine can also attract blowflies if it soils heavy tags.
- Treat all injuries with fly repellent during hot weather. Injuries and even insect bites can invite blowflies.
- Put fly repellent on docking and castration sites on lambs in warm weather. Check the wounds periodically until they have healed. To avoid this problem, dock and castrate early in the spring or later in the fall — just try to avoid hot weather.
- Use fly traps or large electronic bug killers to cut down on the number of flies in the barn and surrounding areas.
- Be especially vigilant during prolonged wet periods in the summer. Warm and moist conditions invite flystrike.

If you have a maggot problem with your whole flock, which is unlikely unless they were attacked by dogs, you can use sheep dip on them.

Nose Bots

The nose bot, *Oestrus ovis*, is a mature fly that is dark gray and about the size of a bee. Both the mature fly and the larvae can cause problems for sheep. The full-grown larvae are thick, yellowish white grubs with dark transverse bands and are about 1 inch (2.5 cm) long. These pests are found primarily in the frontal sinuses of sheep. When deposited by the fly on the edge of the nostril, the grubs are less than 1⁄12 of an inch (0.2 cm) long. They cause irritation as they crawl through the nostrils and sinuses and gradually move up the nasal passages. The resulting inflammation causes a thin secretion that becomes quite thick if infection occurs. These thickened secretions can make it difficult for the sheep to breathe, and it may sneeze frequently. The sheep can become run down because of their lack appetite or from the stress of being so annoyed by flies that they cannot graze in peace.

Sheep put their heads to the ground, stamp, and suddenly run with their heads down to avoid this fly. They often become frantic and press their noses to the ground or against other sheep as the flies attack them. This is usually during the heat of the day, letting up in early morning and late in the afternoon, and is more prevalent in areas with a hot summer.

Scab Mites

Several kinds of parasitic mites produce scab in sheep. *Psoroptes ovis* is the common scab mite. This mite is a pearl-gray color and about 1⁄40 of an inch (0.06 cm) long, with four pairs of brownish legs and sharp pointed brownish mouth parts.

Mites puncture the skin and live on the blood that oozes from the wounds. The skin becomes inflamed and then scabs over with a gray, scaly crust. The wool falls out, leaving large bare areas.

To determine whether mites are present, scrape the outer edge of one of the scabs (mites seek the healthy skin at the edge of a lesion) and put the scrapings on a piece of black paper. In a warm room under bright light, examine the paper with a magnifying glass. The mites become more active when warm and can be seen with the glass.

The common scab mite, often called "mange mite," is still a reportable disease in most states but has been all but eradicated in sheep. However, if mites are found, all sheep must be treated in one session because mites are quite contagious. Clean sheep should not enter infected premises for 30 days.

Other Disorders of Sheep

Parasites may be the most common problems for shepherds, but a number of other disorders and diseases can crop up. Within the following groups of disorders, the dietary disorders are the most common.

Dietary Disorders

Common dietary disorders include bloat and grass tetany.

Bloat

As mentioned in chapter 6, bloat is a form of upset stomach with potentially deadly consequences. Bloat is caused by the inability to adequately expel the gas that is constantly being produced in the rumen.

Bloat is most often caused by sudden changes in diet, especially a change from a dry feed to a lush feed. Bloat is more common in sheep being grazed on legume pastures than in those being grazed on grass or grass–legume mixed pastures. It is also more common when pastures are wet after a rain or from early-morning dew. Barley is the one grain that is commonly associated with bloat in sheep.

The left side of a bloated sheep blows up like a balloon; the sheep goes off feed and acts very uncomfortable. Prevention is the best medicine: Feed dry hay before turning the sheep onto lush pasture.

How to Treat Bloat

If signs of bloat appear, they must be addressed quickly.

1. Get a cup of liquid down the sheep that consists of ½ cup (118.3 mL) of water with ½ cup (118.3 mL) of cooking oil. Add 2 tablespoons (29.6 mL) of baking soda and mix well. For a young animal, ¼ cup (59.2 mL) is probably sufficient. For a full-grown animal, although 1 cup (236.6 mL) is ideal, try to force at least ½ (118.3 mL) of a cup.
2. Place a stick (a piece of doweling works well) in the sheep's mouth as you would a horse's bit. This bit gets the animal to work its mouth quickly, which helps to kick start the belching mechanism.
3. If the animal doesn't begin belching and seems to be getting worse, you can insert a stomach tube to vent the gas.
4. The last resort is to puncture the rumen with a sharp, sterile knife. (We've never had to go beyond the liquid and bit!)

Grass Tetany

Grass tetany is a nutritional disease that results from a deficiency of magnesium. Affected animals (often lactating ewes) avoid the rest of the flock, walk with a stiff and unnatural gait, and quit eating. They may appear nervous, twitch or stagger around, or may frequently lie down and get back up. In the late stages, the animal lies flat on its side and pedals its legs at the air; its breathing becomes labored, which is followed by convulsions and death.

Grass tetany is seen most often in areas that have low levels of magnesium in the soil and is worse if the area also has high levels of potassium or nitrates. Your county Extension agent can advise you on magnesium levels in your area. The disease is less likely to be seen in sheep grazing pasture that has a good mix of legumes in with the grass, because these deep-rooted plants help to bring magnesium up within the soil matrix.

The recommended treatment is 50 to 100 mL of a 50 percent saturated magnesium sulfate solution injected subcutaneously. For prevention, shepherds in areas where grass tetany may be a problem should provide a high-magnesium, trace mineral, salt mixture.

Diseases Caused by Viruses

Several disorders in sheep are caused by viruses. Aside from ovine progressive pneumonia, most of these disorders are not often fatal in full-grown sheep; however, bluetongue can be fatal in lambs, and soremouth has *zoonotic* potential — in other words, it can be transmitted from animals to humans.

Bluetongue

Gnats carry the virus that causes bluetongue. This disorder occurs in warm, wet weather, when the gnats are out. Affected sheep develop a high fever and inflammation of the mucous membranes. Subsequent signs include ulcerations around the lips, tongue, and dental pad and crusty discharges from the nose. Though generally not deadly to mature sheep, it causes birth defects in a high percentage of lambs if the ewes have the disease during pregnancy. Because it is a viral infection, it must run its course, but your veterinarian might prescribe antibiotics to stave off secondary infections.

Ovine Progressive Pneumonia

Any disease can cause chronically thin sheep, but if nutritional deficiencies and parasites have been eliminated as a reason, the next most likely cause is ovine progressive pneumonia (OPP). Ovine progressive pneumonia is a slow virus that is similar to AIDS in humans, taking at least 2 years to manifest its signs. The virus slowly causes progressive lung damage. Ewes gradually lose stamina and body condition and have serious breathing problems, ending in physical weakness and fatal pneumonia.

While there is currently no cure for or vaccine against OPP, there are new tests that make disease control possible. To avoid OPP, all breeding animals must be tested annually, and infected animals must be eliminated. Be sure to purchase only OPP-free breeding stock replacements. Because OPP is transmitted from ewe to lamb primarily through milk, a valuable breeding ewe that has OPP should be isolated from the flock, with her lamb taken immediately at birth and raised on colostrum replacer and lamb milk replacer. This is almost 100 percent effective in preventing OPP in the offspring of these ewes.

Any animals that test positive for OPP should be isolated from the rest of the flock, because it can be transmitted from infected animals via respiratory secretions when animals are confined to crowded quarters. All sheep that test positive do not necessarily come down with the disease. However, once signs appear, the disease is invariably fatal. Any animals that tests positive should be isolated and culled if the flock is to be protected. Ovine progressive pneumonia is another example of a "purchased disease" and certainly underlines the need to be extremely careful when buying the initial flock and any replacement animals. Animals being considered for purchase should be tested before introducing them into the flock to protect your sheep from disease. Request proof that the flock has been tested for OPP.

Soremouth

Soremouth, the scientific word for which is *ecthyma,* is caused by a contagious virus that can be transmitted fairly easily from sheep to humans. If soremouth is suspected, wear gloves while handling your sheep. The symptoms start with pustules and scabs on the mucous membranes — including lips, eyes, and teats — and feet.

A vaccine is available for this disease, but standard veterinary practice is to only vaccinate animals if an actual outbreak has occurred. Animals that are already displaying clinical symptoms are vaccinated with the rest of the flock. The reason for not vaccinating before the disease has shown up is that the current vaccine is a *live* vaccine, meaning that it could cause an outbreak if given to a clean flock.

Disorders Resulting from Bacteria

The good news about these bacterial diseases is that they aren't actually very common. The bad news is that when they hit, they hit hard, often with high mortality rates.

Blackleg

Caused by the bacterium known as *Clostridium chauvoei,* this disease is characterized by swelling in the heavy muscles and some lameness. Initially the disease is accompanied by fever, but by the time the other symptoms become evident, the fever may have broken, and the animal's temperature may be subnormal. Most often, the disease follows an injury, though occasionally it seems to come out of nowhere.

Blackleg is often a fatal disease, but if caught early, treatment with antibiotics may be effective. In areas where *Clostridium* species are prevalent, vaccination is the best option.

Boils

Caused by the bacterium *Corynebacterium ovis,* boils (or caseous lymphadenitis) is seen throughout North America. Once introduced into a flock, it seems to affect most of the members. The bacteria are thought to initially enter the sheep's body through a nick or a cut. Once in, the bacteria start to form an abscess in which they live. These abscesses may be external in the areas where there are lymph nodes, but they can also be internal. The external form often clears up, but the internal form is generally deadly. Boils are thought to spread from one sheep by the bacteria escaping from ruptured boils.

If you notice any kind of abscess, isolate the sheep immediately and have your veterinarian drain the abscess and test it to determine if it is indeed C. *ovis*. If so, the sheep is probably best culled, unless it is a high-value animal. In the case of very valuable animals, if the disease is caught early, the veterinarian may be able to surgically remove all the abscesses. Never try to drain an abscess yourself. When done improperly, drainage just spreads the disease to the rest of the flock and maybe even to people. There is a vaccination available, and when purchasing new sheep, an enzyme-linked immunosorbent assay (ELISA) is available to test for the disease.

Johne's Disease

A chronic, infectious disease with worldwide distribution, Johne's (pronounced *yown-ees*) disease is caused by a hardy bacteria by the name of *Mycobacterium paratuberculosis*. These bacteria are distantly related to those that cause tuberculosis and leprosy in humans. Unlike most pathogenic bacteria, which die pretty quickly when they're out in the environment and not in a host, these bugs have been documented to live up to a year outside of a host.

The main symptom of Johne's disease is unexplainable weight loss, despite reasonable food intake. Weight loss may be accompanied by intermittent diarrhea. This combination of weight loss and diarrhea is easily confused with other diseases.

Although many animals in infected flocks carry the organism, many develop an immune response that permits survival. It can be tested for by running a bacterial culture on fecal matter and by having an ELISA done. Unfortunately, there is no approved vaccine for sheep.

Listeriosis

Although the bacteria that causes listeriosis *(Listeria monocytogenes)* is a relatively common soil microorganism, the disease seems to only be associated with feeding spoiled silage. If feeding silage, make sure that it is properly ensiled and unspoiled. There are no effective treatments for this disease, and when an outbreak occurs, the mortality rate is high. Symptoms include disoriented, circular walking; facial paralysis and drooping ears; lowered eyelids and depression; and abortion in late-term ewes.

Malignant Edema

A disease that results from a wound that's become infected by one of several *Clostridium* species, malignant edema is usually fatal. If a wound infected by *Clostridium* is caught early, then antibiotics may be effective. Clostridia are environmental bacteria that normally reside in soil and manure. The best approach to infection with these bacteria is keeping wounds clean and maintaining injured animals in a clean environment — like on a healthy grass pasture — instead of in a manure-contaminated yard. Vaccines are available.

Pinkeye

Infectious keratitis, known commonly as pinkeye, is another infection caused by a chlamydial organism. This disease can be quite contagious. Infected eyes become red, inflamed, and watery; if the disease is allowed to progress, the eyes become opaque and ulcerated, with blindness being the eventual outcome. Your veterinarian can supply an ointment or a powder to put in the eyes of infected animals.

Scrapie

Scrapie, like mad cow disease (bovine spongiform encephalopathy), chronic wasting disease in deer and elk, and Creutzfeldt-Jacob disease in humans, is caused by something called a *prion*. A prion is like an abnormal subparticle of a protein molecule. The science and understanding of prions is a new field, so there's lots to learn. Scientists aren't sure if these diseases are directly or indirectly related, and they're not sure if they can be transmitted between species.

Of all the diseases discussed here, scrapie has been recognized for the longest period — almost 200 years. Despite the fact that the disease has been around for so long, little is understood about the causative prions and how the disease works. There is a slew of symptoms for scrapie, ranging from exaggerated movements to weight loss, itching, and scraping and biting wool on the sides and legs. There is no known cure for the disease, but there is a Voluntary Scrapie Flock Certification Program managed by the USDA's Animal and Plant Inspection Service (see Resources). When purchasing sheep, look for flocks that are enrolled in the program

Spider Syndrome

The spider syndrome is a recessive, genetic disorder that seems to particularly affect Suffolk and Hampshire sheep, although it may be seen in some of the other black-faced breeds as well, probably as a result of crossbreeding with Suffolks. The disorder manifests itself with skeletal abnormalities, such as splay-leggedness. (Its name comes from the fact that animals suffering from the disorder develop legs that look like spider legs.) In spider syndrome, the growing animal doesn't correctly convert cartilage to bone as it develops.

If purchasing a ram of a black-faced breed, make sure he's been DNA tested as free from the "s" gene that causes the disorder.

Hoof-Related Problems and Care

Many foot diseases can be prevented by proper and periodic hoof trimming, which is most easily done in the spring when hooves are still soft from wet weather and in the fall after the start of the rainy season. The amount of hoof wear depends on whether the soil conditions are mud, sand, or gravel and whether the barn has a dirt or concrete floor. Hooves that are in good shape should ideally be trimmed twice a year, but they may need trimming more than twice a year, especially when the weather is wet for prolonged periods.

Lameness

You can help prevent sheep from becoming lame by:

- Trimming all feet each spring before turning sheep out on new pasture.
- Trimming again at shearing time or later in the year. Untrimmed hooves curl under on the sides, providing pockets for accumulation of moist mud and manure that create an ideal environment for foot-disease germs.
- Maintaining dry bedding during winter.
- Keeping sheep away from marshy pastures during wet months.
- Changing location of feeding sites occasionally to prevent accumulation of manure and formation of muddy areas.
- Having a footbath arrangement.

Footbath. Having a footbath is one of the most important aspects of treating foot disorders and avoiding lameness in sheep. If you run the sheep through a trough of plain water first, it keeps the bacterial bath clean longer. Be sure that sheep have water elsewhere and are not thirsty, so they do not drink any of the footbath.

Feet should be trimmed before the footbath, not just to allow better penetration but because the footbath chemicals hardens the hooves and makes them more difficult to trim. Remember to disinfect the knife after every hoof so you do not needlessly spread germs.

Zinc sulfate has become the solution of choice for footbaths. It has at least a tenfold greater benefit if the animals are allowed to stand in the solution for 1 hour on two occasions spaced about a week apart. Do *not* use other chemicals in the footbath, such as formalin or copper sulfate, because you could severely burn or "pickle" the sheep's feet.

Trim nonlimping sheep first, put them in the footbath, and then turn them into clean pasture (that is, pasture that hasn't been grazed for at least a week). Next, bathe the feet of limping sheep and keep them in a dry area if possible, treating them regularly every 5 to 7 days, or have them walk through the bath on the way to daily feeding.

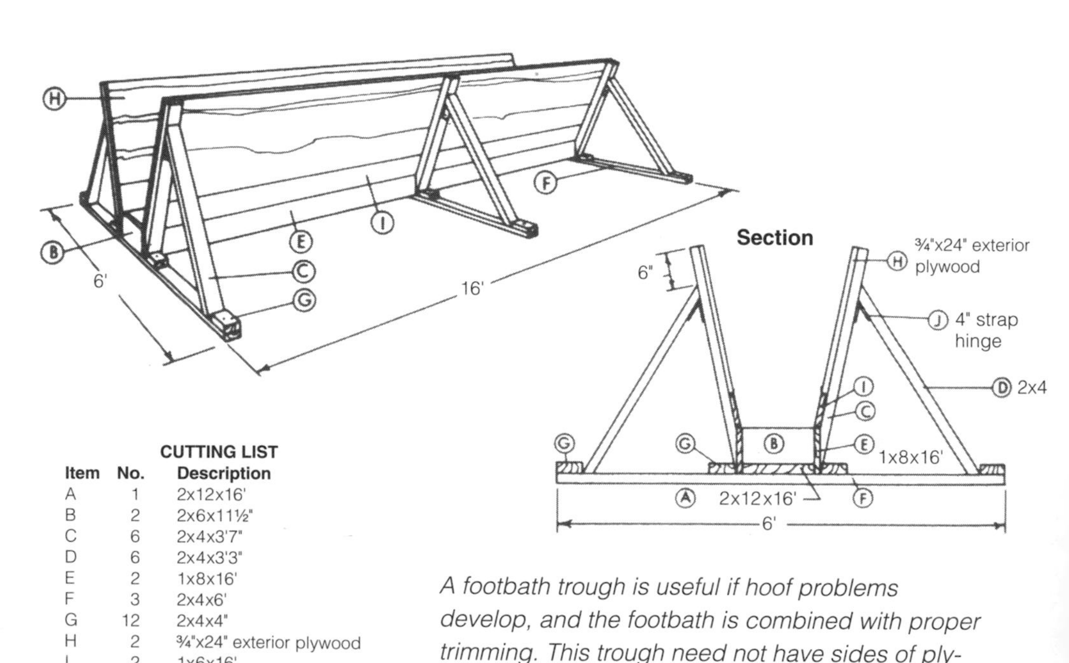

CUTTING LIST

Item	No.	Description
A	1	2x12x16'
B	2	2x6x11½"
C	6	2x4x3'7"
D	6	2x4x3'3"
E	2	1x8x16'
F	3	2x4x6'
G	12	2x4x4"
H	2	¾"x24" exterior plywood
I	2	1x6x16'
J	6	4" strap hinges
K	12	⅜"x3½" bolts

A footbath trough is useful if hoof problems develop, and the footbath is combined with proper trimming. This trough need not have sides of plywood. Shiplap or 1x6s can be used. A gate is needed at the exit to keep sheep standing in the bath for the required time. (Midwest Plan Service)

You can construct an alternative to the standard footbath trough in the following manner:

1. Use a 4 x 8 sheet of ½-inch (1.3 cm) waterproof sheathing plywood.
2. Nail 2 x 4s around the edge.
3. Caulk the edges to make it watertight.

A temporary pen around the perimeter completes the unit. The size of the pen can vary, but it should hold upward of eight to ten animals or more. Footbaths, in conjunction with other good husbandry practices, such as vaccination and proper foot trimming, can stave off the foot disorders that can lead to lameness in sheep.

Check limping sheep. When you notice a sheep limping, try to discover the reason. Notice which foot is being favored, then catch the sheep and trim all four hooves if they need it. Do the sore one last to avoid possibly spreading infection.

Foot Examination in Limping Sheep

1. Look for a lump of mud, a stone, or a sharp splinter caught between the claws of the hoof that seems to be sore.
2. If there is nothing there, check the gland. Sheep have a deep gland between the two toes of each foot, with a small opening on the top of the front of the hoof. This can be readily seen if you look for it. (The gland's secretion is waxy and has a faint, strange odor that is said to scent the grass and reinforce the herding instinct.) When these glands become plugged with mud, the secretion is retained and lameness occurs.
3. Squeeze the gland. Sometimes a fairly large blob of waxy substance pops out. If this was the problem, then the sheep should improve.
4. If there is no evidence of a plugged gland or a foreign object, try to determine if a hoof disease is present. You have to have a clear idea of what a normal hoof looks like before you can spot a diseased one. If you're not familiar with sheep hooves, compare the sore one with another foot.
5. If everything else looks good, check any injury (including cuts) that might be causing the problem. We once checked a limping ewe and found that she'd cut her udder and was getting hit with a flystrike!

Hoof trimming. Using a hoof knife or jackknife, trim the hoof back to the level of the foot pad so that the sheep can stand firmly and squarely on both claws. The purpose of trimming, other than to prevent lameness, is to give a good flat surface on the bottom of the hoof and make sure that both pads are evenly flat. To do this, trim off the excess horn so that it is level with the sole and does not protrude too far in front. If there are still pockets where mud or manure can gather, scrape these out with the point of your knife or the hook on the end of the hoof knife, and trim the hoof back a little farther. Notice the shape of the hooves on your half-grown lambs for the ideal.

Hoof knives are sold in two sizes — large for cows and smaller for sheep. In dry weather, when feet are drier and harder to trim, hoof shears can be useful.

Among the best hoof shears are Swiss-manufactured pruning shears (Felco-2, available from Premier [see Resources]), which have come into routine use in many large commercial flocks. The curved blades have less tendency to slip on tough, dry hooves. These shears are slightly more expensive than the traditional Burdizzo shears but are more than worth the extra money. The specially tempered blades are thin and very sharp, requiring less than 25 percent of the "squeeze" power needed with the more traditional shears. Because of their sharpness and ease of use, exercise caution when first using these, because it is very easy to overtrim the hoof or cut your hand, even if you are experienced. *Always* wear a leather glove and arm protection on the opposite hand when trimming hooves to avoid accidental injury if the sheep kicks.

Foot Rot

Sheep raisers once thought that foot rot was a spontaneous disease of wet weather. But it is actually a bacterial disease, and according to Ann Wells, a veterinarian whose specialty is sheep and goats, it is "one of the biggest disease problems for the sheep industry."

Two different bacteria contribute to foot rot. The first, *Fusobacterium necrophorum*, is always present in soil, but by itself won't cause any problems. When the second bacterium, *Dichelobacter nodosus* is also present, the disease can run rampant through a flock. The good thing is that *D. nodosus* can only live out of a sheep's foot for about 2 weeks, so by keeping sheep off a pasture for 2 weeks the organisms will die off in the soil, but the bad thing is that some sheep can act as carriers of foot rot, constantly reinfecting the soil that other sheep walk on. Quarantining new animals for 2 weeks is the best way to stop foot rot from being introduced to a foot rot–free flock.

D. nodosus is an *anaerobe*, which means it grows in an oxygen-free environment, deep in the hoof tissue. This is why hoof trimming is an important part of foot rot treatment, so that dead tissue is removed and oxygen can enter.

Foot rot starts with a reddening of the skin between the claws of the hoof. Odor is faint or absent in the beginning but becomes noticeable as the infection worsens. Infection begins in the soft horny tissue between the hoof, or on the ball of the heel, and spreads to the inner hoof wall. As the disease progresses, the surface of the tissue between the underrun horn has a slimy appearance. The horny tissue of the claws becomes partly detached, and the separation of the hoof wall from the underlying tissue lets the claw become misshapen and deformed. There is relatively little soft-tissue swelling.

In severe infections, it is often more practical to dispose of the most seriously affected animals and concentrate treatment on the milder cases. Use the following steps to treat for foot rot:

1. Trim the hooves, removing as much of the affected part as possible. Disinfect the knife after each hoof, and burn the hoof trimmings.
2. After trimming, have sheep walk through a footbath prepared with zinc sulfate.
3. Hold the sheep on a dry yard or pasture for 24 hours, if possible, after the footbath.
4. Vaccinate (see the section on vaccines and other biologicals on page 202).

Foot Scald

Foot scald is sometimes mistaken for foot rot. It involves the soft tissue between and above the toes and the "heel," usually only on one foot. Hooves with this disorder have inflamed, moist tissue and sometimes open sores. Foot scald is very similar to athlete's foot in humans.

Treating Foot Scald

Treat foot scald by trimming hooves and then spraying them with hydrogen peroxide. If there is no improvement, treat with a zinc sulfate footbath — the same kind as for foot rot. If you do not have footbath facilities, you can do the following:

1. Purchase special booties, place them on the sheep's feet, and fill them with the footbath solution. Alternatively, you can use a large fruit-juice can filled with 2 inches (5.1 cm) of the footbath solution.
2. Soak the affected foot for 5 minutes.
3. Repeat if necessary.

Causes of the disorder are dampness, wet pastures, prolonged walking in mud, or abrasion caused by dirt or foreign objects lodged between the toes. It occurs primarily during wet periods, and the condition sometimes improves without treatment in dry weather. Foot scald is a major problem only because it lessens the foot's resistance to more serious disease, such as abscess or foot rot, and causes sheep to eat poorly and not get enough exercise.

Bumble Foot

Bumble foot is a true abscess that occurs within the hoof structure. It usually afflicts only one foot. It is considered infectious but not extremely contagious like foot rot.

The infection causes formation of thick pus, and as internal pressure increases, the sheep becomes more and more lame. Sometimes you can see a swelling above the hoof. When compared with the other foot, the infected foot will be warmer.

The disorder is caused by bacteria in manure and dirt, which enter through cuts or a wound, infecting the soft tissue. There is usually a reddening of the tissue between the toes. This infection may become advanced if not treated and can move into the joints and ligaments. If that occurs, it is almost incurable because it is impossible to reach.

Abscess is dangerous in pregnant ewes as they will fail to graze, be slow to feed on grain, and not get enough exercise, which can bring on pregnancy toxemia. Insufficient nutrition also leads to low birth weight of lambs and having insufficient milk for them.

Unless pressure is released by an incision, the abscess may eventually break and discharge pus. When it is opened or breaks, squeeze out the pus and treat with antiseptic.

Drugs for Sheep

Good shepherds are prepared for emergencies by having a supply of standard medicines on hand. Such medications include:

- Bloat medication
- Antibiotics
- Anthelmintics (worm medications)
- Propylene glycol for pregnancy toxemia
- Calcium phosphate or other treatment for milk fever
- Disinfectants
- Mineral oil for constipation
- Dextrose solution

- Footbath preparation
- Uterine boluses
- Clean or sterile equipment (like syringes and needles)

Discuss with your veterinarian which antibiotics, anthelmintics (worm medications), and vaccinations to use. If the vet is close by or if you have easy access to a farm-supply store that sells a wide variety of antibiotics, you probably don't need to keep any on hand. Before you administer an antibiotic, it's a good idea to have the veterinarian run culture and sensitivity testing if the illness isn't critical. This tells you which antibiotic will be most effective against the bacterium that's causing the problem. For example, if a ewe's udder is hard and hot from a mastitis infection, take a milk sample to the veterinarian. Usually within 48 hours, the vet will call back with a specific recommendation for which drug to use. On the other hand, if you don't live near a veterinarian or a good supply store, you'll need to keep a broad-spectrum antibiotic around.

Iodine and hydrogen peroxide are both good disinfectants for treating wounds. We clean first with the peroxide, and then coat the wound area generously with iodine.

How to Administer Medications

Follow label directions for dose and type of administration (as well as withdrawal days before slaughter). The types of administration of certain drugs are as follows:

- Oral, such as boluses (large pills) given with a bolus gun or with capsule forceps
- Oral powder, such as vitamins, placed well back on the tongue for treatment of an individual animal or mixed with feed, minerals, drinking water for general treatment of the whole flock
- Oral liquid, given as drench with a syringe or in the drinking water
- Oral pastes, easy-to-use products that are sold with an applicator for smearing on the sheep's tongue (several worming medications are sold as paste)
- Spray-on and sprinkle-on products, such as pinkeye spray or insecticides, including maggot or screwworm bombs
- Dips, which are also common for insecticides when treating a large number of animals; a tank full of solution is prepared, and the animals are forced to "swim" through the tank
- Pour-ons, such as iodine on newborn lamb navel, disinfectant on minor wounds, and certain insecticides

- Subcutaneous injection, medication given just under the skin
- Intradermal injection, medication given into the skin
- Intramuscular injection, liquid such as antibiotics injected into heavy muscle
- Intramammary injection, administration of fluid or ointment such as mastitis drugs through the teat opening
- Intraperitoneal injection, liquid given through right flank into the abdominal cavity; such administration should be done by a veterinarian
- Intraruminal injection, administering fluid into the rumen on the left side, as for bloat remedy if it is too late to be given by mouth; should be done by a veterinarian
- Intravenous injection, fluid administered into a vein; is best done by a veterinarian or very experienced producer
- Intranasal, spraying of vaccine up the nasal cavity
- Uterine boluses to prevent infection after an assisted lambing

Injections

Sterile procedures must be maintained to avoid serious infections. Use only clean and sterile syringes (boiled for at least 30 minutes if new, sterile, disposable syringes are not being used) and sharp, sterile, *disposable* needles. Needles can be boiled, but this causes them to become dull. Dull needles are one of the most common causes of injection site infections because they force dirt, grease, and bacteria through the skin. Storing needles in alcohol can also cause the points to be blunted because they strike against the side of the container.

Disposable plastic syringes are inexpensive and can be ordered from veterinary-supply catalogs. In some states, they can be obtained from your local drugstore or purchased at farm-supply stores.

If you are withdrawing doses for several sheep, protect the contents from contamination by sanitizing the top of the vial with disinfectant as above, then inserting a sterile needle *that you will leave in the bottle*. Fill the syringe, leave the needle in the bottle, and attach a separate needle to the syringe for injection. For the next dose, detach the used needle, fill the syringe with the needle left in the vial (leaving the needle in the vial), and reattach a new or disinfected needle. In this way, you protect the medication from contamination and can save the balance of the contents through the dating period. While this is true of an inactivated vaccine, you cannot save a live vaccine. Once opened and exposed to air, live vaccines become unstable and can't be stored for later use.

How to Fill a Syringe with Medication

1. Clean the top of the vial with a disinfectant.
2. Swirl or shake the bottle to thoroughly mix the contents without causing undue bubbles.
3. While holding the vial upside down, pull the syringe plunger back to approximately the volume of drug to be removed, insert the needle into the center of the vial stopper, and depress the plunger forcing the air into the vial. (This prevents creating a vacuum in the vial and difficult removal of the dose.)
4. Withdraw a greater volume of drug than needed, then express the excess drug back into the vial to remove any air bubbles that may form in the syringe.

Once the needle is filled with medication, do not let it touch anything or it will no longer be sterile. If possible, have a helper hold the sheep or hand you the necessary medicine and equipment.

An alcohol swabbing of the skin prior to injection gives the impression that the skin has been sterilized, but this is not really the case. It takes approximately 6 to 8 minutes for alcohol to kill common disease-causing germs. The alcohol swab mechanically removes the majority of the skin bacteria contained in the body oils. Simple wetting of the skin with alcohol or most other disinfectants merely puts bacteria in the solution where they can be more readily picked up by the needle and carried into the injection site. Paula's veterinarian says that injections placed in dry, "clean" skin (that is, free from excessive grease, manure, and so on) result in fewer injection-site contaminations.

For the same reason, you should avoid, if at all possible, injecting wet sheep. Routine vaccinations should always be scheduled when the weather is dry.

Protect drugs from freezing and from heat. Many medicines require temperatures above freezing, and below 50°F (10°C). Read the label on each medication for storage directions. Many antibiotics require refrigeration. Check the expiration date on the package.

Read the dosage instructions carefully and follow them, or use the medication according to the advice of your veterinarian. On some drugs, there is not much leeway between the effective dose and the overdose that could be fatal or harmful.

Subcutaneous Injection

The medication usually should be at body temperature, especially with young lambs. It can be given in the neck, but the preferred place is in the loose, hairless skin behind and below the armpit (axillary space), to the rear of the elbow, over the chest wall. Be careful not to inject into the armpit — this can happen if the injection is made too far forward. The armpit is actually a large cavity underlying the entire shoulder blade area, crossed by the major artery, vein, and nerves that serve the front leg. Some vaccines are highly irritating, and if injected into the axillary space, could cause a severe reaction and lameness.

A dose of more than 10 mL is best distributed among several sites instead of all in one place (use an even lower cutoff for lambs).

Do not inject near a joint or in areas with more than a small amount of fat under the skin. With this type of injection, veins are usually not a problem, but if you want to make sure you are not in a vein, the plunger can be pulled out very slightly before injecting. If it draws out blood, try another spot. Medication for subcutaneous use should never be injected into a muscle.

The preferred location for giving a subcutaneous injection is in the loose, hairless skin behind and below the axillary space (armpit), to the rear of the elbow, over the chest wall.

Subcutaneous Injection Technique

1. Pinch a fold of loose skin.
2. Insert the needle into the space under the skin, holding the needle parallel to the body surface.
3. Push the plunger.
4. Rub the area afterward to distribute the medication and hasten absorption.

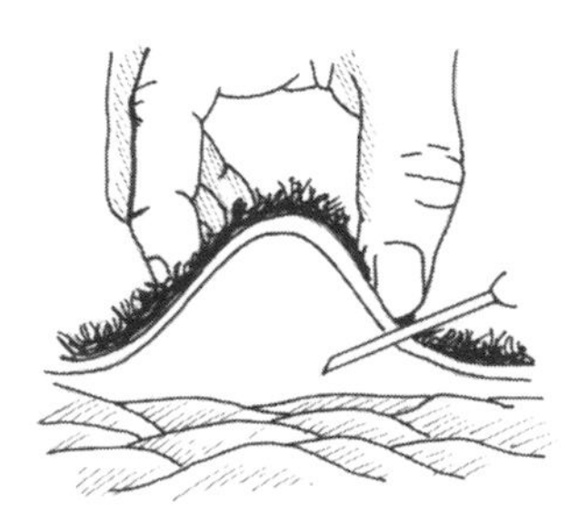

Intradermal (or Intracutaneous) Injection

Intradermal injections are rarely used. The needle is inserted so close to the surface that it can be seen through the outer layer of skin, in a site the same as for subcutaneous administration. Injection is made slowly while the needle is being drawn out to distribute the dose along the needle's course.

Intramuscular Injection

Sites for intramuscular injections include the neck or the thigh. If possible, have an experienced person demonstrate this method so you can see the exact place that avoids nerves and the best cuts of meat.

With an assistant holding the sheep still, thrust the needle quickly into the muscle. To be sure the needle is not in a blood vessel, use the technique described for subcutaneous injection. It is usually best not to inject more than 10 mL of medication into any one spot.

Intramuscular shots are the most common form of injection. The best sites for administration are either the top of the neck, about 6 inches (15 cm) down from the center of the back, or the top of the thigh where it joins the hip, about 8 inches (20 cm) down from the center of the back. When giving an intramuscular injection, quickly thrust the needle into the muscle (inset). *Make sure you haven't accidentally entered a vein by pulling back slightly on the plunger first; if blood enters the syringe, the needle is in a vein; withdraw the needle and reinsert.*

Method for Intramammary Injection

1. Remove the cap of the infusion tube, and gently insert the tip into the teat canal. Do not remove the cap until you are ready to use it. This avoids bacterial or fungal contamination, which could complicate an already-serious condition.
2. Squeeze the medication into the teat.
3. Massage the dose upward toward the base of the udder.

Again, the importance of cleanliness cannot be overemphasized! Most udder infections can be treated with antibiotics, but unsanitary infusion techniques could introduce fungi and molds that are not sensitive to the antibiotic, resulting in a totally untreatable condition.

Intramammary Injection

The tips of udder antibiotic syringes are designed for cattle and are difficult to use in sheep, though some come with a fine-tip design that works well for sheep. Cleanliness is paramount when infusing the udder. First, milk out the affected side of the udder as completely as possible. Afterward, wash your hands and the udder thoroughly, then carefully disinfect the teats several times a few minutes apart. A solution of ½ laundry bleach and ½ water works. Dry the end of the teat with a clean paper towel to avoid injecting germs when you insert the medication.

Intraperitoneal Injection

Injection by the intraperitoneal method should only be done by a person who is familiar with aseptic technique and anatomy. Complications (peritonitis) are common after this procedure. It is easier if one person holds the sheep, straddling it just in front of the shoulders.

Clip the wool from the right flank, in the shallow triangular depression below the spine, between the last rib and the point of the hip bone. Medication injected into the center of this depression goes into the peritoneal (abdominal) cavity. Scrub the injection site with soap, rinse, dry, and disinfect the skin with iodine.

Medication should be at sheep body temperature. A sterile 25-mL or 50-mL syringe and a sterile 16-gauge needle are required. Disinfect the bottle stop-

Antibiotic Use: Prevention Is Better Than Cure

The availability of antibiotics should not encourage improper sanitation practices or "fire-engine" treatment of diseases that can be prevented through proper management and vaccination. There is concern that improper use of antibiotics can give rise to new strains of drug-resistant bacteria that pose a threat to both humans and animals. Physicians and veterinarians have both noticed that antibiotics that were once effective at low doses must now be given at much higher doses to both humans and animals to accomplish their purpose.

per before withdrawing the medication, and use a separate sterile needle to give the medication to reduce the possibility of introducing the infection into the body cavity. Hold the needle perpendicular to the skin, pointed toward the center of the body. Quickly insert the full length of the needle, and inject the medication. If it does not inject easily, the needle may be clogged with tissue or may be in the wrong place. If so, withdraw the needle, replace it with a new one, and try again. Rub the injection site with disinfectant afterward.

Antibiotics

Antibiotic is the general term for a group of products that either kill or seriously impair bacterial growth. They are effective against many bacteria, but are useless against viruses.

Antibiotics are only effective when present in adequate concentration. Giving a low concentration (below recommended levels) or discontinuing treatment too soon may fail to kill the more resistant bacteria present in the infection. This could result in a relapse of the condition or, more seriously, a chronic infection, which could be difficult or impossible to treat if the bacteria develop resistance to the antibiotic.

Care must be exercised to ensure that antibiotics, as well as other drugs, are properly used and not overused. Mastitis and certain respiratory diseases are among the few examples of disorders for which there are no vaccines. While management practices can minimize the occurrence of these diseases, antibiotics are needed once the infection is established.

Certain forms of antibiotics can upset normal body functions. Some may "sterilize the gut" (that is, kill the beneficial bacteria that both aid in digestion and compete with harmful bacteria and fungi), making animals susceptible to enteric upsets and infections. Many shepherds give yogurt (which contains cultures of beneficial digestive bacteria) to a lamb after antibiotic therapy or illness to reestablish the "friendly" bacteria.

Antibiotics are often used when they are of no benefit whatsoever, as in the case of disease caused by viruses. When the exact cause of sickness is unknown, there is a temptation to give a shot, usually a wide-spectrum antibiotic, to see if it helps. Ideally, any illness should have an accurate diagnosis to see if any antiserum or vaccination is available and what, if any, antibiotic treatment would be effective.

Parasite Preparations

A wide variety of products are available for parasite control. The best product is one that gives broad-spectrum control. The best policy is to use parasite preparations *only* when really necessary. The products that are based on the "avermectin" chemical family currently give the broadest control, and include a product called Ivomec (Merck, Rahway, NJ), which contains ivermectin.

Vaccines and Other Biologicals

Certain immunizing agents are intended solely for disease protection. They have little if any effect in treating the disease. These agents are proteins that are called antigens, and they only stimulate the sheep's immune system to produce protection against a particular disease. It should be well understood that vaccination and immunization are not the same thing, because administration of the antigen by vaccination results in immunization only if the sheep's immune system is normal and functioning. Vaccination must be accomplished well ahead of the period in which disease exposure may occur, because it usually takes up to 1 month for maximum immunity to develop. Very low levels of protection are observed at 2 to 3 weeks after vaccination, and with a few kinds it can take up to 45 days after the last dose of some vaccines for maximum protection.

Immunizing agents fall into one of four classes and all are commonly called vaccines:

- **Antisera.** Often called serums or antitoxins, are derived from the serum of hyperimmune animals, which are those that have received multiple doses of vaccine to confer a high specific antibody level against the particular disease.

- **Bacterins.** Contain killed bacteria and/or fractions of the bacterial cell.
- **Toxoids.** Contain the inactivated toxins produced by bacteria, usually clostridial organisms such as those that cause tetanus and enterotoxemia.
- **Vaccines.** Derived from viral agents.

Antisera

When an antiserum is injected, we are only "borrowing" antibodies produced in another animal to confer temporary, or passive, immunity. The period of immunity usually lasts from 10 to 21 days. This type of vaccine is used to protect animals for a short time when disease is present in the herd and to treat infected animals. Antisera can be used for young lambs.

On rare occasions, antiserum may sometimes be administered along with a vaccine to give immediate protection while the animal's own active immunity is developing. Check with your veterinarian before administering antiserum and vaccine together, because in some instances the hyperimmune serum can neutralize the vaccine.

Bacterins

Bacterins are suspensions of bacteria that were grown in culture media and then chemically or heat killed. These vaccines are not capable of producing disease and can be used without danger of spreading disease. The bacteria used in the production of the various bacterins are highly antigenic strains isolated from animals that have succumbed to the particular disease.

Bacterins are often suggested as an aid in establishing immunity to specific diseases. Always follow the manufacturer's label. Most bacterins require a priming injection, followed by a booster in 1 to 4 weeks. More often than not, very little immunity is obtained after the priming injection — actual protection is obtained after the booster shot. Bacterins do not confer long-lasting immunity. The animal is usually protected for a maximum of 6 months to a year between boosters.

Toxoids

Solutions of inactivated toxins make a vaccine known as a toxoid, which is derived from bacteria that cause disease by producing toxins that enter the bloodstream and cause severe tissue or nerve damage (such as tetanus, enterotoxemia, and blackleg). Since it is the toxin produced by the bacteria and not the bacteria itself that causes disease, toxoids stimulate the sheep to produce neutralizing antibodies against the toxin, thereby protecting against their deadly effect.

Vaccines

A vaccine is a modified live or killed biological preparation that, when administered to the animal, stimulates the immune system to build its own protective antibodies. Modified live vaccines contain strains of the virus that cannot cause the disease but still retain the immune-stimulating potential. With few exceptions, modified live virus vaccines produce greater and longer-lasting protection than the inactivated (killed) virus vaccines.

Immunizing Shots

Store all immunizing supplies in a cool place, but do not allow them to freeze. Purchase vaccines from a reputable source because if they are not properly stored or transported before you buy them, they may be worthless.

Vaccines can be applied in a number of ways — scratching the skin, injecting subcutaneously, injecting intramuscularly, and spraying into the nasal cavity. You must follow the directions of the manufacturer or veterinarian very carefully regarding both the dosage and the manner of administration. Ewes and lambs are ordinarily vaccinated on the side of the breast bone (lower chest wall behind the elbow) or the side of the neck. Do not inject vaccines into the armpit under any circumstances.

What shots are necessary? There is no hard-and-fast rule. Guidelines depend on your area, the climate, the type of operation, the prevalence of nearby sheep flocks, the purchasing of new animals, and the conditions under which the sheep must be raised.

Unapproved Drugs

Unfortunately, we do not have as many options in approved drugs as other countries, such as Australia and New Zealand, which have larger sheep industries. It is not practical for U.S. drug companies to spend time and fantastic amounts of money to get approval of a medication for sheep, which are considered a "minor species."

However, veterinarians can prescribe any drug that can be legally obtained for *extralabel* use (that is, use for which the drug has not been specifically approved by the FDA) if a legitimate client-patient-veterinarian relationship exists. This is a crucial legal point. If a sheep producer uses an unapproved drug without a veterinarian's prescription, such use is not only illegal, but the producer will be held legally and financially responsible if residues are produced and detected. Using these drugs without your veterinarian's guidance could also prove deadly to your sheep, as not all drugs can be safely interchanged between species. Your veterinarian can prescribe any needed medications and give advice on the proper dispensing, dosing, and withdrawal times.

Actually, many unapproved drugs, especially wormers, are used commonly. There is much pressure on the government to make it easier for a drug, which is already approved for one livestock species, to become approved for another species. This would take much less investment by the manufacturers.

EIGHT

Problems of Rams, Ewes, and Lambs

This chapter covers sexual and reproductive problems of rams and ewes, as well as problems that tend to be gender and age specific. As is true with the health problems discussed in chapter 7, good husbandry and management techniques will minimize or eliminate most of these troubles before they begin.

Rams tend to have far fewer problems than do ewes but, because of the rams' influence on the breeding flock, their problems can have far-reaching effects.

Without doubt, newborns are the most vulnerable class of livestock on any farm. Studying the section on lamb problems can help you protect your babies.

Problems with Rams

Your ram is far more than 50 percent of your flock. He contributes genetic material to each of your lambs, so his health is critical to production.

Epididymitis

In recent times, awareness of ram epididymitis, a disease caused by one of several organisms that damage sperm-producing tissues, has increased greatly. The infection is well under way and contagious before being noticed during a physical examination. Signs can include swelling of the epididymis (located at the base of the testis) and the presence of hard, lumpy tissue, and these signs show that the disease is far advanced. It is transmitted from one ram to another

SHEPHERD STORY

Thirty years ago, Phil LaRocca started a 200-acre organic vineyard and winery in Northern California. Phil ran his vineyard organically, and he incorporated a few sheep to help control vegetation among the vines and so his family would have good meat that they'd raised themselves.

"At first, we'd purchase a small number of weaned lambs in the spring and butcher them in the fall. Then the kids showed sheep in 4-H, and we started a small flock. We ate our lambs, traded some to neighbors, and sold a few.

From those auspicious beginnings, Phil has developed a flock of about three hundred animals and is paving a new path for certified organic wool and lamb with the bulk of his flock run on a second farm. "We're hoping that our status as certified organic producers will provide us with a strong marketing niche. Organic cotton has been doing really well for a while now, so it seems like the demand for organic wool should be there. We're talking with some children's clothing manufacturers who've been using organic cotton, and they seem interested in our product."

Certified organic production may not make sense for everyone. Phil says it cost him about $700 for the initial inspection, and he has to commit to paying 0.5 percent of his gross sales to the certifying organization, but he thinks it will pay off in the long run. "I did have the advantage that I was getting two farms certified and am now able to sell certified organic wine as well as lamb and wool. We're trying to start something here — start an industry — so I'm willing to bear those costs."

To raise certified organic livestock, the animals can eat only certified organic feed, and the use of chemical wormers and medications is prohibited. "The key to our approach is pasture: The animals run through a number of different paddocks and are moved every 21 days on average. Controlling the timing of the moves helps to control parasites, and we feel that healthy pastures give us healthy animals."

via the ewes during breeding season. In some cases, the ewe may become infected, resulting in abortions, stillbirths, and weak lambs.

Enzyme-linked immunosorbent assay (ELISA) is available to accurately test for the disease. Demand a negative ELISA on any ram you're thinking of purchasing! Accurate testing also enables producers to identify diseased rams so they can be readily identified and culled. Treatment is generally unsuccessful, but if caught early through testing, when the only indication is the presence of white blood cells in the semen, then high doses of antibiotics, such as tetracycline and streptomycin, may be effective. Although this can be a way to save a valuable animal, it requires isolation and extensive monitoring. Also, a new vaccine is available that may be worth investigating if you live in an area where the disease is prevalent.

Orchitis

Inflammation of one or both of the testicles, known as *orchitis*, may be caused by several species of bacteria. If just one testicle is affected, the ram can still reproduce after treatment with antibiotics (oxytetracycline or procaine penicillin are the two most commonly used), but if both testicles are infected, he will be sterile. The symptoms are swelling, pain, and heat in the affected testicle.

Pizzle Rot

Combine a high-protein diet with the bacteria *Corynebacterium renale*, and your ram may have trouble with pizzle rot. C. *renale* is among the flora that inhabit the skin of the *prepuce*, or foreskin, of the ram's penis. Normally, C. *renale* causes no problems, but when the ram is on a high-protein diet, he passes more urea in his urine, and bacteria break down the urea into ammonia. The ammonia "burns" the area, causing ulceration. At the least, it's painful for the ram; at the worst, it causes all kinds of secondary problems, from flystrike to death.

Prevention is fairly easy — don't feed a diet that's too high in protein. For example, don't feed rams straight legume hay; feed a grass–legume mixed hay. Minor cases of pizzle rot can be treated with antibiotic ointments. Advanced cases may require antibiotic injections and/or surgery.

Ulcerative Dermatosis

Ulcerative dermatosis is a form of venereal disease that can be transmitted from the ram to the ewes. This disease is caused by a virus and is uncommon, but it can be mistaken for pizzle rot because the early symptoms are similar. It occurs most often in fall and winter and gets started when an irritant (like snow, ice, or cockle burrs) damages the skin and allows the virus to enter. If a ram develops an infection, he should be isolated. The disease eventually runs its course, but if the ram introduces it to the ewes, it will circulate through the flock forever.

Disorders in Ewes

Ewes can suffer from their own unique set of health problems. This is especially true during preganancy and lactation, when their bodies are under tremendous stress.

Abortion

Following are several of the many causes of abortion:

- Moldy feed
- Injury
- Chlamydia
- Vibriosis
- Toxoplasmosis
- Salmonellosis

Feed that contains mold spores can infect and destroy the placenta. This in turn cuts off nourishment to the fetus.

Injury is often a cause of abortion, such as when a ram is running with the pregnant ewes and bumps them away from hay or feed. Narrow doorways, where sheep rush through for feed, are dangerous, especially as the ewes become large as pregnancy progresses. Dog attacks nearly always cause abortions among the ewes that have been injured or chased.

Disease is another obvious cause of abortion, and these diseases are more well defined than ever. There is now a combination vaccine for enzootic abortion and vibriosis (EAE-vibrio). This vaccine can be administered 2 weeks before breeding, with a booster shot given each year 2 weeks before breeding.

Enzootic Abortion of Ewes

Enzootic abortion of ewes (EAE) is caused by *Chlamydia psittaci*, which causes late-term abortions, stillbirths, and weak lambs. *C. psittaci* is not the same species of *Chlamydia* that causes respiratory diseases, epididymitis in rams, conjunctivitis (pinkeye), or arthritis in sheep; however, in flocks where *C. psittaci* is enzootic, other problems, such as pneumonia, often arise.

C. psittaci is spread to susceptible ewes by contact with aborting ewes, infected fetal membranes, uterine discharges, or a dead fetus. Susceptible ewes thus infected will most likely abort their next lamb unless they become infected early during gestation, in which case it would happen with the current pregnancy. When the organism is first introduced to a clean flock, abortions can reach as high as 60 percent, but after a year of two, they drop to about 5 percent.

Vaccinations are available, but they cannot help during an outbreak. Since the signs of vibriosis and EAE are similar, laboratory analysis must be used to identify the exact cause of an abortion. The EAE–vibrio vaccine protects from both.

During an outbreak of EAE, giving tetracycline (300–500 mg per head per day) is helpful in bringing the situation under control. This should be started immediately if more than one ewe aborts and continued until test results come back. Also, because EAE is contagious, aborting ewes should be removed and strictly quarantined for 30 days. Bedding from quarantine pens should be burned.

Vibriosis

Campylobacter fetus is the bacteria that causes vibriosis. These bacteria, which live in the gallbladder and intestine of the ewe, invade the uterus, placenta, and fetus during late pregnancy. When C. *fetus* is introduced to a clean flock, 90 percent of the ewes lose their lambs, but once that has happened, the ewes become immune to the bacteria and won't abort again. However, they can be carriers that contaminate the feed and water, thereby infecting other ewes.

Ewes can be vaccinated at the beginning of the breeding season and then again about 90 days later. Once vaccinated, they require one annual booster shot. Your veterinarian may recommend tetracycline or a penicillin preparation for treating ewes in a newly infected flock.

Milking Considerations for Ewes That Have Had Abortion or Stillbirth

When a ewe loses the lamb in the last few weeks of pregnancy or has a stillbirth and there is no orphan to graft on her, she should be milked out on the third day and again in a week if she has a full udder. If the lamb was born dead due to a difficult birth rather than disease, the first milking should be done at once and the colostrum frozen for future use. With a tame and docile ewe, you may want to continue the milkings for a while and then taper off, saving all the milk for future bummer lambs.

Toxoplasmosis

Toxoplasmosis is caused by a microscopic protozoan (coccidium) whose natural host is a cat. Other species, including sheep and humans, can become an unnatural host to this organism when it "gets lost" in its normal migratory route. This parasite invades many tissues, causing infections in the brain, eyes, uterus, fetal membranes, and the fetus itself. Clinical signs are consistent with the particular tissue that has been damaged. Abortion and stillbirth are most commonly observed. Infection occurs when cats defecate, leaving the infectious organisms on hay, grain, and other food consumed by sheep. Ground-up grain is a common target because it constitutes a ready-made litter box. There is no vaccine or effective treatment for sheep, though farm cats can be treated. Stray cats should be removed immediately. Strict sanitation; clean, uncontaminated water; dry, protected storage of hay and grain; and off-the-ground feeding troughs reduce the incidence and spread of disease.

Salmonellosis

Several species of *Salmonella* bacteria can cause abortion. The good news is that the disease is pretty rare, but the bad news is that when it strikes, it's highly contagious.

Affected ewes develop a high fever and severe diarrhea and tend to go off feed. Most abort, and some may die from the disease. Those that give birth don't have milk to raise the lambs, and the lambs are generally weak and die.

There is no vaccine against *salmonellosis*, and no approved antibiotics, although your veterinarian may be able to provide an extralabel antibiotic (see Unapproved Drugs in chapter 7). The infection often results from poor sanitation, so keep feed and water clean.

Mastitis

An infection that results in inflammation of the udder, mastitis usually affects only one side. It can be caused by one of several species of bacteria, or sometimes by yeast. It can be acute or chronic.

In the acute form, the ewe runs a high fever (105–107°F; 40.6–41.7°C) and usually goes off her feed. One side of her udder is hot, swollen, and painful. She will limp, carrying one hind leg as far from the udder as possible, and not want the lamb to nurse. The milk can become thick and flaky, full of curds, or watery. Early detection and prompt treatment can minimize udder scarring. One type of acute mastitis results in gangrene. The udder becomes almost blue and is cold to the touch. This type of mastitis is critical, and requires treatment with intramuscular or intravenous (IV) antibiotics in order to save the ewe. The ewe may also require IV fluids.

Chronic, or subclinical, mastitis may be undetected, showing up at the ewe's next lambing when she has milk in only half of her udder and the other half is hard. Mild cases may be caused by bruises. Bruising can results from large lambs, especially near weaning, that bump their mothers with great zest as they nurse (sometimes lifting her hind end right off the ground) or twins who pummel her simultaneously. Mild mastitis has fewer symptoms, and the ewe may just wean the lambs by refusing to let them nurse.

Sheep people say that mastitis is governed by Murphy's Law. The severity and incidence of mastitis is directly proportional to the value of the ewe, her lack of age, the number of lambs she delivered, the severity of the weather, and how busy you are at lambing time.

Take a milk sample to your veterinarian as soon as you suspect mastitis. The sample can be examined under a microscope, and culture and sensitivity tests can be done. The affected half of the udder is not likely to be able to produce milk again unless mastitis is noticed early and treated promptly. Ewes that are treated aggressively at the first signs of mastitis have a good chance of cure.

Mastitis Treatment

Veterinarians often prescribe injectable antibiotics for mastitis. Your veterinarian may also advise you to use a teat medication for cows, and although the applicator is large and inconvenient, it can be used with sheep. The product is squirted directly into the teat. Infected sides should be milked out as completely and as often as possible, and the milk should be destroyed. Combination treatment drugs are available for both acute and chronic mastitis, and these agents are effective against several of the causative bacteria.

Milk Fever

Because so much calcium is needed to form the bones and teeth of the developing lambs and for milk production, the ewe may suddenly be unable to supply all of the calcium that's needed. This deficiency may be caused simply by a calcium deficiency in her diet, or it could be triggered by a metabolic disturbance. Milk fever (also known as lambing sickness and hypocalcemia) is serious and can be fatal in a short time. The ewe's lack of sufficient calcium more often becomes apparent *after* lambing, but it can manifest in the 24 hours that precede birth. (If milk fever comes on before lambing, it is easily confused with pregnancy toxemia.) Abrupt changes of feed; a period without feed; or a sudden, drastic change in the weather near the end of pregnancy seems to increase the likelihood of milk fever.

Milk fever is a true medical emergency in which the ewe's life is in jeopardy. Several commercial veterinary preparations for treatment are sold in sheep-supply mail-order catalogs, including a paste that can be given orally during the early onset of milk fever. But never give the paste once the ewe is down, because if she can't swallow correctly, the paste will enter her lungs and kill her very quickly. Once the ewe is down, IV treatment is the method of choice and works very quickly in restoring the ewe.

This type of treatment should not be attempted by a novice. Have your veterinarian come out as quickly as possible. If veterinary assistance is not available, inject the ewe subcutaneously with 75 to 100 mL (given in five places) of calcium gluconate. This can be purchased at a drugstore; tell them it's for livestock use. Always keep calcium gluconate on hand if you don't have a veterinarian nearby. Subcutaneous injections act more slowly but have less of a chance of causing cardiac arrest (heart failure) than IV injection, making it safer in the hands of a novice. Calcium gluconate can also be given intraperitoneally, in the paunch on the right side of the ewe. Lay the ewe on her left side, and inject into the right side if she has already lambed. The intraperitoneal method is faster than subcutaneous injection and safer than intravenous injection.

The dramatic improvement that results from the calcium is simply amazing. Paula has seen it in ewes, and I've seen it in cows. In each case, the mom gets up only minutes after treatment. Maintenance therapy for milk fever involves giving some paste twice a day for 3 days after the onset. Some ewes tend to have this problem with each lambing, so keep an eye on these ewes in the future.

Signs of Milk Fever

The onset of milk fever is sudden, and the disease progresses rapidly. The first signs are excitability, muscle tremors, and a stilted gait, and these signs are followed by staggering, rapid breathing, staring eyes, and dullness. The ewe then lies down and is unable to rise. In the final stages, she slips into a coma and then dies. Although this disorder is called a "fever," the temperature remains normal or subnormal, and the ears become very cold. To be successful, treatment should start before the ewe is down; however, if she is down but not comatose, there is a chance for recovery.

Pregnancy Toxemia

Pregnancy toxemia, also known as *ketosis*, is highly fatal if not treated or if the ewe does not lamb right away. It usually occurs in the last week or so of pregnancy and often in twin- or triplet-carrying ewes. It can be reliably diagnosed by using a urine test strip that is available from a veterinarian, farm supplier, or even a drugstore. However, this disease can usually be recognized just by its symptoms.

The problem is caused by an excessive buildup of *ketones* (which are the by-products of fat metabolism) in the bloodstream. The disease tends to cause a vicious cycle, because multiple or large fetuses require high amounts of nutrients, but as they grow and take up more abdominal room, the ewe's ability to consume sufficient feed to support both her and her offspring is drastically reduced. The disease is not a "thin ewe" or blood sugar problem, but one of insufficient energy (calorie) intake. When the ewe consumes less energy than she needs to sustain herself and the lambs, she begins to use stored body fat to provide this energy. If the ewe is breaking down significant levels of body fat, she may reach the point that ketones are being produced faster than her body can excrete them. When this occurs, the ketones build up to toxic levels, and pregnancy toxemia occurs. This disorder often afflicts ewes that are on a high-

Pregnancy Toxemia or Milk Fever?

It can be difficult to tell the difference between pregnancy toxemia and milk fever. Pregnancy toxemia can be accurately diagnosed by test strips, but it can be a complicating factor in milk fever. So, a diagnosis of pregnancy toxemia does not rule out milk fever. You can make an intelligent guess by reviewing the circumstances.

If the disorder happens before lambing and there is any possibility that the ewe may not have been fed properly in the last month, it's probably toxemia. If in doubt, call your veterinarian immediately, as milk fever is a life-or-death situation.

If it occurs after lambing, the ewe is providing good milk for twins or triplets, and she has had adequate feed with molasses, the problem is more likely to be milk fever, possibly with a trace of pregnancy toxemia as a complication. Most mail-order veterinary suppliers sell commercial preparations for milk fever that also contain dextrose or other ingredients for pregnancy toxemia, so they can be used to treat either or both.

fiber (hay), low-energy (no grain) diet and ewes that are too fat early in pregnancy. Stress and forced activity also demand energy, which can contribute to the problem or actually trigger the toxemia in borderline cases of inadequate nutrition. Simply stated, prevention requires *calories*. To determine which ewes are not obtaining sufficient nutrition, use the test kits as a precautionary measure — especially on ewes that you expect to have multiple lambs.

The symptoms to watch for include sleepy-looking, dopey-acting, dull-eyed ewes that are weak in the legs and have sweet, acetonic-smelling breath (it smells like model-airplane glue!). They usually refuse to eat, are unable to rise, grind their teeth, and breathe rapidly. Recovery is unlikely if treatment is delayed too long.

Luckily, if caught early, it's easy to treat. Four ounces (118.3 mL) of propylene glycol (this is *not* the type of glycol used in antifreeze, which is poisonous), 4 ounces (118.3 mL) of glycerin diluted with warm water, or a commercial preparation for treating pregnancy toxemia can be given orally twice a day. To prevent relapse, continue treatment for 4 days even if the ewe seems to have recovered.

Keep propylene glycol (or the commercial medication) on hand before lambing for prompt treatment of any suspected cases. Once a full-blown case has occurred, it might already be too late for treatment to be effective, and a cesarean section will be required to save the ewe. The lamb will be lost unless the ewe is very close to normal lambing time. Subclinical pregnancy toxemia (also confirmed by the ketosis test) is a more mild form of the same disease and is characterized by a weakened ewe that may produce a small or a dead lamb.

Retained Afterbirth

In almost all cases, the afterbirth comes out normally, usually within the first hour after the lamb has been born depending somewhat on the ewe's activity. If the afterbirth is partially hanging out, do not attempt to pull it out, as this might cause straining and prolapse or some other injury. A veterinarian does not consider the afterbirth to be truly retained until at least 6 hours have passed since the birth, so you can allow some time to pass before starting to worry. Some ewes eat the afterbirth (or if you have dogs that have access to the pasture where sheep are lambing, they'll relish this choice morsel), so you may think it is retained, even though it has passed.

Physical removal of the afterbirth is best done by a veterinarian, who can differentiate between the maternal and the fetal cotyledons. Manually removing the placenta sooner than 48 hours after the birth is usually not advised, and in the meantime the veterinarian may prescribe a drug (oxytocin) to help the ewe expel it. If allowed to remain in the ewe, a retained placenta results in uterine infection and sterility. Nutritional deficiencies, particularly selenium deficiency, seem to play a role in retaining the afterbirth.

Vaginal Prolapse

Prolapse of the vagina is most common before lambing, but it occasionally follows a difficult labor. The vaginal lining is seen as a red mass protruding from the genital opening. Do not delay treatment, for it will worsen progressively and become more difficult to repair. Early detection is important. Even though this problem occurs infrequently, be on the watch for it.

If the lining is just barely protruding, confine the ewe in a crate that elevates her hind end, thus decreasing the pressure. Leave her head out to eat and feed her mostly on grain plus some green feed (like grass, weeds, or apples). Avoid ground-up grain or rolled oats. The dust might cause coughing and aggravate the problem.

A more certain solution, which also should be started as soon as possible to increase the chance for success, is to use a *prolapse retainer* (otherwise known as a ewe-bearing retainer), which is a flat, plastic tongue, or you can make your own prolapse loop.

1. Use 1⁄16-inch (1.02 cm) wire.
2. Cover only the loop part with soft rubber tubing as a cushion.
3. Slip the rubber tubing onto the wire before making the final bends.
4. Bend as shown.
5. Disinfect the loop before use.

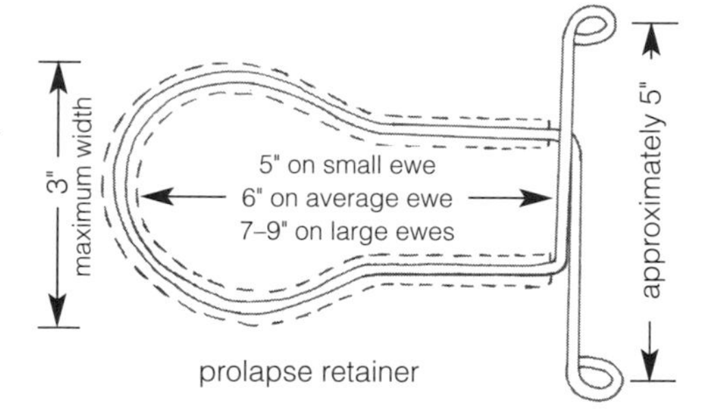

prolapse retainer

Before attempting to insert the prolapse retainer, wash the prolapse and elevate the ewe's hind end considerably. You can tie the ends of a length of rope to each hind leg and loop the rope up over the top of a post, or have the rope short enough to just go around your neck, so that the ewe is raised by it. You can hold her steady in this position if she is on her back, with her hind end raised and steadied by a bale of hay.

To use the retainer or loop to replace the protrusion:

1. Cinch a rope or belt around her middle so that she cannot strain after you replace the prolapse. Tie a 1⁄4-inch (4.1 cm) rope securely around her flank in front of the udder, but not so tightly that she can't lie down

or get up. The rope has to be removed when she goes into labor. Sometimes the ewe stops straining after a couple of days as the swelling goes down.

2. Wash your hands, and disinfect the loop if you have not done this already.
3. Wash the prolapse with cold (not hot), antiseptic water, or put both mild soap and antiseptic in the water to help contract it.
4. Watch out for a flood of urine as you gently replace the vaginal lining. Its bulging may have blocked the opening to the urinary tract. If prolonged, this blockage can be fatal.
5. Replace the lining, using lubricant if necessary, and gently press out all the creases. This is much easier with the hind end elevated than it would be if the ewe were lying flat. Even holding her on her back, with her shoulders on the ground and her hindquarters up against your knee, relieves much of the pressure on the replaced vagina.
6. Holding the vagina in place with one hand, insert the prolapse loop straight in, flat, and horizontally. If you have made a loop from the pattern given, it should be long enough that the forward end contacts the cervix.
7. The loop is held in place by tying it to clumps of wool or by sutures if the ewe has been sheared or closely crotched. There is also a new prolapse harness (see Resources) that can be used with the prolapse loop to hold it in place better than tying to the fleece.
8. Give a shot of a broad-spectrum antibiotic to avoid infection.

The ewe can lamb while wearing the loop or retainer, or you can remove it as she goes into labor. It is safer to leave it in place and try lambing that way, so that prolapse doesn't recur with ejection of the lamb.

Mark this ewe for culling, because prolapse causes permanent damage and might happen again. Since the disorder could be a genetic weakness, it is best not to keep any of her lambs for breeding.

Suturing the vagina to hold it in is often the best approach if the prolapse occurs after lambing, but it is less than ideal before lambing because the sutures must be removed. **Suturing the vagina can be done in the following manner:**

1. We have successfully used dental floss, a curved needle, and pliers to get a good grip on the needle and pull it through.
2. To ensure safety, use only one deep stitch at the top of the vaginal opening and one across the bottom.
3. Insert the needle from right to left at the top, then bring it down and insert it from left to right at the bottom.

4. Knot the two ends together on the right side. The advantage of sewing this way, rather than crossing the stitches across the center of the opening, is that you can tell when the lamb is coming. There is room for the feet and nose to present themselves, allowing you time to cut the stitches.

If the ewe is beyond 143 days of gestation, your veterinarian can prescribe or give medication, such as dexamethasone or oxytocin, to start labor. After lambing, replacement and suturing can be done.

One factor is now known to be particularly important in preventing prolapse. Selenium, which is known to increase lamb survival and prevent white muscle disease in lambs, has been noted to increase muscle tone and help counteract a prolapse tendency in pregnant ewes.

A ewe that has had a selenium-supplemented ration or a selenium–vitamin E injection will usually have an adequate level of plasma selenium and will produce milk with sufficient selenium. Too much selenium is acutely toxic, so a selenium-enriched ration, plus a mineral–salt mix with selenium, plus injections would be a dangerous combination. There are several selenium products available. Check the concentration carefully before use, and follow label instructions. A newborn lamb should never be injected with more than 1.0 mg of selenium.

There is reason to anticipate an increase in selenium deficiency. Increased forage yields are speeding the depletion of selenium in the topsoil, and increased animal stocking per acre in a given area also contributes to the problem.

Wondering about selenium in your area? Check with the county Extension agent, or have your veterinarian do a blood test so that supplementation can be done on an informed basis.

Selenium Supplementation

Experts at the USDA Sheep Station in Dubois, Idaho, suggest that people who live in an area with selenium deficiency should inject ewes with this mineral 1 week before lambing to maintain muscle tone. Because of the potential trauma to ewes, injection could be given earlier along with the last vaccination. The slightest indication of a prolapse calls for an additional injection of selenium, along with the usual prolapse repair measures. While selenium–vitamin E injectables have instructions for intramuscular administration, many experienced sheep veterinarians recommend subcutaneous injection to avoid incidence of muscle damage at the injection site.

Lamb Problems

Although adult sheep can suffer from the diseases discussed in this section, lambs are far more susceptible to them and they are more often deadly in lambs. Lambs are highly vulnerable during the first 10 days of their lives to many ailments. (Chapter 10 has more information on general lambing problems.)

Acidosis

The disorder known as *acidosis*, which is also called acute indigestion or "founder," is caused by excessive production of lactic acid in ruminants that suddenly gorged themselves on grain or other feeds that are high in carbohydrates. It also occurs in feedlot lambs fed high-grain–low-roughage diets. Fermentation of the high-energy diet results in excessive production of lactic acid and causes toxicity. Acidity increases in the rumen until severe digestive upset or death occurs. Manifestation of acidosis can be marked by the sudden death of numerous lambs; this manifestation is similar to that of enterotoxemia, making accurate diagnosis difficult. Signs include inappetence, depression, lameness, coma, and death. Feed that is at least 50 percent roughage (hay and/or pasture) is safe for the lambs. Any shift to a higher percentage of grain should be made very gradually.

Constipation

A constipated lamb usually hunches up while standing and looks uncomfortable. There is no sign of droppings or only a few very hard ones. Sometimes the lamb grinds its teeth, and if the condition is left unchecked, the lamb can go into convulsions and die.

For most cases of constipation, including that caused by unpassed meconium, administer 2 tablespoons (29.6 mL) of vegetable oil or 1 tablespoon (14.8 mL) of castor oil for very small lambs (under 2 weeks old). For lambs as old as 2 months, carefully give ¼ to ½ cup (59.1–118.3 mL) of vegetable oil. The dose may need to be repeated two or three times.

"Pinning" is an external kind of stoppage that is fairly common in very young lambs, usually those younger than 1 week. Since early manure is pasty, it can collect and dry into a mass under the tail, effectively plugging up the lamb. If not noticed and corrected, the lamb will die. (These manure plugs also make a prime target for flystrike.) Clean off the mass with a damp rag or a paper towel, trim off some of the wool if necessary, disinfect the area if it is irritated, and oil it lightly to prevent further sticking. Check the lamb frequently.

This is another good reason to keep mother and lamb in the pen for the first 3 days, so you can easily inspect for this and other problems.

Occasionally, a lamb may have a rare birth defect in which it is born without an anus. This condition often goes undetected for the first few days, until the distended abdomen and discomfort are observed. Quick detection and surgery are the only treatment.

Entropion

Frequently, when a lamb is born, the lower (or sometimes the upper) eyelid, or both, may be rolled inward; this disorder is called *entropion*. When it happens, the eyelashes chafe the eyeball, causing the eye to water constantly, inviting infection and even blindness. Entropion is a hereditary defect that is more prevalent in wooly-faced breeds. Do not keep such a lamb for breeding. Mark it with an ear tag or notch for slaughter.

Inspect each lamb at birth so that the condition can be found at once and corrected. The easiest method for correcting entropian is: Inject 1 mL of penicillin just under the skin beneath the lower eye. The penicillin forms a small bulb of fluid that forces the eyelid down into its correct position, and by the time the body has absorbed the fluid, a small piece of scar tissue has formed that keeps the eyelid in position. Use a 22-gauge needle.

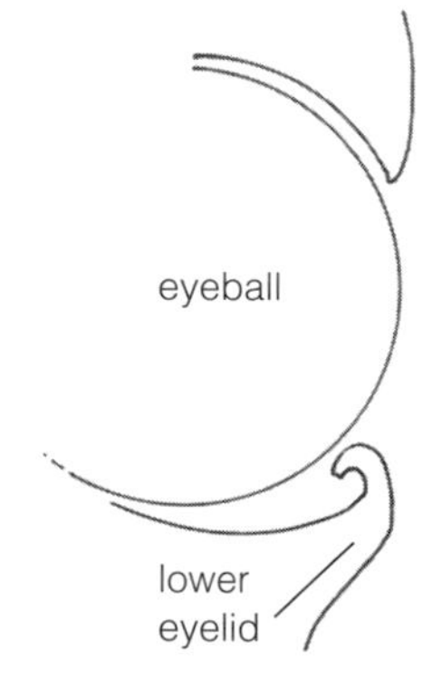

Entropion, showing lower eyelid turned in. Eyelashes will irritate the eyeball. The pocket behind the eyeball becomes infected if the condition is not remedied.

Enterotoxemia

A disease caused by a multiplying of bacteria called *Clostridium perfringens* that often strike your biggest and best lambs — those that eat with the most enthusiasm — is known as *enterotoxemia*, or "overeating disease." This disorder is more common among lambs that are given too much grain and too little roughage (hay) or those that have had an abrupt change in their feed ration. Enterotoxemia sometimes occurs among fairly young lambs that get too much milk from a heavy-milking ewe. Early creep-feeding of *both* hay and grain helps cut down on the incidence.

Specifically, lambs may develop this disease if grain composes more than 60 percent of their ration or if they are brought up to a full feed of 1½ to 2 pounds (0.7 to 0.9 kg) of grain per day too rapidly. Older lambs with a heavy load of tapeworms are especially vulnerable. Wet bedding, chilling, and stress cause variable feed intake, which is conducive to an outbreak.

The immunity provided by the ewe, assuming that the lamb gets a good dose of her colostrum, protects the lamb until about 10 weeks of age. At that point, it should receive its own vaccination. Ideally, a "priming dose" should be given at 9 or 10 weeks of age, with a booster about a month later. If the lamb is kept as breeding stock, it will only need the booster dose in subsequent years, given about 2 weeks prior to lambing.

Enterotoxemia is characterized by sudden death or convulsions and diarrhea. Few lambs live long enough to respond to treatment, but injected antiserum can be tried. Prevention is the only sure thing.

Prevention of Enterotoxemia

Enterotoxemia can be quickly fatal. Thus, prevention is best. Vaccinate your ewes with a multipurpose vaccine during pregnancy that lists clostridium and should be given to ewes 6 to 8 weeks prior to lambing, with a booster dose 2 weeks before lambing. In following years, the ewe only needs the booster shot. The colostral antibodies are passed to the lamb, providing immediate and complete protection against all clostridial diseases, including enterotoxemia and tetanus, the most common ones.

Navel Ill

The term *navel ill* describes infections for any number of organisms that gain entrance to the lamb's body through the umbilical cord shortly after birth. These organisms develop into septicemia, a serious illness in which bacteria invade the bloodstream, usually within a few days.

By treating the umbilical site with strong iodine *as soon as possible* after birth and ensuring that the lamb nurses its mother within the first hour (because the colostrum contains antibodies against many environmental germs), you can minimize the danger of navel ill. A second douse of iodine about 12 hours later is a good practice. Clean bedding in the lamb pen also lessens the chance of infection.

The acute form of navel ill causes a rise in temperature and eliminates the inclination to suck. A thickening can usually be felt around the navel. Death follows quickly.

Tetanus is one of the serious diseases that can be caused by a bacillus that enters through the umbilical site. Certain protection against tetanus is obtained by vaccinating the ewes.

Since navel ill may be caused by various bacteria, it takes a veterinary diagnosis to determine the specific cause and to administer the proper antibiotics. Treatment can consist of IV antibiotics, scour boluses, a tube passed to relieve bloat, and interperitoneal administration of glucose.

Pneumonia

Pneumonia is probably responsible for more lamb deaths than any other single cause (except starvation). In some flocks, this disease can kill 12 to 15 percent of the lamb population. For the most part, pneumonia can be prevented. It is caused by drafts in cold, damp quarters, by overheating with heat lamps followed by exposure to cold, and by exposure to infectious agents. This type of pneumonia is caused by either bacteria or viruses. Proper management is the key to successful prevention of pneumonia. Adequate ventilation in the lambing barn is mandatory. Open-sided barns with burlap bags or the new Tensar (Atlanta, GA) windbreak material to prevent drafts will prevent buildup of ammonia-laden, stagnant air. Use heat lamps no more than necessary, and have jugs with solid bottoms to prevent floor drafts on the newborn.

If pneumonia is a recurring problem in your young lambs, make sure that they are getting adequate amounts of selenium and vitamin E, as marginal levels result in immunosuppression and increased susceptibility to infection.

Mechanical Pneumonia

A type of pneumonia that is caused by foreign body obstruction is known as mechanical pneumonia. Specifically, this type of pneumonia results from fluid (such as excessive birth fluids or milk) or objects entering the lungs. An abnormal birth position or an interruption of the umbilical blood supply to the lamb can result in a respiratory reflex that causes the lamb to attempt to breathe before birth is complete. This causes inhalation of fetal fluids, resulting in mechanical pneumonia. Also, forced bottle-feeding of a lamb with impaired sucking reflex, improper stomach tube–feeding, or improper use of oral medications causes fluid to enter the lungs. There is no known cure for mechanical pneumonia.

Polio (Polioencephalomalacia)

As in humans, polioencephalomalacia — commonly known as polio in sheep is noninfectious and characterized clinically by blindness, depression, incoordination, extreme salivation, coma, and death. The syndrome (which is similar, and possibly secondary to, acidosis) is related to diet. It is often seen in flocks that are moved from severely overgrazed pastures to lush pastures, though the exact mechanism isn't fully understood. It has been shown that the disease is caused by an acute thiamine (vitamin B_1 deficiency and that ruminal contents contain high levels of thiaminase (an enzyme that destroys thiamine).

Empirically, field experience has shown that changing the ingredients in the diet may break this cycle and alleviate the outbreak. In the early stages, treatment with 0.5 gram thiamine hydrochloride stimulates a rapid recovery. Repeat treatments at 2-day intervals as necessary. A lamb that has recovered can contract polio again if the diet remains unchanged.

Scours

Scours (that is, diarrhea) in newborn lambs can be very serious and has many causes. Several kinds of bacteria, some viruses, and overeating all cause scours.

If you are new to lambs, a brief discussion of what normal feces look like during the first few days of life might help you to catch scours early.

1. On the first day, a lamb should pass meconium. *Meconium* is a tarry substance that blocks the anus of a fetus. It usually passes quickly after a newborn nurses the first time, though sometimes it doesn't, resulting in constipation (see the discussion earlier in this chapter).
2. After the meconium phase, the manure is yellow and pasty for about 2 days.
3. It then begins to take on the color of regular sheep feces and begins to firm up. Lambs usually begin passing pelleted manure within about a week to 10 days.

Yellow Scours

One thing that can be confusing for a new shepherd is that one kind of scours is *yellow scours*. Yellow scours is the least serious type, though if left unchecked, it can be deadly. To differentiate yellow scours from normal yellow feces, consider the following:

- Normal yellow feces are pasty and a pale yellow color.
- Normal yellow feces only last for the first day or two.

- Feces associated with yellow scours are runny and have a greenish tinge.

Yellow scours is often associated with overfeeding. It is a common problem with bottle-fed lambs but can also occur in a strong lamb nursing a ewe with an excess of milk. If you are using milk replacer to feed lambs, purchase a good-quality product that is specifically labeled for lambs. The first ingredient listed should be milk! Many cheap milk replacers use soybeans and other plant matter and are very poor substitutes.

If bottle-feeding, substitute a day's feeding with oral electrolytes — give no milk. In a pinch, sport drinks that are sold in convenience stores can be used as an oral electrolyte, or you can make a homemade electrolyte solution (see recipe on page 225). But ultimately it pays to keep some powdered electrolyte on hand that is specifically prepared for livestock. These products (the ones labeled for calves work fine for lambs) contain not only electrolytes but also vitamins, minerals, and energy components.

On the second day, if the fecal matter is returning to a normal consistency, begin feeding milk again, but dilute the lamb's normal ration by giving 50 percent milk and 50 percent water. Scours that result strictly from overeating should resolve by the third day, and you can return to full-strength milk or replacer.

Lambs that are nursing are harder to treat for overeating scours. You can hand-milk the ewe to reduce the amount of available milk, and try to give the lamb a feeding of water or Gatorade so that its appetite is satisfied for a feeding. Scours that continue for a second day (that is, the day after only electrolytes were given) usually indicate that an infection is developing, and the lamb will need treatment for dehydration and infection. You'll need to continue to give the oral solution to replace excessive loss of electrolytes, but start supplying some milk again as well as some form of antibacterial therapy. It's best to not feed electrolytes and milk during the same feeding because the electrolytes can interfere with absorption of the milk's nutrients. What we have found to work best is to feed very small amounts every couple of hours, alternating one feeding of electrolyte with one feeding of milk. (See chapter 10 for a detailed discussion of bottle-feeding.)

White Scours

Infection with *Escherichia coli* usually causes white scours, which are very serious. White scours can result in rapid dehydration, toxemia, and death if not treated immediately. In most cases, this infection is caused by filth, such as poor sanitation, or a lamb sucking on a dirty wool tag from an uncrotched ewe.

Vaccines are available if white scours are a recurring problem in your lambs. The current antibiotic of choice for lamb scours is an oral spectinomycin, but antibiotics are always changing, so before your first lambs arrive, ask your veterinarian to recommend a therapy. A couple of teaspoons of Pepto Bismol or Kaopectate help firm up the stool and forms a protective coating in the intestine in lambs with scours.

Homemade Electrolyte Solution

You can make your own electrolyte solution in a pinch. Here's how:

1 quart (1 L) water
2 ounces (59.1 mL) dextrose (corn syrup)
½ teaspoon (2.5 mL) salt
¼ teaspoon (1.2 mL) bicarbonate of soda

Combine all ingredients. Give this solution for 1 to 2 days. At that point, return to milk feeding, giving smaller quantities than before.

For scours that are not caused by bacteria, it is often helpful to give a few ounces of aloe vera juice to help the digestive system return to normal.

Tetanus

Tail docking and castration can put lambs in danger of tetanus ("lockjaw"). If the ewes have not received a booster of Covexin-8 or a similar product, you should administer 300 to 500 units of tetanus antitoxin at the time of docking or castration. The antitoxin protects the lambs for about 2 weeks while the wounds are healing. Since there is no known cure for tetanus, protection is worth the effort.

Urinary Calculi

A problem of growing ram lambs over 1 month old, castrated or not, is that the salts they normally excrete in their urine can form *urinary calculi,* also known as stones or water belly. These calculi may lodge in the kidney, bladder, or urethra.

Lambs with urinary calculi kick at their stomach, stand with their back arched, switch their tail, and strain to urinate (or dribble urine, frequently with blood in it). Some may recover if the stone is passed soon enough. This blockage of the urinary tract causes pain, colic, and eventually rupture of the urinary system into the body cavity (hence the name "water belly") and death.

If you are watching a lamb that appears to be straining and unable to urinate, put him on a dry floor for a couple of hours. Unless there is a blockage, he will ordinarily urinate in that time. Turn the lamb up, and feel for a small stone that can be worked gently down the urinary passage. Sometimes manipulation of a small catheter tube (from the drugstore) dislodges the stone.

Veterinarians say that nine times out of ten, the plugging is at the outer end of the urethra, so if you can feel the stone right at the end you may be able to dislodge it with gentle pressure. If the passage is cleared and urine spurts out, stop the flow two or three times. It is possible for the bladder to rupture if it is emptied too quickly. If the stone cannot be dislodged, a veterinarian may administer a smooth-muscle relaxer, which has a dilating action that allows the calculi to pass, or remove the stone surgically.

Any of the following can contribute to calculi:

- Low water intake due to cold weather or unpalatable water. Lambs need fresh, warm water during cold weather. Adding salt to the ration and keeping both salt and fresh water in the creep helps. Increasing salt increases urine volume and decreases the incidence of stones; sheep that don't have access to salt do not drink sufficient water, especially in cold weather.
- Ration that has excessive phosphorus and potassium — like beet pulp, wheat bran, and corn fodder — but is low in vitamin A. Correct by adding ground limestone or dicalcium phosphate, 1 or 2 percent of the ration, to make the calcium–phosphate ratio approximately 2:1. Well-formulated lamb feed pellets have this ratio.
- Growing crops by using a heavy fertilizer with high nitrate content. This practice interferes with the carotene roughage that produces vitamin A. Enrichment of the ration with vitamin A counteracts this problem.
- Hard water. This problem can be corrected by adding feed grade ammonium chloride to the ration, approximately ⅕ ounce (6.1 mL) per head per day. This salt is harmless and is found in some pelleted feeds.
- Feeding of *only* pelleted feed. Urinary calculi seldom develop in lambs who receive 20 percent alfalfa.
- Hormonal changes that occur when ram lambs are castrated at less than 4 weeks of age. The absence of testosterone after castration keeps

the urethra from growing to its maximum diameter. If you have a persistent problem with your *wethers* (that is, a lamb that has been castrated before sexual maturity), try castrating after 6 weeks.

- Feeding sorghum-based rations. Cottonseed meal and milo also increase the risk for calculi. Corn and soybean meal are less apt to cause problems.

White Muscle Disease

Selenium again! White muscle disease, also known as "stiff lamb," is caused by a lack of it. If the soil is deficient in this important mineral (as in parts of Montana, Oregon, Michigan, New York, and many other areas), then so is the hay. Hay from localities known to have inadequate selenium should not be fed to ewes after the third month of pregnancy or during lactation, unless it is well supplemented by whole-grain wheat and/or mineralized salt with selenium in it. Supplementation should also include vitamin E.

In areas that are known to be low in selenium, medication should be given to prevent lamb losses. Both oral and injectable products are available and are typically given to the ewe 2 to 4 weeks before lambing, such as when you give the second vaccination against tetanus and enterotoxemia.

Lambs suffering from white muscle disease have difficulty getting up and walking and gradually become affected by muscle paralysis. Though lambs can be treated after birth, once muscle changes occur, they cannot be reversed. Many small, weak lambs or lambs with a stiff neck at birth respond to treatment with selenium.

NINE

FLOCK MANAGEMENT

Good management for reproduction is a key to profitability and peace of mind. Reproduction really goes right through lambing, but I've broken the topic into two chapters: This chapter covers the ram, the ewe, breeding, and pregnancy, and the next chapter covers lambing and early-life management of your flock. The optimal time for lambing varies greatly among geographical areas. The desired lambing time may depend on the availability of pasture, local weather conditions, labor and time restraints, targeted lamb markets, and so on. Choose your lambing time to fit your priorities, and plan to breed about 5 months before you want lambs.

When the cost of hay or grain is a consideration or if you want to minimize labor, than lambing should be timed to take advantage of new pasture growth. Some shepherds living in areas with moderate winters and hot summers may choose to lamb in autumn or early winter to maximize weight gains, knowing that lambs experience very poor weight gain in hot temperatures. Those in the northern areas often begin lambing in March or April to avoid the severe subzero temperatures. People in temperate coastal climates may let the rams run with the ewes all year and let nature take its course, if they have no target date for market lambs. What constitutes "early" or "late" lambing depends on your climate.

Successful Breeding

Many factors can influence how good your breeding season is:

- **Day length.** All sheep are *photosensitive*, meaning that reproductive activity is affected by the length of daylight. In the fall and winter, reproductive activity is highest. Ewes all go into heat and are capable of breeding during these seasons, though some individual ewes and

Reproductive Functions

Here are some rules of thumb for reproductive functions — but remember, all animals are unique individuals, and some don't follow the rules!

- **First estrus:** This generally occurs when animals are at least 6 months of age and weigh ⅔ of their adult weight, though a few breeds are known for coming into estrus as early as 4 to 5 months of age
- **Length of estrous cycle:** The range is 14 to 19 days between cycles, with 17 days being the average; if all ewes have been exposed to the ram for 34 days, they theoretically should have had two estrous cycles in which to breed
- **Length of time standing in heat:** The average is 30 hours, but this can range anywhere from 3 to 73 hours
- **Time of ovulation:** 28 hours after the start of the estrous cycle
- **Length of time the egg remains capable of being fertilized:** 12 to 24 hours

particularly some breeds of ewes can breed year-round, or "out of season." But even within breeds that are known for out-of-season lambing, the ovulation rate is lower during the spring and summer than it is in fall and winter. Thus, by breeding out of season, you may actually have fewer lambs.

- **Temperature.** Although high temperatures don't have as much effect on reproductive activity as day length, it does have some. During hot weather, rams may be infertile. Ewes, even if bred, are likely to miscarry.
- **Age.** Ewe lambs generally begin cycling later than mature ewes, don't tend to have a strong a heat, and do not release as many eggs per ovulation. Reproductive activity may be reduced in aged ewes.
- **Nutrition.** Flushing, or increasing the level of nutrition prior to breeding, increases reproductive activity.
- **General health.** Good health pays off with more lambs born and greater lamb survival.

Advantages of Early and Late Lambing

EARLY LAMBING

- There are fewer parasites on the early grass pasture.
- Ewe lambs born early are more apt to breed as lambs.
- You can sell early lambs by Easter, if creep-fed, and get a better price for early meat lambs.
- You can have all lambs born by the time of the best spring grass.
- There are fewer problems with flies at docking and castrating time.

LATE LAMBING

- It is easy to shear ewes before lambing.
- It avoids the danger of lambing in severe weather.
- Mild weather means fewer chilled lambs.
- Ewes can lamb out on the pasture.
- Less grain is required for lambs, since you have lots of pasture.

SHEPHERD STORY

Although Richard Parry's flock is larger than average, by western range standards, at 1,500 head it's nothing to get excited about. But, Richard's approach to shepherding is something to get excited about. Some of Richard's techniques may not apply to your small flock, but his philosophy should.

After graduating from high school in 1971, Richard became a rancher in his own right because of his father's untimely death. The ranch had already been in his family for several generations, and it was up to him to keep it going.

Richard's dad and granddad before him had operated a conventional range-style sheep outfit. At one time, they ran 3,000 head. Lambing was done in sheds in the winter, then ewes and lambs were moved up to Forest Service allotments to graze in the summer. In spite of hired herders staying full time with the flock, death losses were high. The home place was farmed for winter feed.

Richard started his career with a farm-sized flock, but soon was up to 1,000 head and practicing the same approaches his family had before. "By 1985, I was about ready to chuck it all. We were struggling along, making less money, working harder. But then, I went through my first 'paradigm shift'; I discovered managed grazing. I really began to get excited again, to see hope for our operation. We began by building grazing paddocks on one-half the farm ground.

"Between 1985 and 1990, the paradigm shift continued. I discovered *The Stockman Grassfarmer* magazine, holistic management, and the Ranching for Profit school and group. We were still doing some farming, raising small grains and hay, but the more we learned and the more we shifted our thinking, the more we realized that the farming end of things wasn't profiting us."

Today, Richard's operation has made a big turnaround. He lambs on his

The Ram

The ram contributes to the genetics of all your lambs, so you want to obtain the best ram you can get.

Ordinarily, the "best" ram is a well-grown 2-year-old that was one of either twins or triplets. Being a member of a multiple birth in no way affects the chances of twinning in the ewes he breeds — this is controlled by the number of eggs the ewe drops to be fertilized, which is influenced by genetics and encouraged by flushing. However, his daughters will inherit a genetic inclination toward having twins, especially if their mothers had the same inclination. In other words, both the ewe's and the ram's twinning capabilities will show up in the following generations. The ram also greatly influences other traits, such as conformation and fleece type.

irrigated paddocks in late spring–early summer. "That change lowered both our costs and our labor."

A set-stocked pasture lambing program, which has been practiced for years in New Zealand, has proven to be the most effective method for Richard. The sheep are broken into groups of about fifty ewes for those expecting single births, and smaller groups for ewes expecting twins and triplets. How does Richard know which is which? He uses pregnancy scanning.

Each group is kept in a 15-acre paddock during lambing. As the lambs and their moms form strong bonds, groups are brought together to form larger bands, and timed movement begins. Richard stresses that this system is slow moving initially, so the lambs and ewes are well bonded.

Richard says, "We finally found a system of production that provided us with a temporary advantage — it allowed us to keep going. But operating in a commodity system is a problem. We're always looking for other advantages that will allow us to stay in business and have found two. We're working on direct marketing our lamb. This is important, to get out of the commodity-based system.

"Our latest 'advantage' was a real philosophical breakthrough: We've gotten out of feeding hay in the winter by purchasing a second ranch in a desert area. We move the dry ewes there for the winter, and they rough through really cheaply. We run the entire band with one herder, two Border Collies, and three guardian dogs during the winter. We don't farm at all anymore. In fact, we had an auction and sold all the farm equipment."

Sheep are grazers by nature, and Richard Parry's success comes from his recognition of that fact and his ability to develop a system that mimics nature.

Use a young ram sparingly for breeding. One way to conserve his energy is to separate him from the ewes for several hours during the day, at which time he can be fed and watered and allowed to rest.

One good ram can handle twenty-five to thirty ewes. On a small flock where the ram gets good feed, you can expect about 6 years of use from him, though you don't want him breeding his daughters, granddaughters, and great granddaughters indiscriminately. On open range he may only last for a couple of years.

For a really small flock, it may not make sense to purchase a ram. When we first started our flock, we were breeding for late-spring lambs, and Sherry, from whom we bought the girls, was breeding for early-spring lambs. So, for a nominal fee, we'd use one of her rams to breed our ewes. Sherry always had several rams to make sure she had at least one to breed her stragglers, and we were spared the expense of keeping a ram for the first year or so. Then, during a year when our flock had grown to a pretty large size, we needed a second ram, but rather than purchase it, we worked out a similar arrangement with another nearby shepherd with an even larger flock.

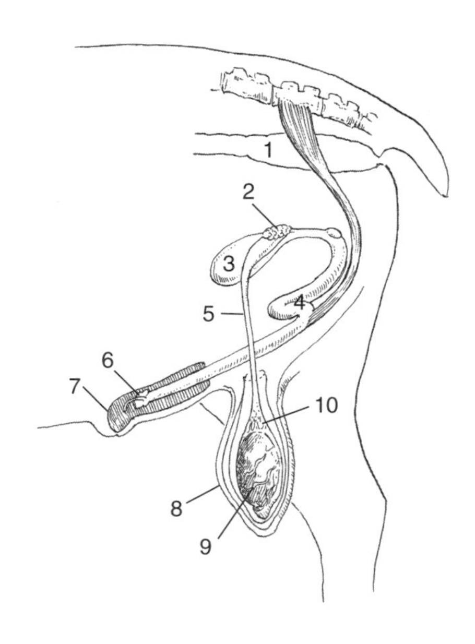

1. Large intestine and rectum
2. Seminal vesicle
3. Bladder
4. Urethra
5. Vas deferens
6. Penis
7. Sheath
8. Scrotum
9. Testicles
10. Epididymis

The male reproductive tract

Artificial Insemination

Artificial insemination (AI) is another option for small flocks. Artificial insemination is a little trickier in sheep than in cattle, and it must be done by a veterinarian or a specially trained AI technician. However, it provides a good ram's semen to a small flock without having to keep a ram.

Interest in using AI in sheep production is increasing. This technique allows breeders to increase genetic diversity in their flocks. For example, Susan Mongold, a breeder of Icelandic sheep, fought her way through several years of red tape to be able to import semen from Iceland. Why? To increase the genetic diversity within her flock. "The Canadian flock, from which we got our animals, originated from two imports of only eighty-eight animals. We wanted the very best genetics to improve our flock," she told me.

Artificial insemination is a procedure that uses a laparoscope (a technique similar the one commonly used on humans with bum knees). Since AI is surgical, ewes should be in good health and should not be overly stressed beforehand. The procedure needs to be synchronized with the estrous cycle, so ewes are often treated with hormones so that they all come into heat at the same time. Artificial insemination is not for everyone. However, it is opening new doors for some shepherds, and as the system evolves, it will benefit more producers.

Considerations and Techniques for AI

Artificial insemination has become extremely common in the cattle industry — particularly the dairy industry, where almost all animals are bred through the use of AI. The sheep industry has not been so quick to take up AI, for a couple of reasons:

- The use of vaginal insemination, which is the method normally used on cows, is easy, and fairly inexpensive, but when done on sheep has generally poor results. The intrauterine method has a high rate of fertilization, but it's expensive, and must be done by a veterinarian, or other highly trained individual, with specialized equipment.
- The value of sheep on an individual basis has not been high enough to merit the same level of interest that has come about in the cattle industry. In other words, the highest value rams in the world might run to several thousand dollars, but top value dairy bulls run several hundred thousand dollars.

Preparing the Ram

If you are buying a new ram or borrowing one, try to obtain him well enough in advance of the breeding season so that he becomes acclimatized to his new home. A week is about the minimum you want him around before he has to "work," and 2 weeks is better. Keep him separated from the flock and on good feed and pasture until breeding time. Remember to change his ration gradually when you first bring him home, and use good judgment in feeding; excess weight results in a lowering of potency and efficiency, so keep him in good condition.

During the breeding season, feed the ram about 1 pound (0.5 kg) of grain per day, so that if he is too intent on the ewes to graze properly, he will still be well nourished. Remember that he needs good feed throughout the breeding season and for a short time thereafter. After all, he's "working" hard!

There are two schools of thought about what to do with a ram after the ewes are bred: The first school says remove him from the flock as soon as breeding is complete, and keep him separated until next breeding season. The second school says leave him with the flock most of the year. Okay, so which approach should you use? That depends on your breed of sheep and your management goals. Ask yourself the following questions:

- Can your breed of sheep breed at a very young age — in other words, could he breed his daughters before you want them bred?
- Do you have a breed that can breed out of season? If so, he may rebreed ewes when you don't want them bred. On the other hand, maybe you want lambs to be born throughout the year.
- Do you have facilities where he can easily be kept separated for long periods? Do you want to deal with a separated critter?

We've always had good luck leaving the ram with the flock all summer. Our ewes dropped their lambs on the pasture in early summer, and since Karakuls don't typically breed too early, the ram was no problem, running with the group. In early fall, when the hours of daylight started to drop and the chances of his breeding the ewes came on, we'd move him to a separate pen until January, when we were ready to breed the ewes. If Karakuls had been a breed known for breeding throughout the year, we would have separated him immediately. (Though Karakuls are known for a longer breeding season than some other breeds.)

Provide a cool, shady place for him in the heat of summer. An elevated body temperature, whether from heat or even an infection, can cause infertility. Semen quality is affected at 80°F (26.7°C) and seriously damaged at 90°F

(32.2°C). Several hours at that temperature may leave him infertile for weeks and ruin any plans you had for early lambing. If your climate is very hot in the summer, shear his scrotum just before the hot weather, run him on pasture in the evening, at night, and in the early morning, but keep him penned in a cool place with fresh water during the heat of the afternoon. (High humidity and temperature can also decrease his sex drive.)

August is generally the beginning of breeding season for January lambing, though many breeds won't begin to breed until September. You can wait until later to turn the ram in with the ewes if you want to start lambing later in the spring. The gestation period is 5 months (148–152 days), so count back from your desired lambing date to determine the best date to introduce the ram.

Ewes are in heat for about 28 hours, with about 16 to 17 days between cycles. So 51 to 60 days with the ram should mate all the ewes, including the yearlings that sometimes come into heat late.

A sense of smell greatly determines a ram's awareness of estrus in the ewes. Some breeds of ram have keener olfactory development than others and can detect early estrus that would go unnoticed by other breeds. Those with the "best noses" for it are Kerry Hill, Hampshire, and Suffolk rams, in that order.

Effect of the Ram on the Ewes

The presence of the ram, especially his scent, has a great effect on estrous activity of the ewes. This stimulus is not as pronounced when the ram is constantly with the ewes as it is when he is placed in an adjoining pasture about 2 weeks in advance of when you would like the breeding season to start. A teaser ram may also be used (see the section on ewes, later in this chapter).

Anyone who has had more than one ram at a time is conscious of the "social" differences seen within a group of rams — one must always assert dominance. Any time rams are reunited after a period of separation, there is the inevitable fighting and head butting until the pecking order is reestablished.

There is actually a wide range of sexual performance among rams. It has been documented that the mating success of dominant rams exceeds that of subordinate ones. This in itself can cause problems, since aggressive potential and ram fertility are *not* necessarily related. If the dominant ram is infertile, then flock conception rates can suffer.

Ram-Marking Harness

To keep track of the ewes that have been bred, you can use a "marking harness," which is used on the ram and is available from many sheep-supply catalogs. The harness holds a marking crayon on the chest of the ram. Ewes are marked with the crayon when they are bred. Inspect the ewes each day, keeping track of the dates so you will know when to expect each one to lamb. Use one color for the first 16 days the ram is with the ewes, then change color for the next 16 days, and so on. If many ewes are being remarked, it means that they are coming back into heat and thus did not become pregnant the previous time he tried to breed them. In those cases, you may have a sterile ram.

If the weather was extremely hot just before or after you turned him in, you can blame it on the heat. But to be safe, you should turn in another ram in case your ram's infertility is not just temporary.

Painted Brisket

Instead of a purchased harness, you can daub marking paint on the ram's brisket (lower chest). Mix the marking paint into a paste with lubricating oil, or even vegetable shortening, using only paints that will wash out of the fleece.

Raising Your Own Ram

One advantage of raising your own ram is that you get to see what he would be like at market age if he were being sold for meat. The older a ram gets, the less you can tell about how he looked as a lamb or how his offspring will look when they are market age.

The way ram lambs are raised can have some effect on their future sexual performance. Studies have shown that rams raised from weaning in an all-male group will show lower levels of sexual performance in later life. When several rams are run with the flock, the dominant ram will breed far more ewes than less dominant rams. If the dominant ram happens to have low fertility, you may be left with unbred ewes.

Recurrent Ram Selection

If you're raising market lambs for meat, you might try a system called "recurrent selection of ram lambs," which consists of keeping the *fastest-growing* ram lambs sired by the *fastest-growing* ram lambs. No, this is not a misprint. Recurrent selection of ram lambs is a way of improving the potential for fast growth in your lamb crop. It involves changing rams fairly frequently and leaves you the problem of disposing of a 2- or 3-year-old ram. If he is a good one, you can probably sell, or trade him, to another shepherd for breeding, or see the discussion on mutton in chapter 11.

The "Battering" Ram

"Battering" rams are not funny and can inflict serious, sometimes permanent, crippling injuries. When you are raising a lamb for a breeding ram, do not pet him or handle him unnecessarily. Never pet him on top of his head — this encourages him to butt. Do not let children play with him, even when he is small. He may hurt them badly, and they can make him playful and dangerous. He will be more prone to butting and becoming a threat if he is familiar with humans than if he is shy or even a little afraid of them.

Leading a ram with one hand under his chin keeps him from getting his head down into butting position. A ram butts from the top of his head, not from the forehead. His head is held so low that as he charges you, he does not see forward well enough to swerve suddenly. A quick step to the right or left avoids the collision.

If you have a ram that already butts at you, try the water cure: a half pail of water in his face when he comes at you. After a few dousings, a water pistol or dose syringe of water in his face usually suffices to reinforce the training. Adding a bit of vinegar to the water makes it even more of a deterrent.

A dangerous ram that is very valuable can be hooded so that he can only see a little downward and backward. He must then be kept apart from other rams, because he is quite helpless.

Strange rams fight when put together. Well-acquainted ones will, too, if they've been separated for a while. They back up and charge at each other with their heads down. Two strong rams, that are both very determined, will continue to butt until their heads are bleeding and one finally staggers to his knees and has a hard time getting up. Rams occasionally kill one another. (Never pen a smaller, younger ram with a larger, dominant one.) Once they have determined which one is boss, they may butt playfully, but will fight no big battles unless they are separated for a time.

To prevent fighting and the possibility of serious injury, you can put them together in a small pen for a few days at first. In a confined area, they can't back up far enough to do any damage.

If no pen is available, you have two options:

- Use a ram shield, which is a piece of leather placed over the ram's face that inhibits frontal vision. For a pretty reasonable price, you'll effectively stop a butting ram without interfering in any of his other functions.
- "Hopple," "yoke," or "clog" the ram — all of which are old European practices.

Hoppling a ram (the modern term would be *hobbling*) was an old system of fastening the ends of a broad leather strap to a foreleg and a hind leg, just above the pastern joints, leaving the legs at about the natural distance apart. This discourages rams from butting each other, or people, because they are unable to charge from any distance and little damage can be done if they can't run. They may stand close and push each other around but will do nothing drastic. Hoppling also keeps them from jumping the fence, which rams sometimes do if ewes are in the adjoining pasture.

Yoking is fastening two rams together, 2 or 3 feet (0.6 or 0.9 m) apart, by bows or straps around their necks, fastened to a light board, like a 2-inch by 3-inch (5.1 cm by 7.6 cm) piece of lumber. Both yoking and hoppling necessitate watching to be sure that the rams do not become entangled.

In *clogging*, you fasten a piece of wood to one foreleg by a leather strap. This slows down and discourages both fighting and fence jumping. Close watching is not necessary.

Ewes

No single ewe has a major impact on your production, but as a collective body, these animals are crucial to success.

Before the breeding season begins, some preparation will make your breeding season more successful. Test your ewes (and the ram) for worms, and worm as necessary. Also check everybody for ticks and other "problems." Trim away any wool tags from around the tail, and trim their feet, because they'll be carrying extra weight during pregnancy and it is important for their feet to be in good condition. By taking care of problems now, you reduce your chances for more serious problems later. For instance, if you eliminate ticks before lambing, none will get on the lambs and you will not have to treat for ticks again.

Seventeen days before you want to start breeding, put your ram in a pasture adjacent to the ewes with a solid fence between them. Research has shown that the sound and scent of the ram bring the ewes into heat earlier.

Some owners of large flocks use a castrated male, known as a *teaser*, to stimulate the onset of estrus in the flock. He is turned out with the ewes about 3 weeks before breeding. Since it always seems that the male lambs make the best pets, this is one way you can keep a pet and feel no guilt for feeding a nonproductive wether!

Never pen the ram next to the ewes before this sensitizing period just prior to breeding. Remember, "absence makes the heart grow fonder." It is the sudden contact with the rams that excites the females.

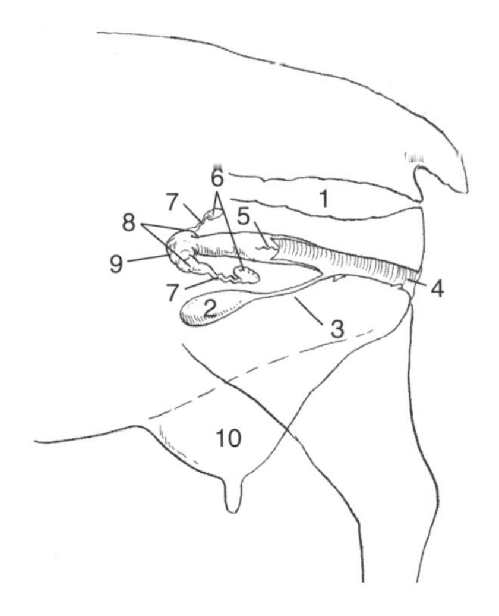

1. Large intestine
2. Bladder
3. Urethra
4. Vagina
5. Cervix
6. Ovary
7. Oviduct
8. Uterine horns
9. Uterus
10. Udder

The female reproductive tract

Vaccines

The vaccine that is most important to both ewes and their lambs is a multipurpose vaccine that is effective for *Clostridium* species of bacteria. Ewes need to be injected twice the first year — the primer shot can be given as early as breeding time or as late as 6 to 8 weeks before lambing, with the booster shot given 2 weeks prior to the calendar lambing date for the flock. For subsequent lambings, ewes require only the booster given 2 weeks before lambing. This protects ewes and the lambs until they are about 10 weeks of age against all of the clostridial diseases, including tetanus.

Other important vaccines include:

- A vaccine that protects against *Chlamydia*, which causes enzootic abortion of ewes and vibriosis. This vaccine is typically administered between 2 weeks and 1 month before breeding.
- A vaccine that protects against some forms of pneumonia and other respiratory viruses; this is typically administered 30 days or less before lambing.

Flushing

Flushing is the practice of placing the ewes on an increasing plane of nutrition, that is, in a slight weight-gain situation, to prepare for breeding. (It is not as effective if the ewes are fat to begin with, and fat ewes may have breeding problems.) Flushing can be accomplished by supplementing the diet with grain, or better pasture, depending on the time of year you are breeding. It is most productive when initiated 17 days before turning in the ram and continued, tapering off gradually, for about 30 days. There seem to be no advantages to starting earlier. This system not only gets the ewes in better physical condition for breeding, it also helps to synchronize them by bringing them into heat at about the same time, which prevents long, strung-out lambing sessions.

Flushing is also a factor in twinning, possibly because with better nourishment the ewes are more likely to drop two ova. The USDA estimates that flushing results in an 18 to 25 percent increase in the number of lambs, and some farmers think it is even more.

You can start with ¼ pound (0.1 kg) of grain a day per ewe and work up to ½ or ¾ pound (0.2 or 0.3 kg) each in the first week. Continue at that quantity for the 17 days of flushing. When you turn in the ram, taper off the extra grain gradually.

The ewes will probably come into heat once during the 17 days of flushing, particularly if you have put the ram in an adjoining pasture. But it's best not to turn in the ram yet — during their second heat, they drop a greater number of eggs and are more likely to twin.

The ewes should not be pastured on heavy stands of red clover, as it contains estrogen and lowers lambing percentages. Other clovers and alfalfa may have a similar effect, though it tends to be weaker in these legumes. Bird's-foot trefoil, another legume, doesn't have this effect at all.

Ewe Lambs

The exception to the flushing would be the ewe lambs, if you decide to breed them. They will not have reached full size by lambing time, so you would not want them to be bred too early in the breeding season. Don't breed them until a month or so after you've begun breeding the mature ewes. Breeding season is shorter for ewe lambs than for mature ewes. Some breeds mature more slowly, like Rambouillet, and some much faster, like Finnsheep, Polypay, and Romanov.

Ewes that breed as lambs are thought to be the most promising, as they show early maturity, which is a key to prolific lambing. Ewe lambs should have attained a weight of 85 to 100 pounds (38.6–45.4 kg) by breeding time, as their later growth will be stunted slightly in comparison with that of unbred lambs. If not well fed, their reproductive life will be shortened, and unless they get a mineral supplement (like trace mineralized salt), they will have teeth problems at an early age.

If replacement ewes are chosen for their ability to breed as lambs, the flock will improve in the capacity for ewe lamb breeding, which can be a sales factor to emphasize when selling breeding stock. Choose your potential replacement ewes from among your twin ewes. Turn these twin ewe lambs in with a ram wearing a marking harness or a paint-marked brisket. The ones that are marked, and presumably bred, can be kept for your own flock. Sell the rest.

Culling

By keeping the best of your ewe lambs and gradually using them to replace older ewes, you should realize more profit. To know which to cull, you need to keep good records (see Sample Ewe Record Chart on pages 244–245), and this necessitates ear tags. Even if you can recognize each of your sheep by name, you are more inclined to keep more accurate, efficient records with tags than without them.

Record the following information in your books: fleece weight, wool condition, lambing record, rejected lambs, milking ability, lamb growth, prolapses, inverted eyelids, any foot problems or udder abnormalities, and any illnesses and how they were treated. With an accurate recorded history of each animal, you know better what to anticipate.

At culling time, review the records and inspect teeth, udders, and feet. The following types of ewes should be culled:

- Ewes with defective udders
- Ewes with a broken mouth (teeth missing)
- Limping sheep that do not respond to regular trimming and footbaths
- Ewes with insufficient milk and slow-growing lambs

Improvements of a flock require rigid culling. Consider age, productivity (including ease of lambing and survivability), and general health. Udders, feet, and teeth are always prime areas for inspection.

Be objective and practical. The runt you tube-fed and bottle-fed may be adorable, but it is not a viable choice for breeding stock.

Feeds

Do not overfeed ewes during the early months of pregnancy. A program of increased feeding must be maintained during late gestation to avoid pregnancy disease and other problems. Overfeeding early in pregnancy can cause ewes to gain excessive weight that may later cause difficulty in lambing.

Have adequate feeder space (approximately 20–24 inches [50.8–61 cm] per ewe) so that all ewes have access to the feed at one time; otherwise, timid or older ewes will get crowded out. If possible, they should be given a free choice mineral–salt mix that contains selenium. This can make it unnecessary to inject selenium prior to lambing (to protect lambs from white muscle disease). *Never* use a mineral mix intended for cattle because it may be fortified with copper at levels that are toxic to sheep. Some geographical areas require selenium supplementation that is above the legal limits available in commercial mineral supplements. Check with your local veterinarian or Extension agent.

Feeding in the Last 4 or 5 Weeks Before Lambing

By the fourth month of pregnancy, ewes need about 4 times as much water as they did before pregnancy. And, since 70 percent of the growth of unborn lambs takes place in this last 5- to 6-week period, the feed must have adequate calories and nutritional balance to support that growth.

During the last month of gestation, the fetus becomes so large that it displaces much of the space previously occupied by the rumen. This necessitates giving feed that is higher in protein and energy, as the ewes have trouble ingesting enough feed to support themselves and the growing lamb if they're fed on low-quality roughage. If they aren't getting enough protein and energy, they use excessive quantities of stored fat reserves, which can lead to pregnancy toxemia. Poor energy supplementation can also result in hypoglycemia (low blood sugar), which mimics the symptoms of pregnancy toxemia. Pregnancy toxemia is *not* necessarily a "thin-ewe" problem.

A good grain mix would be ⅓ oats, ⅓ shelled corn, and ⅓ wheat (for the selenium content). Barley is a good feed, if available. Grain rations can be supplemented to 12 to 15 percent protein content with soybean meal or another protein source. Grain and hay should be given on a regular schedule to avoid the risk for pregnancy disease or enterotoxemia by erratic eating. Approximately 1 pound (0.5 kg) of grain per day (more for larger ewes) is a good rule of thumb.

At this time, watch for droopy ewes — ones going off their feed or standing around in a daze. See chapter 8 for troubles ewes may suffer at this time, including pregnancy toxemia. Exercise and sunlight are valuable to all critters, but especially to pregnant ewes. Lack of exercise contributes to many pregnancy problems. If necessary, force exercise by spreading hay for them in various places on clean parts of pasture, once a day, to get them out and walking around.

The Importance of Proper Feeding in Late Pregnancy

Poor feeding in the last 4 weeks (last 6 to 8 weeks for twinning ewes) leads to:

- Low birth weight
- Low fat reserve in newborn lambs, resulting in more deaths from chilling and exposure
- Low wool production from those lambs as adults
- Increased chances of pregnancy toxemia
- Shortened gestation period, with some born slightly premature
- Ewes slower to come into milk, and less milk
- Production of "tender" layer (break) in the ewe's fleece; this weakness causes the fibers to break with the slightest pull and decreases the wool value

Excessive feeding can result in excessive growth of the lambs and an overweight condition in the ewe, which can lead to serious lambing problems.

The Ketone Test

One way to be sure that your prolific ewes — those carrying twins or triplets — are getting enough nutrition (energy) is to check for ketones in the urine. Better to avoid pregnancy toxemia (ketosis) than to be forced to treat it later as an emergency.

Ewes that are not getting enough feed to meet their energy (caloric) requirements will use reserve body fat. When fat cells are converted into energy, waste products called ketones are created. Pregnancy disease results when the ketones are produced faster than they can be excreted. They rise to toxic levels in the bloodstream, which can be easily detected in the urine. A simple test kit for ketones, available at a pharmacy, can be used to identify ewes with caloric deficiencies. Use the ketone test results to separate the ewes that need extra feed, thus avoiding underweight or dead lambs and pregnancy toxemia problems.

Year 2000 Sample Ewe Record Chart

EWE NAME OR NUMBER	NUMBER OF LIVE LAMBS BORN (DATE)	AVERAGE WEANING WEIGHT*	NUMBER OF WEANED LAMBS (DATE)
e.g., 101	2 (5/13)	48	2 (7/19)
102	2 (5/19)	39	1 (7/19)

Records are crucial for good management. Add dates where appropriate. (Weight in pounds.)*

WOOL CONDITION	WOOL WEIGHT* (DATE)	VACCINATIONS, HOOF TRIMMINGS, ILLNESSES, OTHER INFORMATION (DATE)
excellent	*7 (4/28)*	*trimmed hooves (4/28)*
excellent	*7.5 (4/28)*	*trimmed hooves (4/28); 1 lamb died of scours (5/22)*

TEN

LAMBING

Lambing is both hard work and rewarding, but the hard work needs to be emphasized. And if you choose to lamb in the winter, the work will be even harder.

Preparation for Lambing

Certain husbandry practices done a few weeks before lambing can be very helpful when the time for lambing arrives. These practices — shearing, crotching, and facing — can help keep a clean environment for the newborn lamb and remove obstacles that could make your job harder.

Shearing

If the weather is mild and you do your own shearing and can ensure gentleness, ewes can be sheared up to 3 or 4 weeks before lambing. See chapter 11 for information on shearing. There are some advantages in having ewes sheared before lambing.

- No dirty, germ-laden wool tags for lambs to suck
- Clean udder makes it easier for lambs to find teats
- Fewer germs in contact with the lamb as it emerges at birth
- Easier to assist at lambing, if necessary
- Easier to spot an impending prolapse (see Vaginal Prolapse, page 216)
- Easier to predict lambing time by ewe's appearance
- Ewe less apt to lie on her lamb in pen

- Shorn ewes require less space in the barn, at feeders, and in lambing pens
- Shorn ewes aren't as apt to sweat in jugs and contract pneumonia
- Shorn ewes seek shelter in bad weather

Crotching

Actually, the first five advantages of shearing before lambing are gained also by *crotching* (sometimes called crutching or tagging), which is trimming wool from the crotch and udder and a few inches forward of the udder on the stomach. Only about 4 or 5 ounces (118.3 or 147.9 mL) of low-value wool are removed, and this can be washed for spinning or sold with the fleece.

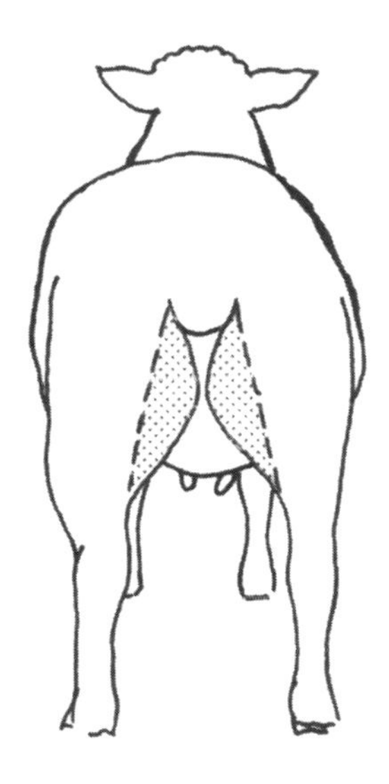

The shaded area indicates area for crotching.

Facing

Another practice of value before lambing, or before the ewe is turned out of the lambing pen, is *facing*, or wigging (trimming the wool off the ewe's face). Facing is often done at the same time as crotching and has several purposes. In closed-faced sheep (sheep with heavy wool around the eyes and cheeks), it has the following advantages:

- Avoids wool blindness
- Enables the ewe to locate and watch her lamb more easily
- Helps prevent the accumulation of hay chaff and burrs in the wool while eating hay at the bunk

The Lambing Process

As the time approaches for actual lambing, the lamb will drop, giving the ewe a sway-backed, sunken appearance in front of the hip bones, and a restless attitude. Dropping is much more noticeable if she has been sheared. She will often pick out her spot to lamb, and lie down apart from the rest of the sheep, sometimes pawing the ground before lying down. Too much lying around, without any observable cud chewing, may be an early sign of the droopiness caused by toxemia (see chapter 8). The ewe should normally have made a bag — or her udder should obviously be filling with milk — by now.

Sometimes, a ewe that is really close-up to lambing will try to mother-up, or take lambs, from a ewe that's had her lambs. This happened to us one spring: One of our ewes had triplets, and another came along and stole one. We, none the wiser, thought that one had twins and the other had a single — until the next day when the thief had two of her own. In this case it worked out okay, and she raised both her twins, and the one she'd stolen the previous day, but sometimes a ewe won't pay any attention to her own lambs after she's confiscated somebody else's.

Feeding habits change when a ewe is close-up. She may refuse a grain feeding just before lambing, or she may walk away from a feeder and plop out a lamb.

Ewes, especially those that are huge with twins or triplets, start grunting several days before lambing, as they lie down or get up. The vulva relaxes and often is a little pinker than before, but should not be protruding or red, which could be the beginning of a prolapse (see chapter 8).

Some ewes may have a mucous discharge, which can be clear or slightly bloody, starting about 2 days before they actually show go into labor. They may also discharge mucus for up to a week after. (If the discharge is yellow or pussy looking, then she has an infection.)

Labor is beginning when the ewe lies down with her nose pointed up, then strains and grunts. Early in labor, the water bag appears. It looks like a balloon protruding from the vulva and is a dark, bluish red. How long labor lasts varies from animal to animal, but is largely influenced by age. First-time ewes usually take significantly longer than older ewes.

Helping Out

Most lambs are born without assistance, and without trouble, but there are times when you're going to need to help out, and it's always a quandary to know when to help and when to butt out. You want to give the mama time to expel the lamb herself if she can, but not wait until she has stopped trying. A good rule of thumb is to allow a half hour to an hour after the water bag breaks, or up to 2 hours of labor, before you jump in. Wait a little while longer for first-time ewes, say up to 3 hours.

If a ewe has been at it for a long time and is showing no sign of action, get her up and walking. If she's in a small jug, move her out so she can get exercise.

When to Assist with Lambing

As a general rule, let the ewe go on her own until:

- The lamb's one front leg and nose are both showing, but the other front leg is nowhere in sight.
- There are two right or two left legs showing (mixed up twins).
- The lamb is showing but the ewe isn't making progress.
- The ewe is obviously becoming weak and tired, and nothing seems to be changing.
- She has been in obvious labor for a couple of hours, with no sign of change.

The size of the pelvic opening is usually large enough for the lamb's body to come out if it is in the normal position, with the front legs and the head coming first. If it is not in this position, delivery is seldom possible without some repositioning of the lamb or veterinary assistance. Even a single lamb can achieve many abnormal positions and since there are often two or more lambs being born, the situation can become even more complicated.

If you do need to assist, prepare for the event in the following manner:

1. Wash your hands and arms; we always washed well with warm, soapy water first, and then swabbed our hands and wrists with iodine if we weren't using obstetrical gloves (which can be purchased from farm-supply stores or from your veterinarian).
2. Wash off the ewe.
3. Lubricate one hand with an antiseptic lubricant, or mineral oil, and slip it in gently.
4. Try to find out the position of the lamb (during lambing season, keep your fingernails short).

Actually, getting in there is something that's nice to have a mentor for the first time you try it, though if there's nobody around to help, you'll have to go for it on your own. The first step is to try to identify the lamb's legs and position. Make sure that the legs you're feeling all belong to the same lamb. In twin births, it's easy to get their legs mixed up.

The front legs are muscular above the knees and bend at the knees in the same way that the foot (pastern) joint bends, with the knuckle pointing forward. The hind legs have a prominent tendon and bend the opposite way from the back foot. The hind legs also have a sharper knuckle that points backward. If you have a small lamb, catch it and feel the difference between its front and back legs. The legs should be aligned so that the tops of the legs are on top. If the legs are bottom side up, then the lamb is upside down in the womb and needs to be turned.

When repositioning a lamb to change an abnormal position, avoid breaking the umbilical cord. When the umbilical cord breaks, the lamb will attempt to breathe, and this causes it to suck amniotic fluid into its lungs. Rarely do these lambs live through birth — most often they drown in the birth canal. The odd one that survives birth will suffer from mechanical pneumonia and often dies within a day.

When helping, time your pulling to coordinate with the ewe's labor contractions. If she is tired and has stopped trying, she will usually start again when you start pulling on the lamb.

Ringwomb

Ringwomb, or the failure of the cervix to properly dilate, is a fairly common cause of problems during lambing. In normal circumstances, the cervix softens and dilates concurrently with the uterine contractions, but in ringwomb the contractions start without softening and dilation of the cervix. There are several possible causes for ringwomb, including infections, hypocalcemia, and high concentrations of estrogenic compounds in feeds.

Ringwomb will generally require a veterinarian's involvement, with cesarean delivery being the most common approach. If you can't get a veterinarian out, you may be able to stretch the cervix manually with your fingers.

Lamb Positions

The good news is that the overwhelming majority of lambs come in the normal, front-feet-first position, and require no help. But the following section reviews the problems you may encounter, and provides some guidance on how to deal with them if you do.

Normal Birth

The nose and both front feet are presented, and the lamb's back is toward the ewe's back. It should start to come out a half hour to an hour after the ewe has passed the water bag. This is the most common position (thankfully), and she should need no help unless the lamb is large or has a large head or shoulder (see next position).

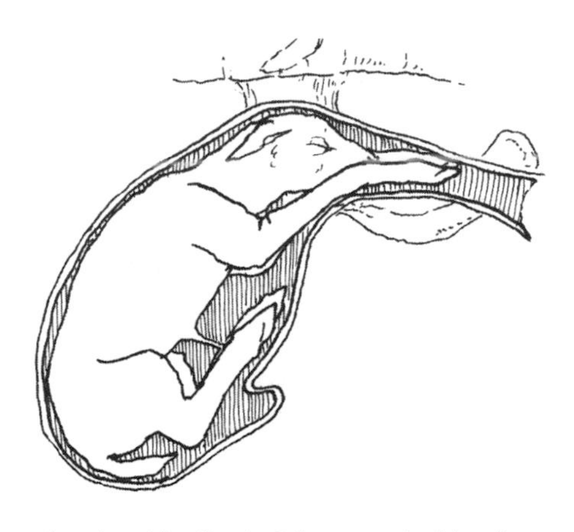
In a healthy flock, 95 percent of lambs should present normally, with few assists required.

Large Head or Shoulders (Tight Delivery). Even with the lamb in normal position, if it's extra large or the ewe has a small pelvic opening, the ewe may need some help. Sometimes the shoulders are large and are stopped by the pelvic opening. Use a gentle outward and downward pulling action. Pull to the left or the right, so the shoulders go through at more of an angle and thus more easily.

Occasionally, the head is large or may be swollen if the ewe has been in labor for quite a while. Assist by pushing the skin of the vulva back over the head. When the lamb is halfway out (past the rib cage), the mother can usually expel it by herself, unless she already is exhausted.

When the head is extra large, draw out one leg a little more than the other while working the vulva back past the top of the lamb's head. Once the head is through, you can extend the other leg completely and pull out the lamb by its legs and neck. If both of the legs are pulled out together, the thickest part of the legs comes right beside the head, making delivery more difficult for both the ewe and you.

Use mineral oil or an antiseptic lubricant with a difficult, large lamb. Use a lamb-puller placed over the lamb's head so that the top of the noose is behind the ears and the bottom of the snare is in the lamb's mouth.

Pulling gently from side to side is more helpful than only pulling outward and downward, as in normal delivery. Gentle pulling on the head as well as the legs is better than pulling on the legs only.

A lambing snare, which can be purchased or made at home, and a couple of soft ropes are essential for saving lambs that have malpresented. The snare goes over the head and in the mouth to keep the head from slipping backward or sideways while you work, and a soft rope secures each leg. Pull in time with the ewe's contractions.

Front Half of Lamb Out, Hips Stuck. A ewe that is weary from labor may need a little help if the lamb's hips get stuck. While pulling gently on the lamb, swing it a bit from side to side. If this doesn't make it slip out easily, give it about a quarter turn while pulling. A small ewe giving birth to a large lamb often needs this kind of assistance.

Head Coming Out Before One or Both Legs

In this case, either one or both of the legs are bent back.

One leg back. To change this position to a normal birth position, attach a snare-cord, or a lamb-puller, behind the ears and inside the mouth and a second cord to the one leg that is coming out. Then push the head and the protruding leg back enough to enable you to bring the retained leg forward so you can pull the lamb out in normal position. The cord on the head is important, for the head may drop out of the pelvic girdle, making it difficult to get it back in again.

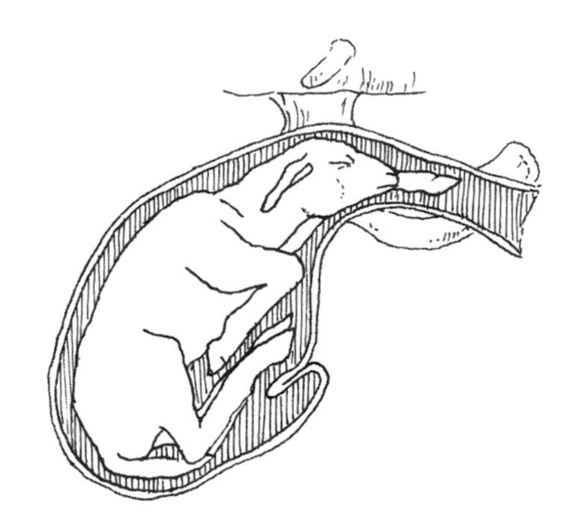

One leg back

Set the ewe so she's lying on the side that has the leg out properly and the turned-back leg is on top. This makes it a little easier to get the turned-back leg into the correct position. Usually, once you've got that leg straightened out, the birth proceeds quickly.

Two Legs Back. To correct this position, attach the lamb-puller onto the head. Try to bring one leg down into position, then the other, without pushing the head back any farther than necessary. Attach a noose of cord onto each leg as you get them out, then pull the lamb.

If your hand cannot pass the head to reach the legs, elevate the ewe's hind end, which gives you more space. A hay bale, or metal garbage can, can be used as a prop. With the snare over the lamb's head, push the head back until you are able to reach past it and bring the front legs forward, one at a time. Put the ewe back into the normal reclining position, start the head and legs through pelvic arch, and pull gently downward.

Both Legs Presented, with Head Turned Back

This is one of the most difficult malpresentations to deal with and often requires a cesarian section, or surgical removal, to get the lamb out. The head may be turned back to one side along the lamb's body or down between its front legs. If the front legs are showing, slip a noose of heavy cord over each front leg, then push the lamb back until you can insert a lubricated hand and feel for the head position. Then bring the head forward into its normal position. With a noose on the legs, you won't lose them. While pulling gently in a downward direction on the legs, guide the head so that it passes through the opening of the pelvic cavity at the same time that the feet emerge on the outside.

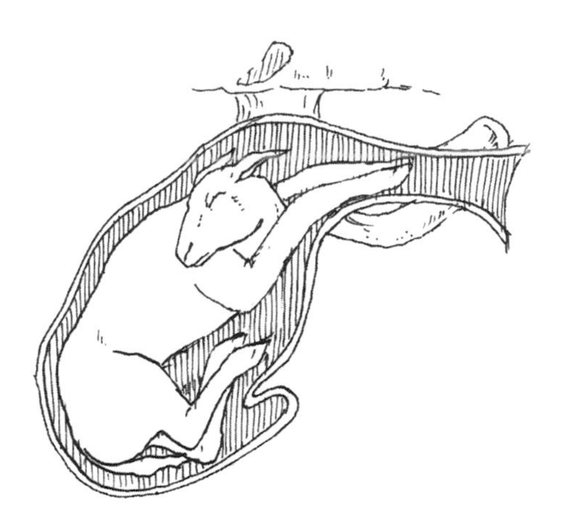

Both legs presented, head turned back

If the head does not come out easily, it is either too large or the lamb may be turned on its back (with its back down toward the ewe's stomach). With cords still attached to the legs, you may have to push it back again and gently give it a half turn so that its legs are pointed down in a normal position, for it will come out easier that way.

If you have a hard time getting a grip on that slippery head to bring it into position, try to get a cord-noose over its lower jaw. Insert your hand with the noose over your fingers, then slip it off onto the chin. Be sure it does not clamp down on any part of the inside of the ewe, as this may tear her tissues. By pulling on the noose that is over the chin, you can more easily guide the head into position.

Hind Feet Coming Out First

When the hind feet are coming out first, pull gently, because the lamb often gets stuck when it's halfway out. When this happens, swing the lamb from side to side while pulling, until the ribs are out, and then pull it out quickly. Sometimes it's easier on the lamb to twist it a half turn, so that its back is toward the ewe's stomach, or rotate it a quarter turn while it's being pulled out. Pulling the lamb out quickly is extremely important because the umbilical cord is pinched

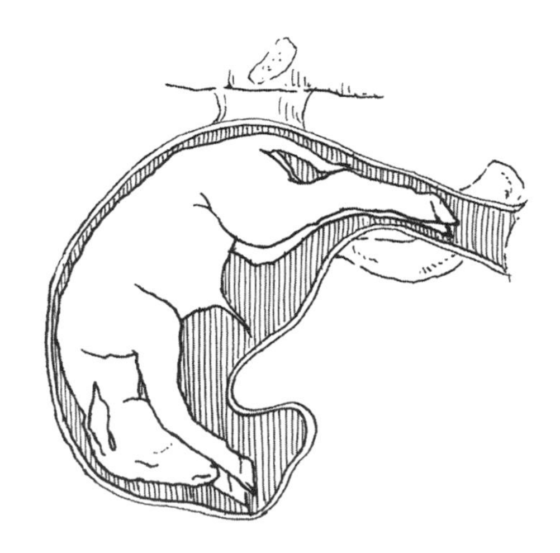

Hind feet first

once the lamb is half out and the lamb will inhale mucus if it tries to breathe. You must also immediately wipe off the mucus that covers the lamb's nose to prevent it from suffocating.

Breech Birth

In the breech position, the lamb is presented backward, with its tail toward the pelvic opening and the hind legs pointed away from the pelvic opening. It is generally easier to get the lamb into position for delivery with its back feet coming out first, but once you get it started out in this position, speed is important. The lamb will try to breathe as soon as the umbilical cord is pinched or broken, so it may suffocate in mucus if things take too long. Wipe off its nose as soon as it pops out.

Do the following if you decide to rectify the breech position before delivery:

1. Slightly elevate the ewe's hind end (use the hay bale or metal garbage can again).
2. Push the lamb forward in the womb.
3. It will be a tight squeeze, but reach in and slip your hand under the lamb's rear.
4. Take the hind legs, one at a time, flex them, and bring each foot around into the birth canal.

When the legs are protruding, you can pull gently until the rear end appears; then grip both the legs and the hindquarters if possible, and pull downward, not straight out.

If the ewe is obviously too exhausted to labor any more, try to determine if there is another lamb still inside her. If not, go ahead and give her a penicillin shot or insert an antibiotic uterine bolus to prevent infection.

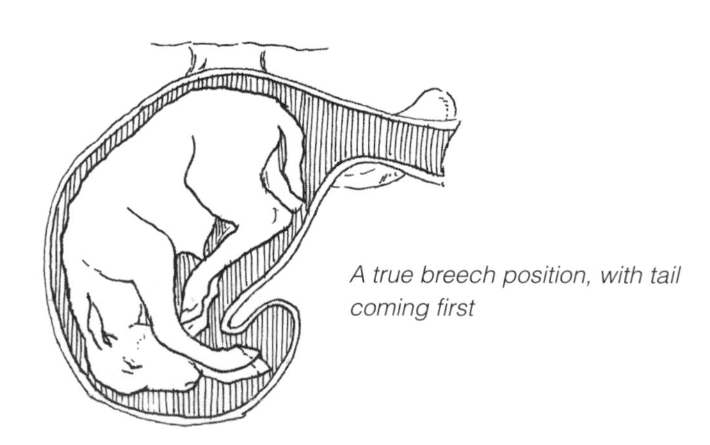

A true breech position, with tail coming first

Lamb Lying Crosswise

Sometimes a lamb lies across the pelvic opening, and only the back can be felt. If you push the lamb back a little, you can feel which direction is which. It can usually be pulled out easier hind feet first, especially if these are closer to the opening. If you do turn it around to deliver in the normal position, the head will have to be pulled around. If it is also upside down, it will need to be turned a half turn to come out easily.

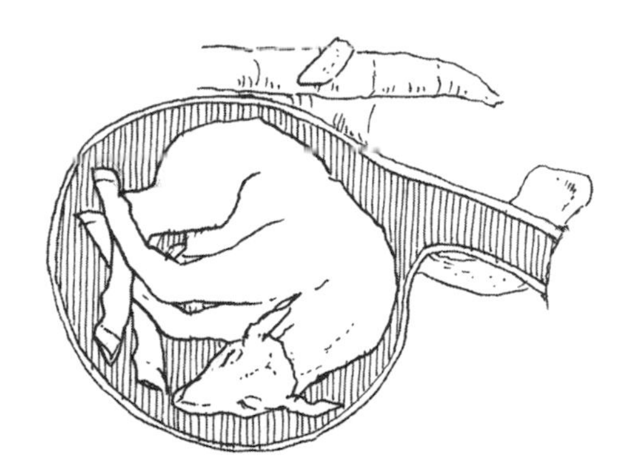

Lamb on its side

All Four Legs Presented at Once

If the hind legs can be reached as easily as the front, deliver by the hind legs so you don't have to reposition the head. If you choose the front legs, the head also must be maneuvered into the correct birthing position along with the legs. Attach cords to the legs before pushing back to position the head.

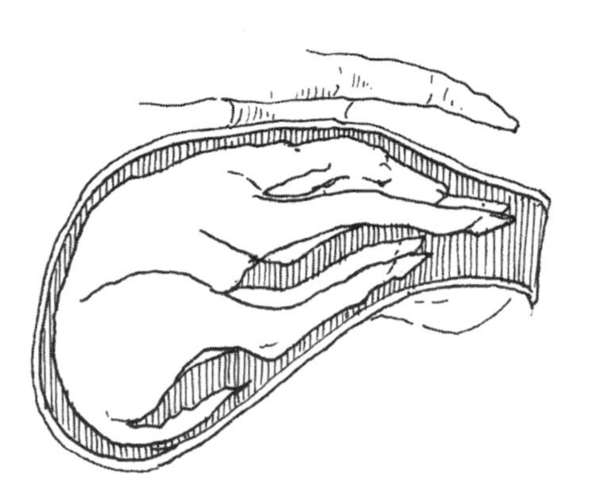

Four legs at once

Twins Coming Out Together

When you have too many feet in the birth canal, try to sort them out, tying strings on the two front legs of the same lamb and tracing the legs back to the body to make sure it is the same lamb, then position the head before pulling. Push the second lamb back a little to make room for delivery of the first one.

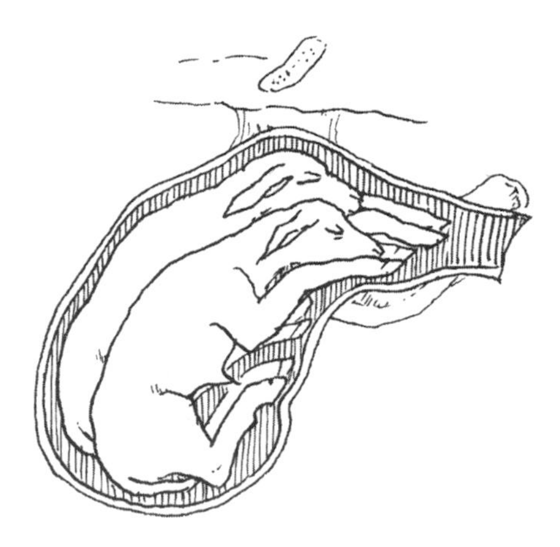

Twins coming out together

Twins, One Coming Out Backward

When twins are coming out together, one and sometimes both may be reversed. It is often easier to first pull out the one that is reversed. If both are reversed, pull the lamb that is closer to the opening. Very rarely, the head of one twin is presented between the forelegs of the other twin, a confusing situation.

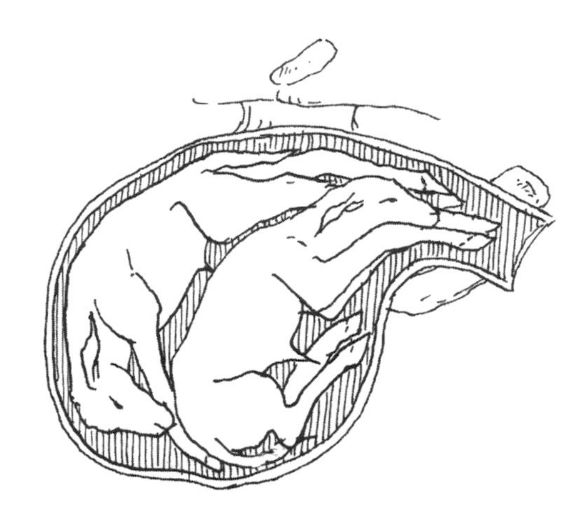

Twins, one backward

When to Call the Veterinarian

If a ewe is obviously in distress, has labored for more than an hour and has made no progress, or you cannot get the lamb into proper position for delivery, then it's time to call the veterinarian. You're paying for this doctor's services, so be sure to learn all you can. They don't ordinarily explain things unless you ask questions and show an interest.

If a lamb is dead in a ewe and so large it can't be pulled out, a veterinarian may have to dismember the lamb to remove it.

After Lambing

If you are there when the lamb is born and mom isn't getting right after licking it down, wipe the mucus off its nose, place it at the ewe's head quickly so she can identify it as her own, and clean it off the rest of the way. (Now is the time to graft on an orphan or triplet that needs a foster mother. See page 267.) A lamb that's having difficulty breathing from excess mucus in the throat and lungs benefits from a quick swing. Grasp it firmly by the hind legs and swing it aggressively in an arc several times — centrifugal force will expel the mucus. Be sure that you have a good grip on the lamb to avoid throwing it out of the barn, and be sure that its head will clear the ground and all obstacles!

If the ewe is exhausted by a difficult labor, then place the lamb at the ewe's nose so she can begin bonding. Help dry off the lamb with old towels (old flannel sheets cut up into rags also work really well) so that it doesn't get too cold from being wet too long. Don't remove the lamb from the mom's sight if you can help it, as this can disrupt bonding. Overuse of a heat lamp to dry the lamb can result in a chill after removal and can predispose the lamb to pneumonia.

How to Deal with the Umbilical Cord

In advance, make sure you have a 7 percent tincture of iodine solution in a small, wide-mouth, plastic jar.

1. Snip the umbilical cord to just about 2 inches (5 cm) long, with either *dull* scissors or a *dull* knife; a dull instrument is preferable because it reduces the chance of bleeding.
2. Hold the lamb so that the umbilical cord hangs into the container and is submerged in the iodine solution.
3. Press the container against the lamb's belly.
4. Turn the lamb up so that the entire cord and the surrounding area are covered.

Iodine should be applied as soon as possible after birth, because many bacteria can enter via the navel. The iodine penetrates the cord, disinfects it, and assists in drying. Avoid spilling the iodine on the lamb or applying it excessively — it has a strong odor that may mask the lamb's natural odor and cause the ewe to reject it. As an extra precaution against infection, you can treat the cord with iodine again in 12 hours.

If the cord is not cut to the proper length, some ewes try to nibble at the navel and can injure the lamb. One year Paula had an excited ewe that chewed the tail off her newborn lamb, nibbling it as if it were an umbilical cord. She put a band on the tail, above where the ewe had chewed it, dunked it in iodine, and left for a few minutes. When she got back, the ewe had had a second lamb, and she chomped the tail off that one, too! She then proceeded to lick the lambs off like normal and was a wonderful mother. (If this had happened with more than one ewe, Paula would have suspected a nutritional deficiency, for these abnormal behaviors are often a sign of nutritional problems.)

Feeding the Lamb

The lamb is best fed on Mama, but at times, you need to help get the process going. When the ewe stands up, she'll nudge the lamb toward her udder with her nose. The lamb is born with the instinct to look for her teats and is drawn to the smell of the waxy secretion of the mammary pouch gland in her groin. If the udder or teats are dirty with mud or manure, a swab with a weak chlorine bleach solution before the lamb nurses will clean things up and help prevent intestinal infection in the lamb.

Let the lamb nurse by itself if it will, but do not let more than a half hour to an hour pass without it nursing, as the colostrum (the ewe's first milk after lambing) provides not only warmth and energy but also antibodies to the common disease organisms in the sheep's environment. We usually opt not to interfere for about 20 minutes after the lamb is up on all fours and looking for a teat. If after 20 minutes it hasn't found the teat, is trying to nurse but doesn't seem to be getting any milk, or the mom won't let it nurse, we intercede. (This is easy with some ewes, which will stand really still until you and the lambs get things settled, but it can be a real pain with others that just want to wander around in circles and carry on the whole time.) Occasionally, the ewe will not allow the lamb to nurse because she is nervous, has a tender or sensitive udder, or is rejecting the lamb. If the udder appears sensitive, it is often because it is tightly inflated with milk.

Remember my rule of thumb when you're trying to decide whether the lamb is actually getting any milk: A lamb that is getting milk will have its little tail whipping back and forth like a metronome at full speed. When a lamb is getting milk, its body fills out quickly, its skin folds start to disappear, and its little stomach becomes tight. When a lamb is a few hours old and is crying continuously or has a cold mouth, it is not nursing.

If the problem appears to be nothing more than a flighty ewe, but the lamb is still strong and trying to grab a teat, then restrain the ewe and allow the lamb to nurse. The ewe can be restrained with a head gate, a halter, or by pushing her into a corner and leaning your weight against her. (See inset on page 269.)

If the lamb is getting on a teat but doesn't seem to be getting milk, you probably need to unplug the ewe. I say "unplug" and mean it literally: The end of the teat has been protected over the past several weeks by a little waxy plug, which is sometimes hard for a lamb to displace, especially if it's a little weak to begin with. After you've broken free the plug, strip the teats of several squirts of colostrum by massaging down the teat between your thumb and index finger.

Sometimes lambs are a little dumb and need you to help them find the teat in the first place. I've seen newborns try to nurse the front knee, wool on a tail, or other odd spots. Grab the lamb, and force its mouth over the teat while you massage the teat to get a few squirts of milk in the lamb's mouth. Usually, once it gets those first couple of squirts, it settles down to business with no additional assistance. But you'll occasionally find a really slow one that you may have to help for longer.

You may encounter a lazy lamb that, for no apparent reason, does not want to nurse the ewe but will take a bottle with enthusiasm. These lambs can be maddeningly frustrating and can tax both your nerves and your patience with regard to how long you are willing to wait to see if it will begin nursing the ewe. We call these lambs "volunteer bummers."

If for some reason, Mom can't feed it any colostrum (no milk, bag hard with mastitis, too many previous lambs), then you'll need to feed it like a bummer lamb (see the section on orphans later in this chapter).

You have made a great contribution to the colostral protection of the newborn lamb if you have previously vaccinated the ewe (twice) with a vaccine to protect against tetanus, enterotoxemia, and the other common clostridial diseases. These antibodies are absorbed by the mammary gland from the ewe's bloodstream and are incorporated into the colostrum so that they protect the newborn lamb until it starts to manufacture its own antibodies. The small intestine of the newborn lamb possesses a very temporary ability to absorb these large molecular antibodies from the colostrum. This ability to absorb antibodies decreases by the hour and becomes almost nonexistent by 16 to 18 hours of life. Colostrum is high in vitamins and protein and is a mild laxative, which can assist in passing the fetal dung (meconium, the black, tarry substance that is passed shortly after the lamb nurses for the first time).

The longer a lamb has to survive without colostrum, the fewer antibodies it has the opportunity to absorb and the less chance it has of survival if it develops problems. A weak lamb or one of low birth weight can be lost because of a delay in nursing.

When a ewe has too much milk, her udder becomes too full and the teats become enlarged. To rectify this situation, milk out a bit of this colostrum and freeze it in small containers for emergency use. Ice-cube containers or small resealable freezer bags are good options because they allow you to thaw small

---

Be on the Lookout for Starvation

Loss due to lack of colostral antibodies is not the same as loss due to starvation, which occurs from receiving no milk at all. A strong lamb can sometimes survive for a day or more without getting any milk but will become weaker all the time. Many lamb deaths that are attributed to disease are actually due to starvation. Lambs often die having not uttered a sound or indicated that they were starving. Always make sure that the lambs are actually nursing, and always recheck the ewe to make sure she is continuing to give milk for the first few days.

---

quantities as needed. Solidly frozen colostrum will keep for a year or more if it is well wrapped. When saving and freezing colostrum, you should have a combination of colostrum milked from several ewes, for they do not all produce the same broad spectrum of disease-fighting antibodies. Cow or goat colostrum can be stored and used in emergencies.

Thaw frozen colostrum at room temperature or in lukewarm water. Never use hot water or a microwave oven to thaw colostrum because it can denature and destroy the antibodies, rendering the colostrum worthless.

Molasses and Feed for Mama

Ewes are often thirsty after giving birth. We offer the ewe a large bucket of warm water (not hot) that contains half a cup of stock molasses. It is important to have it warmed, as the ewe may be reluctant to drink very cold water, and this can result in lowered milk production. Offer good hay, but no grain the first day, as it could promote more milk than a tiny lamb could use. If the ewe has twins or triplets, however, and seems short of milk, grain feeding should start that first day.

Multiple Lambs

Twins, triplets, and larger "litters" require vigilance to assure that all the lambs are claimed by the ewe and that they all get their share of colostrum. If the ewe does not have plenty of milk for all the lambs, increase her grain consumption gradually. Unless she shows some reluctance about the molasses, continue offering it in lukewarm water during the time that she is in the jug with the lambs.

If you have multiple lambs that are crying a lot, they are probably not getting enough milk. If not all of them are crying, assist the hungry ones by holding them to their mother. If she is short on milk, give a supplemental bottle. When a ewe does not have enough milk for multiple lambs, you can still let them all nurse her, but supplement one or all of them with a couple of bottle-feedings a day. Give 2-ounce (59.1 mL) feedings the first couple of days, and increase to 4 to 6 ounces (118.3–177.4 mL) by the third and fourth day. Continue this process, gradually increasing the quantity as they grow, for as long as the ewe's milk is not adequate. See chapter 6 for the newborn lamb milk formula to feed for the first 2 days, then gradually change to lamb milk replacer (not calf milk replacer).

Marking Lambs

A brightly colored, small nylon dog collar or a collar made of yarn is a convenient way to flag any lambs that need special observation; these collars really make them stand out in the mob.

If you have more than two or three ewes, which should produce two to six lambs, the lambs are best identified by ear tags. This way, you can keep records of lamb parentage, date of birth, and growth, and it will be easier to decide which sheep to keep for your flock and which to sell. With identification tags also on your ewes, you can be certain which lambs are hers, even after they are weaned.

Livestock-supply catalogs sell a variety of tags. Some are metal, and some are plastic and come in a variety of colors. Some can be marked with an indelible marker, and others come with a preinscribed combination of numbers and letters (your name if you wish) that you specify during ordering. The different colors of the plastic tags can be used to identify sex, whether twins or singles, the month born, and so on.

Use of Ear Tags

Some tags are a self-clinching type, while others need a hole to be punched. Tags should be applied while the lamb is still penned with its ewe. Never use large, heavy, cow tags on adult sheep. Similarly, tags intended for mature sheep are often too heavy for a lamb's ear to support. If you're using the small, metal lamb tag, insert it onto the ear approximately half the length of the tag to leave growing room for the maturing ear.

Problems with Newborn Lambs

As well as making sure the lambs are getting adequate feed, there are some other problems to watch for. Hypothermia is one of the most common problems facing newborns, and it doesn't have to be all that cold for hypothermia to occur. In fact, a lamb can suffer from starvation hypothermia on a fairly warm, sunny day.

Another fairly common problem is "weak" lambs. Weak lambs often result if the ewe had a long, difficult delivery.

Hypothermia

Guard newborns and young lambs against hypothermia, which is implicated in about half of all lamb deaths. Hypothermia has two basic causes: exposure and starvation. As it implies, exposure hypothermia is primarily a result of extremely cold temperatures or cold temperatures mixed with drafts. This can kill wet lambs within the first few hours of birth. Starvation hypothermia can occur in lambs from 4 and 5 hours old to a couple of days old.

Once dry and fed, lambs can withstand quite low temperatures, but due to a large ratio of skin area to body weight, wet or hungry lambs can chill quite quickly. A hypothermic lamb will appear stiff and be unable to rise. Its tongue and mouth will feel cold to the touch. You must warm it immediately with an outside heat source, because it has lost its ability to control its temperature. Wrapping it in a towel or blanket will *not* suffice.

There are several methods of warming lambs. Water warming is probably the best choice for very cold lambs, with air warming a close runner-up. Some people use infrared lamps, but these are probably the least desirable method because they can seriously burn the lamb and can cause fires if used in the lambing shed. If you plan to lamb in winter, consider buying, or building, a lamb-warming box. But if you'll normally lamb in the spring, then you can probably get away with bringing the occasional cold lamb in the house. Warm it in a big cardboard box — or do like we did, and pick up an old playpen (yeah, you remember those prisons we were subjected to as toddlers) at a flea market. Your box or playpen can be set near a woodstove or a heat vent or in front of the oven (with the door open) to warm the air and the lamb. A blow drier also helps, but again, be careful not to burn the lamb.

If you are dealing with a slightly older lamb that's become severely hypothermic from starvation, then it will need an injection of glucose. This happens to lambs that don't get fed within an hour or so after birth. They need the energy in that first hit of milk very badly. If they don't get it, their body temperature begins to fall and they begin to "feed" off the glycogen, or sugar, reserves in their body. Without the glucose injection, the lamb will die during warming.

The use of plastic "lamb coats" in cold weather can be beneficial because they help the lamb retain a great deal of body heat. A newborn lamb appears wrinkly because there is very little body fat under the skin. It takes approximately 3 to 5 days to build up that fat layer under the skin, which acts as natural insulation. When a lamb coat is used to help the lamb retain body heat, the energy that would be used to keep it warm is converted to body fat. This can be especially beneficial to twins and triplets with marginal milk intake.

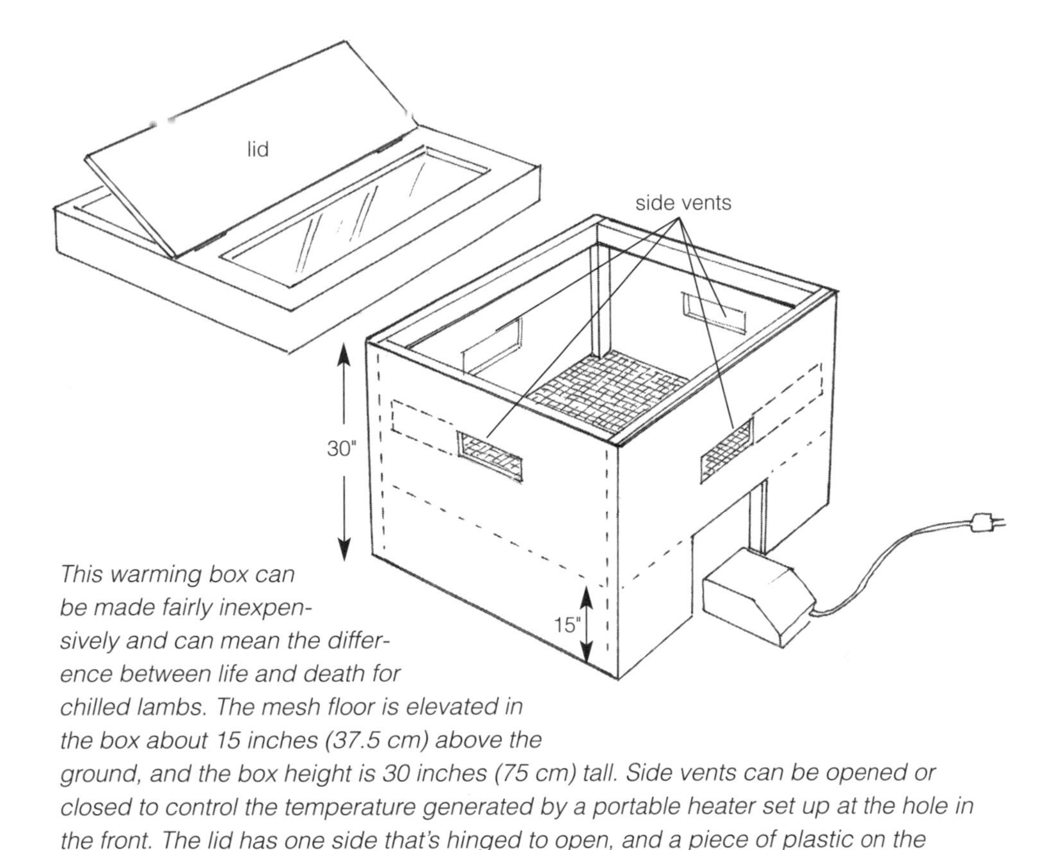

This warming box can be made fairly inexpensively and can mean the difference between life and death for chilled lambs. The mesh floor is elevated in the box about 15 inches (37.5 cm) above the ground, and the box height is 30 inches (75 cm) tall. Side vents can be opened or closed to control the temperature generated by a portable heater set up at the hole in the front. The lid has one side that's hinged to open, and a piece of plastic on the other side for easy viewing.

Warming Up a Frozen Lamb

The warm-water method is probably the best method of warming a "frozen" lamb. This is one that is really, really cold. Submerge it up to its neck in water that is quite warm to the touch. It will begin to struggle, but keep it immersed for several minutes. Dry it well, and place it in a warm environment until it has totally recovered. Feed it 1 to 2 ounces (29.6–59.1 mL) of warm colostrum or milk replacer as soon as it can take it. If you are experienced, force-feeding with a stomach tube after removal from the water and drying speeds up recovery.

Weak Lambs

A lamb that has been weakened by a protracted or difficult birth may be suffering from anoxia (lack of oxygen) or have fluid in its lungs. The first few minutes are critical. If it gurgles with the first breaths or has trouble breathing, swing it as discussed previously. Two or three swings normally get things going. Be sure that you have a firm grasp on the lamb (the lamb will be slick) and that there are no obstructions in the path of your swing.

It is not essential for the first feeding to be colostrum, but make sure the lamb does receive colostrum during the first few hours of life. The lamb's ability to absorb the antibodies in the colostrum drops rapidly from birth to approximately 16 hours of life. For a very weak lamb, you may have to give the first feeding from a baby bottle with the nipple hole enlarged to about the size of a pinhead or use a stomach tube to feed. Give 2 ounces (59.1 mL) of warmed colostrum to give the lamb strength. Do not force the lamb — if it has no sucking impulse, the milk will go into its lungs and cause death. Often, a weak lamb can get up on its feet after just one bottle-feeding (or stomach tube–feeding) and be ready to nurse from its mother without further assistance.

Lamb Resuscitation

If the heart is beating but the lamb is still not breathing, artificial respiration is mandatory.

1. Grasp the lamb by the nose so that your thumb and fingers are slightly above the surface of the its nostrils.
2. Inflate the lungs by blowing *gently* into the lamb's nostrils until you see the chest expand. Release the pressure, and gently push on the lamb's chest to express the air.
3. Repeat the procedure until the lamb begins to breathe.

Exercise caution — don't blow as though you're blowing up a balloon. A lamb's lungs are quite small and can be ruptured by too much pressure. If your attempts are still unsuccessful, sometimes a cold-water shock treatment will do the trick. Dunk the lamb in cold water, such as in a drinking trough. The shock may cause the lamb to gasp and start to breathe. Sometimes a finger inserted gently down the throat will stimulate the coughing reflex and get things going. Then, make sure the lamb is warmed and gets to nurse.

Stomach Tube Emergency Feeding

Several sources (see Resources) provide stomach tubes specifically designed to safely feed severely weak lambs with no sucking impulse. If you need one quickly and there is no time to order, get a male catheter tube from the drugstore and use it with a rubber ear syringe or a 60-mL hypodermic syringe. The tube should be about 14 to 16 inches (35–40 cm) long. (Check the length by holding it against the lamb.) The tube should be kept in warm sterile solution; when it's wet it slips in more easily.

Ensuring that the tube is in the stomach and not the lungs is the single most important step in stomach tube–feeding. (See the box on page 266 for instructions.) Injecting liquid into the lungs would kill the lamb. In slipping the tube into the stomach, you should be able to feel the tube as it goes down if you put your thumb and finger along the left side of the neck and pass the tube with the other hand. If the tube is incorrectly passed in the trachea, you can't feel the tube going through the neck. A tube into the lungs usually elicits a cough, but don't depend on that as a sign. If you think the tube isn't in the correct position, hold a wet finger at the protruding end. If the finger feels cool from moving air, the tube is in the lungs instead of the stomach. Remove it, and try again. Another way to check for proper position is to blow gently on the tube. If it is in the stomach, the lamb's abdomen expands. The air will escape when you release the pressure, and the abdomen will flatten again. If the tube is in the lungs, the air will escape past the tube and up the trachea without this "ballooning" effect.

Check Eyelids

Check the lamb's eyelids to see if they appear to be turned in so that the eyelashes would irritate the eye. This can cause serious trouble and blindness if not corrected, and the sooner it is noticed the easier it is to remedy (see Entropion in chapter 8).

Orphan Lambs

Orphan, or bummer, lambs can result from the death of the ewe, abandonment, rejection, or loss of milk production before the lamb has reached weaning age. A ewe may disown one or all of her lambs, sometimes for reasons known only to her.

Inserting the Stomach Tube: One-Person Method

It is easier for two people to feed the lamb with the stomach tube, but it can be done by one person if the syringe is filled in advance with 2 ounces (59.1 mL) of warm colostrum (or warmed canned milk, undiluted, for this feeding only). Keep the syringe within reach, and then do the following:

1. On a table, hold the lamb's body with your left forearm and with its feet toward you. The lamb's head, neck, and back should be in a straight line if you are looking down from above, but the head should be at a 90-degree angle to the neck.
2. Use the fingers of your left hand to open the lamb's mouth to insert the tube, which should be sterile and warm, if possible.
3. Insert the tube slowly over the lamb's tongue, back into its throat, giving it time to swallow.
4. Push the tube down the lamb's neck and into the stomach. Having checked the tube length previously, you should know about how much of it should stick out. Stop pushing when the end is in the stomach area. The average insertion distance is 11 or 12 inches (28 or 30.5 cm). You cannot insert it too far, but it is important to insert it far enough.
5. Confirm that the position is correct with the wet-finger or blowing test.
6. Insert the end of the catheter tube into the syringe filled with warmed milk, and slowly squeeze the milk into the lamb's stomach.
7. Withdraw the tube quickly to prevent dripping into the lungs on the way out.

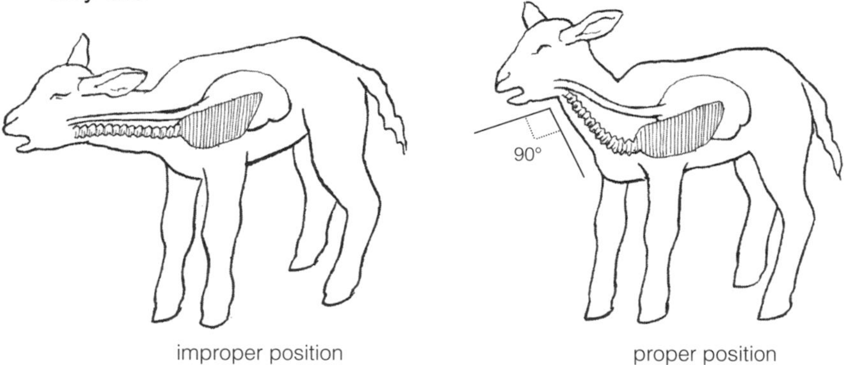

If the lamb's head is held out straight, the trachea is open and there is considerable risk of inserting the tube into the lungs. It's better, although more difficult, to keep the head at a 90-degree angle when inserting the tube into the stomach.

Common reasons for **lamb rejection** include:

- The ewe may have a painful or sensitive udder from mastitis or an overabundance of milk.
- She may have delivered one lamb in one location, then moved and delivered the other, forgetting about the first.
- Some ewes cannot count to two; they may be willing to accept twins, but as long as they have one, they are happy and do not seek out the other.
- The lamb may have wandered off before the ewe has had a chance to lick it off and become bonded to it.
- She may have sore or chapped teats, or the lamb may have sharp teeth.
- Because of a difficult lambing, she may be exhausted and not interested in her lamb.
- The lamb may be chilled and abandoned as dead.
- The ewe may have the new-mother syndrome: Young, first-time moms may be nervous, flighty, confused, or just frightened by the lamb.
- The lambs may have been swapped: If two ewes lamb near each other at the same time, one ewe occasionally adopts and bonds to the other's lamb, but the other ewe does not accept the first ewe's lamb.

If the ewe rejects the lamb *after* it starts to nurse, not before, check the udder for sensitivity and check the lamb's teeth. A little filing with an emery board can remedy sharp teeth. Don't file too much, or the teeth will become sore and the lamb won't nurse, which puts you right back where you started. Apply Bag Balm to the ewe's teats if they are sore or lacerated by sharp teeth. Keep her tied where the lamb can nurse until she accepts it.

Grafting

Sometimes a bummer lamb (either a true orphan or a rejected lamb) can be grafted onto another ewe. Grafting is getting the ewe to accept another lamb as her own, but the process can be complicated and isn't always successful.

Scientific evidence suggests that vaginal stimulation during parturition plays a large role in the ewe's instinct to accept the lamb, which could explain why grafting lambs is more successful the closer it's done to delivery. This could also explain why some "easy lambers" simply walk away from a newborn lamb as if its birth were just a minor occurrence.

Once a ewe rejects a lamb, it is hard to fool her into accepting it. All methods fall into two major categories: the mental, or "brainwashing," techniques in which you attempt to change their hardheaded opinion, or the physical, or "fool-the-sense-of-smell," method.

Encourage grafting in any way you can. There are a number of things to try, such as:

- Use fetal fluids from the ewe onto which the lamb is to be grafted (either its mother or another ewe) and smear over the lamb; this is a tried-and-true method of grafting.
- Rub the lamb with a little water with molasses in it to encourage the ewe to lick the lamb.
- Use an "adoption coat" or "fostering coat" (see Resources), which is a cotton, stockinet tubing applied like a lamb coat. When stretched over an accepted lamb for a few hours, it will absorb the smell and can then be turned inside out and stretched over the lamb you wish to graft. (Shepherd's tip: If you have a heavy-milking ewe with a single lamb, slip a coat on her lamb to have a fostering coat ready to use if needed.)
- Daub her nose and the lamb's rear end with a strong scent-masking agent, which is made for this purpose, or with a dab of petroleum jelly. Since the ewe identifies the lamb primarily by smelling its rear end, sometimes menthol, vanilla, or even an unscented room deodorant on her nose and the lamb's rear, will suffice.
- If it's a case of the "new-mother jitters" or the ewe is high strung and not very tame, a tranquilizer can sometimes work wonders to calm her.
- An old-timer's method is to tie a dog near the pen. Its presence is supposed to foster the mothering instinct. But sometimes this makes the ewe so fierce that she will butt the lamb if she can't reach the dog.
- Another method, which is not actually as cruel as it sounds, is to flick the tips of the ewe's ears with a switch until she becomes so rattled that she urinates from the stress. She may then accept the lamb.
- Immerse the lamb to be grafted on the ewe and the lamb she has accepted in a saturated salt solution to even out the scent.

Forcible Acceptance

If all else fails in your grafting attempts, then it is time to get tough. One solution is to pen or tie the ewe in such a way that she cannot hurt the lamb and it can nurse regularly in safety. You may need to tie her hind legs together temporarily so she can't keep moving and preventing the lamb from nursing. Without the mother's guidance and encouragement, you may need to help the lamb nurse by holding the ewe and pushing the lamb to the right place.

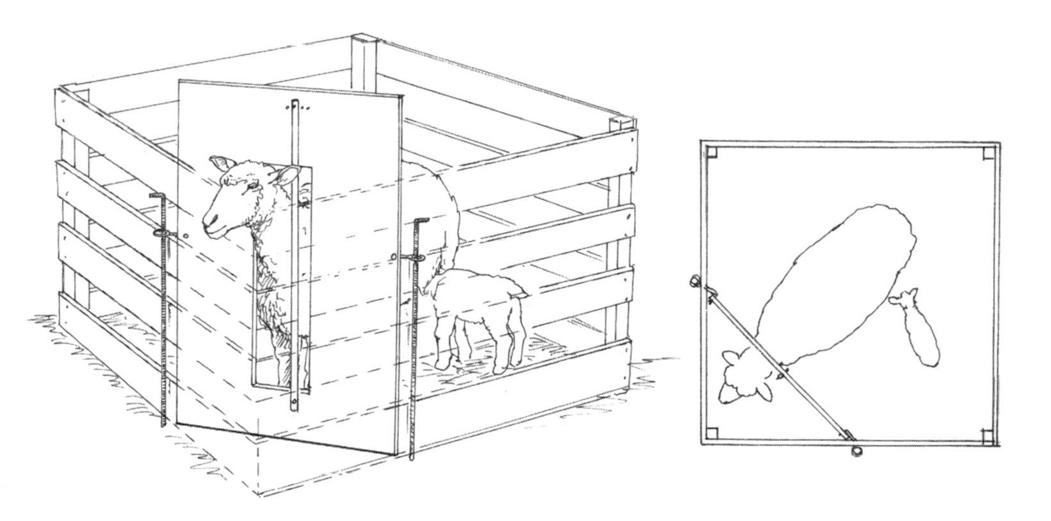

A small pen with a head gate can be used for adopting an orphan lamb onto a ewe or for chores like crotching. While in the head gate, the ewe is free to lie down or stand up, and to eat or drink from feed or water placed in the corner of the pen. (This design comes from MidWest Plan Service [see Resources].)

If the ewe is a hard-core case, a ewe stanchion, which is designed to limit the ewe's movement but still allows her to lay down, get up, and eat, could be necessary. A less elaborate one can be improvised in the corner of the lambing pen. Make sure that the ewe has room to lie down and has plenty of hay and water in front of her. Use molasses in the water, as you would for any ewe that has just lambed. It may take from 1 to 5 days before the ewe is completely resigned to accepting the lamb.

Caution: Exercise care and judgment in the size of the lamb that you are attempting to graft. An orphan lamb that is a week old may be so aggressive at nursing that it will frighten the ewe. Also, if there is a significant difference in age and size between two lambs placed on a ewe, the weaker or younger lamb may not be able to compete, and its growth can be stunted or it may be starved out completely.

The most likely situation is when the ewe has twins and rejects just one of them. Spraying the rear end of both lambs with a confusing scent (like a room deodorant) is the easiest thing to try at first, and most often it works.

If the ewe starts showing any hostility *at all* toward one of her twins, don't wait until she starts butting it — take action right away. The most reliable way is to tie her up. The sooner you stop her from comparing the smell of the two lambs, the sooner she will accept the rejected one. If the ewe has to be tied up, however, be sure she gets water often, for it may be difficult to leave it in front of her.

Grafting an Orphan on a Different Ewe

If another ewe goes into labor and you think she may deliver only one lamb, you might choose to graft the rejected lamb on that ewe, for she may be more willing than the ewe that has proven her desire to reject something.

Have a bucket of warm water and an empty bucket ready. Keep the rejected lamb nearby, and watch the lambing. If you are fortunate enough to catch the water bag, put its contents into the empty bucket, which makes everything much easier.

Follow these steps for grafting:

1. As the ewe delivers her own lamb, dunk the rejected lamb into the water-bag liquid. Or, if you didn't catch the water bag, immerse the lamb into the warm water up to its head.
2. Rub the two lambs together, especially the tops of the head and the rear ends.
3. Present them both to the ewe's nose; usually, she will lick them and claim them both.

Don't neglect the newborn when you are working with the orphan — the new lamb's nose must be licked off by its mother or wiped off by you so that it can breathe. If the mother delivers twins, you may have to take the reject back. Dry it off, and keep trying to get its mother to take it (or bottle-feed it yourself).

If the substitute mother seems to accept a grafted lamb that is much older than her newborn, hobble the orphan's legs so that it doesn't get up and run around too much at first. Let the newborn lamb have the first chance to nurse. If your orphan is a few days old, it doesn't really need the colostrum, and shouldn't get too much of it at one time. Actually, to do this trick, the orphan should be less than a week old, as an older lamb would surely cheat the new lamb out of its share of the milk. In any event, both lambs will have to be supervised carefully.

One worthwhile practice is to save the water bag from a ewe and freeze it in pint quantities. You can thaw this and pour it over a ewe's nose and onto the lamb you want to make her accept. This is not always successful, but worth trying.

Giving an Orphan to a Ewe That Has Lost Her Lamb

When a ewe delivers a dead lamb and you have a young orphan that needs a mother, dunk the lamb in warm water containing a little bit of salt and some molasses. Dip your hand in the warm water and wet its head. By the time she licks off the salt and molasses, she usually has adopted the lamb. If it is a lamb that is several days old and does not need the colostrum as much as a newborn, this gives you an opportunity to milk out and freeze some of the valuable fluid.

In all this talk about grafting an orphan onto a ewe, I haven't mentioned the old way of the "dead lamb's skin." In that method, if a lamb is born dead or dies soon after birth, it was skinned and the skin was fastened like a coat over the orphan. Skinning a dead lamb is not simple unless you are already adept at it. The process is messy and unsanitary because you may not know why the lamb is dead and could be transferring germs and disease.

A cleaner method is to rub a damp towel over the dead lamb, and then rub the towel on the orphan. Before doing this, wash the orphan with warm water, giving special attention to washing the rear end, which is the first place that the ewe sniffs in determining whether the lamb is hers.

I remember a postage stamp issued some years ago, showing a ewe with a lamb. She appeared to be sniffing its head. Sheep raisers laughed, as it was not the usual end for her to be sniffing.

The fewer sheep you raise, the less chance there is that another ewe will be lambing when you need a substitute mother. So if a lamb's mother has died, has no milk, has been incapacitated by pregnancy disease or calcium deficiency, or just outright rejects her baby, you have a bottle lamb.

Bottle Lamb

This is one of the greatest pleasures (and biggest headaches) of sheep raising. Even if the ewe is weakened by a hard labor and/or has no milk, she should be allowed to clean the lamb as much as possible; she will claim the lamb even if she cannot nurse it, and even as a bottle lamb it can stay with her. If the ewe does not lick the mucus off the lamb's nose, wipe it off, dry the lamb, and put iodine on its navel at once. Feed the lamb as directed in chapter 6.

Nature automatically regulates the amount of milk that nursing ewes can give per feeding — small amounts, but often. It is important to control the volume of milk that bottle lambs consume during each feeding. It is tempting to overfeed a bummer lamb — it is so cute and learns quickly how to beg in an irresistible manner. A yellow semipasty diarrhea is the first sign of overfeeding. If this occurs, substitute plain water or an oral electrolyte solution, such as

Gatorade, for one feeding because the lamb needs the fluid but not the nutrients (see chapter 8 for a detailed discussion).

Overfeeding is more common during the first week or two of life than it is later on. Starting orphans on lamb milk replacer (after colostrum) that is prepared with twice the label recommendation of water helps eliminate the problem, but remember — this is a rare exception to the "always-read-and-follow-the label" doctrine that I strongly advocate. Gradually increase the concentration of milk powder in the solution so that it's at full strength about the time the lamb is a week or so old. At the first sign of yellow stools, reduce the concentration slightly for a day or so and then gradually bring it back up.

As bummer lambs get older, their need for water increases, especially if they are beginning to eat grain from the creep-feeder. If they have not yet learned to drink from the water tank, they will attempt to quench their thirst with milk, which is the equivalent of you attempting to quench you thirst on a hot day with a milkshake! Substitute an occasional feeding with plain water, or add some water to their milk to give them extra volume. Judging the need for water in a bummer lamb requires experience and development of a sixth sense. When you are feeding bummers, common sense and observation are your best allies.

The true bummer is a lamb whose mother either dries up or doesn't have enough milk, and the lamb is forced to sneak or "bum" off other ewes. If this occurs before the lamb is 3 to 4 weeks of age, the lamb may lose weight, become skinny, or even starve. If the lamb is big enough or smart enough, it

Suggested Feeding Schedule for an Orphan

AGE	AMOUNT
1–2 days	2–3 ounces (59.1–88.7 mL), six times a day, with colostrum
3–4 days	3–5 ounces (88.7–147.9 mL), six times a day (gradually changing over to lamb milk replacer)
5–14 days	4–6 ounces (118.3–177.4 mL), four times a day, and start with leafy alfalfa and crushed grain or pelleted creep feed
15–21 days	6–8 ounces (177.4–236.6 mL), four times a day, along with grain and hay
22–35 days	Slowly change to 1 pint (0.47 L), three times a day; after the lamb is 3 months old, feed whole grain and alfalfa or pelleted alfalfa containing 25 percent grain, but change rations *very* gradually

Hot Flashes

After bottle-feeding the lamb, if you happen to be holding it on your lap (sometimes an irresistible thing to do), you may notice that it suddenly feels very hot or flushed approximately 5 to 10 minutes after feeding. This hot flash is not actually a sudden increase in body temperature, but rather an acute dilatation of the capillaries of the skin, which releases a short burst of body heat. These hot flashes usually last only a minute or so. Do not become alarmed, as it is a known physiologic phenomenon of sheep (and cats). The mechanisms and reasons for it are poorly understood.

will figure out how to bum from the other ewes in the flock without getting caught. It usually sneaks up behind a ewe just after grain is fed when the ewe's attention is focused on competing for her share or when her head is thrust into the hay feeder. Most ewes become less protective of their lambs and hence less particular about who may be nursing as the lambs get older. Bummer lambs seem to seek out these ewes and can often be seen nursing from behind, between the legs.

Orphan Feeding, Cafeteria Style

If you raise Finnsheep or another breed that gives birth to "litters," or if you have a flock that's large enough to have quite a few orphans, you might want to look into cafeteria-style feeders. There are several commercially made multiple-nipple, cafeteria feeders available; you can order them through sheep suppliers or farm stores. The lambs should be taught to nurse from a bottle on warmed milk replacer, then changed to the milk-feeder.

With this system, the lambs have constant access to milk, and they can suck it out of the feeder as they want it. The milk formula is usually fed cold to reduce the chance of overeating and to reduce bacterial contamination when it is left standing all day. The milk-feeder should be cleaned, disinfected, and supplied with fresh milk daily. For such use, the replacer should be one that stays in suspension well.

Care of Baby Lambs

At this point, I'm assuming that your lambs are up and going. Over the next few months, there are a number of things that need to be done.

Vaccination

If pneumonia is a problem, vaccinate newborn lambs with an intranasal vaccination, as opposed to relying entirely on the passing of immunity through the mother's colostrum. It should also be given its own vaccination for tetanus, enterotoxemia, and other clostridial diseases by the age of 10 weeks.

Docking

If you're lambing in a barn, tails should be docked (removed) before the lambs are turned out. Lambs on pasture are usually easily caught for docking when they're 2 or 3 days old, but if you let them get any older, they become difficult to catch because by 4 days of age they start running really fast. Docking is also much less traumatic on the lamb when it's only 2 or 3 days old and the tail is still small. Sheep of most breeds are born with long tails, and these can accumulate large amounts of manure in the wool, attracting flies and then maggots (flystrike). In other words, the tail can serve as a general source of filth, interfering with breeding, lambing, and shearing.

There are many ways to remove tails, and some are better than others. Docking can be done by a number of methods:

- Cutting with a dull knife (a sharp knife causes more bleeding)
- A knife and hammer over a wooden block
- A hot electric chisel or clamp (this cauterizes the wound to lessen bleeding)
- A Burdizzo emasculator and knife (which crushes the ends of the blood vessels)
- An Elastrator, which applies a small strong rubber ring to cut off the circulation, causing the tail to drop off in a couple of weeks

This is one thing on which Paula and I agree completely: We both favor the Elastrator because it minimizes shock and eliminates bleeding problems. The Elastrator is very economical in terms of supplies and equipment needed and is the easiest method for the beginning shepherd to learn to use. The main disadvantage if you have not vaccinated is the risk for tetanus (but if you haven't vaccinated, tetanus is a risk with all methods).

The Burdizzo emasculator is quick. If the procedure if finished with a mattress suture in the skin, it is almost bloodless and wound healing is nearly around the same time that the tail would have dropped off if you had used the

Elastrator. The disadvantage of the Burdizzo method is the need for more expensive equipment, a suture procedure, and greater operator skill.

What length is best for the tail? It is stylish among purebred producers to cut tails off at the base, leaving practically no tail stub. However stylish "no-tail" docking might be in the showring, the damage it causes to the tissues surrounding the anus predisposes ewes to rectal prolapse. A farm producer should leave 1½ inches (3.8 cm) from the body. As you lift the tail, you will notice two flaps of skin that attach from the underside of the tail to the area on each side of the rectum. The band or cut should be placed just at or slightly past where the skin attaches to the tail (on the tail, not the skin). This leaves enough tail to serve as a cover and prevents damage to the muscle structure that could weaken the area and add to the risk for prolapse later on.

Whatever procedure you use, be clean. The Elastrator rubber rings should be stored in a small, wide-mouth jar of alcohol, disinfectant, or mild bleach solution to keep them sterile and to disinfect your fingers when you reach for one. Dip the Elastrator pliers, Burdizzo emasculator, and/or knife in the disinfectant, too. With the Elastrator, the tail falls off in 1 to 3 weeks, but after 3 days it can be cut off, on the body side close to the band. Dunk the stump in 7 percent iodine. If you live in an area where tetanus is prevalent in the soil, then give a shot of tetanus antisera at docking.

Castration

Castration also can be done early, as soon as the testicles have descended into the scrotum. An emasculator can be used for castration, so there is no wound and thus no opening to attract flies. This is important in late lambing and warm weather. Similar to tail docking, castration by using an emasculator is a bloodless procedure, as it crushes the spermatic cord and arteries. There is also less pain, less setback to the lamb's growth, and no danger of infection. Check to see that the testicles have descended into scrotum, then clamp the emasculator onto the neck of the scrotum, where it joins the body, separately on each testicle cord. Because of the high cost of this well-made piece of equipment, you may not want to buy one for use on a few sheep. You might borrow it from a neighbor who has more animals or buy it in partnership with another sheep raiser. After the emasculator is used, the testicles will atrophy in about 30 to 40 days.

The Elastrator also can be used for castration, when the lamb is about 10 days old and when the testicles have descended into the scrotum. These special pliers stretch the rubber ring so you can pull the scrotum through it, being sure both testicles are down. When the pliers are removed, the ring tightens where it is applied, around the end of the scrotum where it attaches to the body. The device cuts off the blood supply so that the testicles wither within 20 to 30 days. There is no internal hemorrhage or shock, and risk for infection is slight. If you have problems with infection, douse the band with iodine after about a week. In hot weather, you can spray it with fly repellent. The Elastrator is the method we use for castrating. When you first apply the band, the lambs run around like mad for a minute or so, but then they don't seem to be bothered by it at all.

Cryptorchid or Short Scrotum

The Elastrator ring can also be used as a means of sterilization. Extensive tests in Australia have shown that animals sterilized with an Elastrator ring gain weight faster, get to market faster, and have less fat and more lean meat than castrated or uncastrated males.

The rubber Elastrator ring is used on the scrotum, but the testes are pushed back up into the body cavity. The increased heat on the sperm results in sterilization. While the male hormones are still present to increase weight gain with

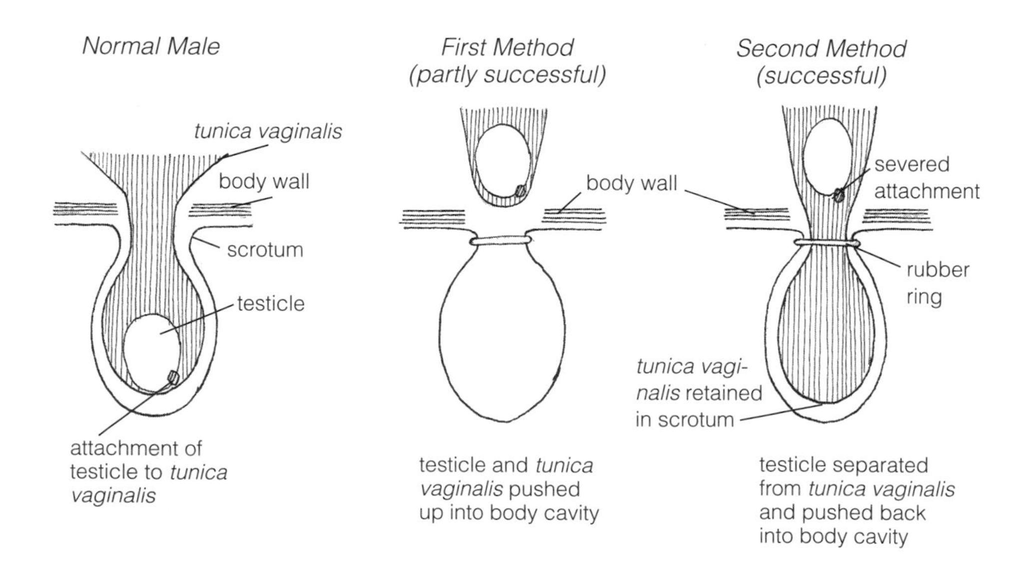

Methods used to induce cryptorchidism (Shepherd *magazine, December 1973)*

more lean meat, the animal shows little or no sex activity. This method is used at about 4 weeks of age, and the animal is called a cryptorchid (meaning "hidden testicles").

Although the body heat results in functional sterilization, do not use the elastrator ring to create "teaser" rams. After an extended period, some develop a Sertoli cell tumor in the retained testicles, which produces abnormal amounts of female hormone that can cause feminization.

Should You Castrate?

There are reasons for not castrating, for example, if you plan to market the animal for meat at 5 or 6 months of age or are thinking of keeping or selling it as a breeding ram. Castrated lambs grow faster than ewe lambs, but uncastrated rams grow faster than both of them and the meat is leaner. So, if you have early lambs and plan on selling the rams for meat at 5 months of age (before breeding season), you can forego castration. However, if you plan to sell to a packinghouse, you will be penalized for not castrating. If you intend to keep the ram for longer than 6 months before slaughter, castration is desirable.

ELEVEN

PRODUCTS AND MARKETING

The key to profit is to make good use of all the potential sources of income connected with your sheep. This requires good planning and good management.

If you've opted to raise purebred sheep, then publicize the superior traits of your breed to boost sales of final products, such as wool or meat, and to make money by selling breeding stock. If you're raising a heritage breed, capitalize on that by explaining the importance of maintaining diversity in the gene pool.

For both profit and pleasure, make use of all the by-products that you can. Think about all the types of products or enterprises that you might be able to spin-off from your sheep project.

Merchandising to Reach Your Market

It is useless to have a good product if no one knows about it. There are many ways to find the buyers who want your nice fleeces or good meat, and the method to use is the one that is most convenient for your situation.

You may opt to market your products through a cooperative effort with other farmers or go it alone. You may choose to market through conventional agricultural marketing systems, like sale barns or wool pools, but will net more cash flow from items you sell directly to consumers.

The easiest way to market fleece is to contact the nearest place that teaches spinning and weaving and leave your name with them or talk to their students. Taking samples of fleece could trigger an immediate response if what you show is of excellent quality.

Many shepherds who market directly publish a newsletter or host a festival or picnic at their farms. As Paula says, "I attended a Shearing Day Festival in Michigan, and it was more fun than a carnival, with an auction of a prize fleece, shearing and spinning demonstrations, spinning contests, and food booths. They had a booth with barbequed lamburgers — most delicious and popular."

SHEPHERD STORY

Lisa Merian was born and raised on a New York State dairy farm where she now raises sheep and goats with her mother and runs a successful, wool-based business. Her business grew out of her interest in fiber arts. "I was fortunate; while I was still in high school, I was encouraged to do independent study in fiber arts. I took classes that were offered at nearby colleges and art schools. I was given the opportunity to study under different people, including Paula Simmons — and I still take classes."

Lisa sells fleeces, rovings, art pieces, and commission work — for example, special-order hand-knit sweaters or felted art wall hangings. "I sell quite a bit of my wool to other fiber artists. Hand-dying has become my forte, and that's helped build my business."

Fleece quality is a big issue in Lisa's enterprise. Intensive skirting and keeping the sheep in clean surroundings are the two strategies she uses to achieve quality.

"Some people rely on sheep coats, but I found they didn't work well for me. The sheep seemed to get caught up in them or tear them off. They were high maintenance.

We keep the flock on good, well-drained paddocks and keep the barn and barnyard clean. Then, before I card the skirted fleece, I check it carefully by hand."

"Developing a wool business takes time. If you want to go that route, keep it simple. I learned the hard way: Bigger isn't always better. And be frugal. Aim for having everything free and clear.

"I've worked out of a room in the house for years, but now we're renovating part of the barn for a shop. We can afford it now."

There are lots of ways to build a business. One approach that's worked well for Lisa is teaching fiber arts, by appointment from the farm, or by working with an area arts council in taking a program into schools around her region. Students, from preschool through high school, have taken classes she's offered, and many times their parents and teachers become clients.

Many enterprising sheep raisers are adding to their income by buying and selling (merchandising) complementary products, such as sheep coats, electric fencing supplies, and certain veterinary products that are otherwise not easily located. Think about what *you* would like to buy and have had trouble locating — maybe this would be a good product to make available to other sheep people.

When you develop a following of customers who appreciate the quality of your products, you won't need signs. If you sell high-quality wool, meat, and other products, you will have regular customers, and they will tell their friends.

Easy Ways to Advertise

If you are on a well-traveled road, try putting up a sign advertising RAW WOOL AVAILABLE FOR HAND-SPINNERS or NOW TAKING ORDERS FOR LOCKER LAMBS. Computer-generated handouts, with tear-off phone numbers on the bottom, hung in local businesses often yield results with little to no up-front investment. Brochures can be prepared to hand out to prospective customers at fairs and farmers' markets. Be sure that signs and brochures are neat and give clear contact instructions, including directions to your farm if you sell from there.

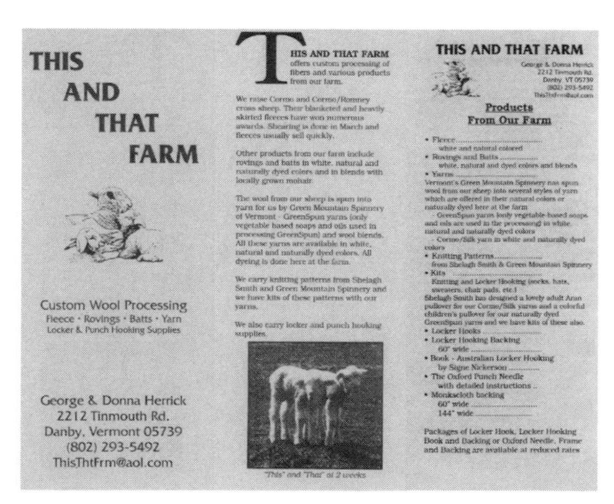

Wool

Wool is truly a remarkable substance! It's very warm, and it both repels water and wicks away perspiration. It's strong, yet elastic, which means it can be spun into a variety of yarns, from very fine to bulky.

Some farm flocks specialize in wool-type sheep, others primarily in meat-type sheep, but generally the trend is toward all-purpose breeds, which are fairly good for both meat and wool production.

For many aspiring shepherds, part of their desire to get a flock of sheep relates directly to knitting, spinning, weaving, and other fiber-arts endeavors.

If this includes you, then handling and caring for wool becomes an important consideration, whichever type of sheep you choose.

Did You Know?

One pound (0.5 kg) of wool can make 10 miles (16.1 km) of yarn!

Wool Production

Wool is produced by *follicles*, which are cells that are located in the skin. There are two types of follicles: primary and secondary. Skin is actually pretty complex stuff. It's made up of two main layers — the epidermis, which is the thin, surface layer, and the dermis, which is the thicker, deeper layer. Skin can contain hundreds of sweat and sebaceous glands, dozens of blood vessels, and thousands of nerve endings per square inch of surface area. The follicle cells are located in the upper layer, or epidermis, but as fibers develop, they push down into the dermis, encased in a tunnel of epidermal cells.

Primary follicles can produce three kinds of fiber (true wool fibers, medullated fibers, and kemp fibers), whereas secondary follicles only produce true wool fibers. The medullated, or med, fibers are hairlike fibers that are as long as the true wool fibers but that lack the elasticity and crimp (or waviness) of true wool. The kemp fibers are coarse and typically shed out with the seasons. Kemp fibers don't take dyes well, but kemp is important in the production of certain types of fabrics, like true tweeds, and in the production of carpet wool.

By 60 days of gestation, a lamb has primary follicles and the fiber that grows from them covering all its skin. Shortly after, the secondary follicles begin to form and are fairly complete by about 90 days after conception, but they don't swing into full fiber production until about 2 weeks after the lamb is born.

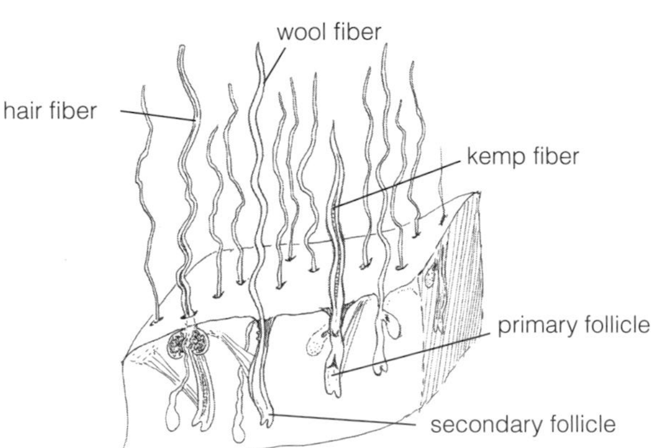

A cross-section of sheepskin, showing the three types of fiber: wool, hair, and kemp. All new fiber growth originates in the skin.

At birth, lambs have more med and kemp fibers because the primary follicles are producing but the secondary ones haven't started yet. When the secondary follicles begin to produce fiber, the proportion of wool fibers increases. The switch to secondary follicle production is evident if you keep an eye on the lambs and is really pronounced in certain breeds, like the Karakul. Coarse-wool and hair breeds, like the Navajo–Churro and the Wiltshire Horn, don't have as high of a proportion of secondary follicles, and therefore tend to have more med and kemp fibers in their fleece. Fine-wool breeds, like the Merino, have a really high proportion of secondary follicles, thereby producing a higher quality, true wool that's both dense and fine.

Heredity determines the wool type, but its quality and strength depend on the health and nutrition of the sheep during each year of fleece growth. One serious illness can easily result in tender, brittle wool, with a weak portion in every fiber of the whole fleece.

The Importance of Nutrition in the Beauty of Wool

The beauty, luster, elasticity, and strength of the wool suffer if the sheep's diet is deficient in protein, vitamins, and minerals. Mixed-grain rations usually have the protein content marked on the labels, and feed stores stock various sheep pelleted rations that are convenient to use. Pasture, grain, and hay provide vitamins, and hay from sunny areas is reported to have a higher vitamin content. Vitamin supplements are available and make a significant difference in the health of older ewes. A handful of brewer's yeast tablets fed daily is an inexpensive way to extend a ewe's life span.

Fiber Structure

All fibers generally have a similar structure. An outer layer, called the *cuticle* provides a protective coating for the inner layer, called the *cortex*. In fine wools, the cuticle and the cortex are basically all there is, but in medium to coarse wools, med fibers, and kemp fibers, the cortex surrounds a central layer, known as the *medulla*. As fibers move from medium wool to kemp, the medulla occupies a greater part of the fiber, with up to 60 percent of a kemp fiber's diameter taken up by medulla.

The cuticle is actually made of flattened cells that overlap each other like the scales of a fish. The scales of fine wool are coarser, while long wools are

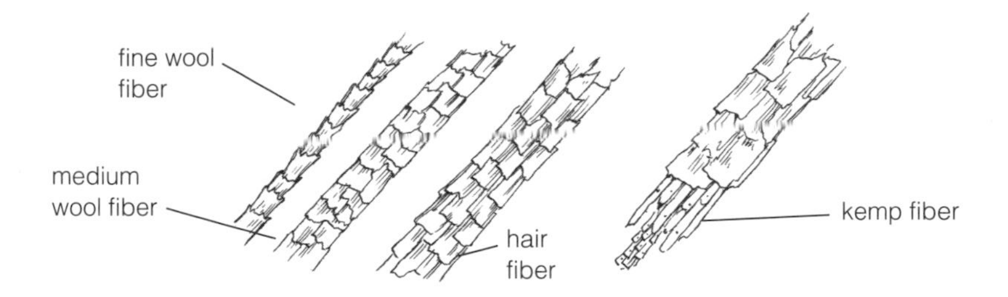

Wool, hair, and kemp fibers have an outer cuticle that is made up of scales and a water-resistant sheath coating that covers the scales called the elasticum. The edges of the scales are more pronounced on fine wools; they are smoothter on coarser wools, which gives coarser wool a more lustrous character. All three types of fiber have a cortex, though only hair and kemp fibers have a medulla in the center of the fiber. The cortex gives fiber its elasticity and durability. The medulla reduces the crimp of the fiber and may interfere with the fiber's ability to take up dye.

smoother; this is why many long wools are more lustrous than the fine wools. The cortex is what gives wool its elasticity, strength, and durability. The medulla is a spongy material.

Like human hair, fibers outside the body are dead. All growth takes place in the skin, at the bulbous end of the root, and forces new wool up and out of the skin. Wool is made primarily of amino acids that link together to form a protein called *keratin*. The protein molecules of wool, unlike those of human hair, are combined in a unique coiled structure that provides the amazing elasticity that's common to wool fibers. This structure is also porous, which allows wool to absorb up to 18 percent of its weight in moisture without becoming damp and up to 50 percent or its weight before becoming saturated.

Evaluating Wool

Commercial buyers purchase wool by the pound (0.5 kg), which is put in a bag, or bale, that can weigh anywhere from 150 to 1,000 pounds (68–453.6 kg). These buyers evaluate the *clip*, or the season's yield of wool, from a flock or group of flocks on the basis of several characteristics, depending on the wool's intended use. Although these criteria for evaluating wool may not have much effect on a small-flock owner with no great amount of wool to sell, the explanations may be of interest. Following are some of the criteria that a wool buyer considers.

- **Yield, or the percentage of clean wool fibers by weight after a raw, or grease wool, is cleaned.** Yield is calculated on the basis of a sample that's weighed raw, scoured, dried, and reweighed. Yield differs according to the breed and can vary 10 to 15 percent within a breed. In North American sheep, it typically runs between 45 and 65 percent.

- **Quantity and type of vegetable matter.** A wool clip that has lots of seeds, burrs, twigs, and other vegetable matter in it is much less valuable than one that is free of these contaminants.
- **Average length and variability of staple.** The length of fibers falls into three major classes: staple, French combing, and clothing. Buyers want minimum variability in the length of fibers within a fleece. Good skirting of fleeces (cutting away belly, leg, and other short wool) and minimizing second cuts during shearing provides a more consistent length.
- **Staple strength and position of break.** *Tender wool* has low tensile strength and breaks unevenly, whereas *broken wool* breaks at the same point on most fibers throughout the fleece. Both of these conditions are largely related to health and nutrition. To test for weakness, stretch a small tuft of wool between both hands. Strum it with the index finger of one hand. A sound staple makes a faint, dull, twanging sound and does not tear or break.

This illustration shows the relative fineness and density of wool on a sheep, and typical clean wool yield. **A.** *A #1 tends to be fine and dense wool, and #4 is coarse and thin; #2 and #3 fall somewhere in between. The length of the wool tends to be the opposite of the denseness, so wool around the neck is usually short.* **B.** *A #1 is typically the cleanest wool, and #4 is typically the dirtiest wool, with a high proportion of vegetable matter in the wool.*

• **Color and colored fibers.** Commercial buyers mainly want bright, white wool that can be dyed without bleaching; however, many hand-spinners and some commercial buyers look for naturally colored wools. If you sell wool commercially, colored fibers, regardless of how many, mixed in with a fleece sends it to the colored bag of wool.

• **Crimp.** Again, buyers are interested in consistent quality when it comes to crimp. Wool from one animal that has too much or too little crimp compared with that of the rest of the flock reduces the value of the bag.

• **Fiber diameter.** "Spinning count," "blood grading," and "micron" systems are all approaches used to describe the fiber diameter. Spinning count originally meant that 1 pound (0.5 kg) of a particular type of fleece wool would spin that many "hanks" of wool (a hank is 560 yards [511.8 m]). So, 70s would spin 70 hanks and 60s would spin 60 hanks. The count system usually went only as fine as 80s count, but German Saxony Merino has been known to grade 90s, where 1 ounce (29.6 mL) of the single fibers laid end to end would stretch 100 miles (161 km)! Spinning count is used more often abroad than in the United States and is always expressed in even numbers.

The blood system of grading the fineness of wool originally indicated what fraction of the blood of the sheep was from the Merino breed, which produced the finest-diameter wool. This term no longer relates actually to Merino or part-Merino blood, but qualifies the relative degree of fiber diameter.

The micron system uses a laboratory test to measure the average diameter of the wool fiber and is most often used by commercial buyers who are purchasing large quantities of wool. In the micron system, the larger the number, the coarser the wool.

Sheep Coats

To sell successfully to hand-spinners, fleeces must be free of vegetation. One way to have clean fleeces is to put coats (or sheep blankets or covers as they are called in supply catalogs or "rugs" as they are called in some countries) on your sheep. Sheep coats have been extensively tested, and not only do they increase the quantity of *clean* wool, which is expected, but they also result in longer staples and improved body weight even under harsh conditions. Coats also make shearing easier, partly because the fleece is cleaner. In areas with severe winters, sheep wearing coats can conserve energy, and this benefit shows up in the maximum percentage of increased wool growth and slightly heavier birth weights for lambs born to ewes wearing covers.

One shepherd reported no death losses from coyotes during coat use and thought that perhaps the sound created by the plastic coats as sheep moved or the sight of the different-colored coats warded off the predators.

Cost seemed to be the main factor in making sheep coats impractical. Cotton coats were not durable around barbed wire or brush pasture. Sturdy nylon-based coats were more durable, but made the sheep sweat during warm weather or close confinement. Woven polyethylene sheep coats were found to be the most practical during large-scale tests in Australia. Being woven, they allowed the wool to "breathe," so hot weather was no problem. Because they partially protect wool from rain, the coats minimize fleece rot and skin disease, according to Australian findings (see Resources for suppliers).

Coats are put on the sheep right after shearing. Try putting covers on a few of your sheep if you are wondering how they would affect the wool of your particular breed. Compare with uncovered fleeces after a year.

The patterns shown below can be made from woven-plastic feed sacks, with heavy, wide elastic used for the leg loops. This material resembles the most satisfactory of the commercial variety.

When using these coats for young, growing sheep of a long-wooled breed, check the fit often to be sure that the coat is not becoming too tight. Elastic, rather than fabric, loops for fastening are better for this reason, although the elastic does wear more quickly and will need replacing annually.

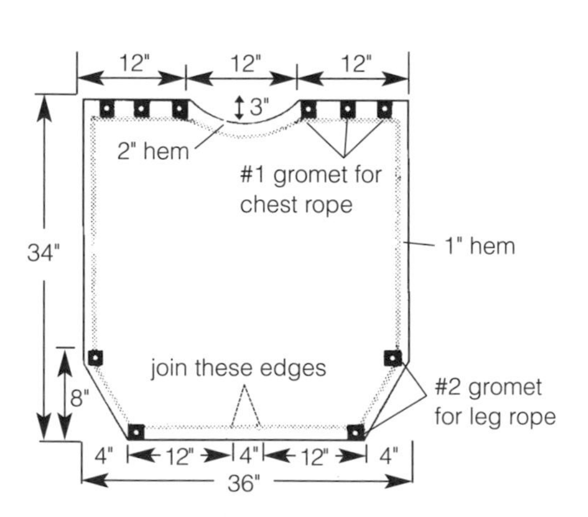

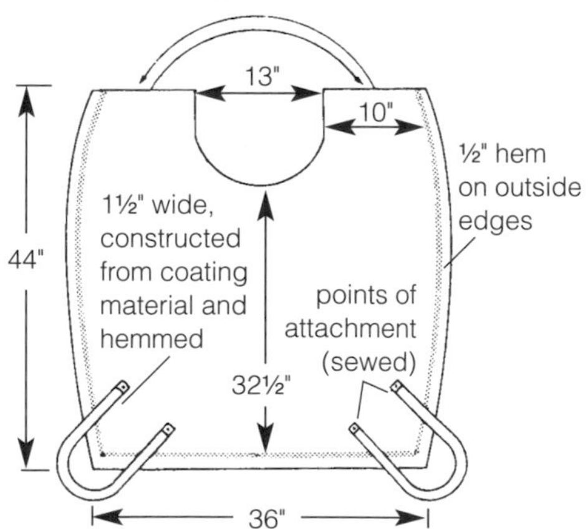

These are two styles of coats to protect sheep, with Number 10 duck or canvas used in most cases. The right pattern can be made in three sizes, with the large having a half-inch (1.3 cm) overlap on the neck flap, and 27-inch (68.6 cm) leg loops. Medium has a 1½-inch (3.8 cm) overlap and 24-inch (61 cm) loops, while the small has a 3-inch (7.6 cm) overlap and 24-inch (61 cm) loops. The loops are 1½-inch (3.8 cm) strips of the coat material, hemmed. Left pattern has grommets used for the chest and leg ropes.

SHEPHERD STORY

Some 4-H projects have a way of becoming lifetime projects, and that's just what happened to Darrin Day. Darrin grew up on a diary farm but showing dairy cattle is really intensive, so his parents bought him a small flock of sheep for his 4-H project. He's had sheep ever since.

Today Darrin's sheep provide aesthetic value for his main enterprise, which is operating a bed and breakfast in Sturgeon Bay, Wisconsin, with partner, Bryon Groeschl. "There are shepherds and farmers who run a B&B, and inn keepers who keep some sheep. I'm in the latter group," he told me with a laugh. "The sheep aren't a profit center, but they're great advertising."

"We normally keep about a dozen sheep in our flock. The flock produces wool, and we sell lambs, but the main thing they're here for now is petting!

"Some visitors come back every year during a certain time to be around the sheep. They may come back for our shearing weekend, or they may come for lambing."

Running a B&B with eight suites and two cabins is a lot of work in its own right, so Darrin doesn't want to have to spend too much time caring for his flock. He's found that Border Leicesters are a good breed for his situation.

"The Border Leicesters are friendly, easy keepers. They're excellent mothers. The ewes lamb on pasture in late spring, and require little or no attention. They get no supplemental grain, and the lambs make great gains right off the grass."

Darrin says that running a B&B is a wonderful way to make a living, but he emphasizes that it's not for everyone. "It's almost like running a dairy farm — you have to be here all the time. And you need patience; 99 percent of your visitors are just great, but you do get 1 percent that are really difficult."

Shearing

Shearing is a major job that has to be done every year on most breeds of sheep. (If you're not interested in wool production and don't want to have to shear, look at the hair breeds). In Australia, scientists are working on a chemical that causes sheep to shed their wool. The chemical is a naturally occurring protein that's administered as an injection. After the injection, the sheep are covered with a hair net that collects the fleece. The protein causes the fleece to break off near the skin surface, and a day later, the sheep begins to grow a new fleece. Although not yet available in North America, it may some day make shearing a much less onerous task for everyone involved — including the sheep.

In the meantime, shearing can be done manually with good-old hand shears or with electric clippers. Either way, shearing is a skill that takes practice to perfect and requires good endurance. Professional shearers make it look really easy and can shear a sheep in just a few minutes, but they've had lots of practice.

Areas where there are still lots of small flocks often have professional shearers available. We've sheared our own sheep and have hired one of these people, and my current recommendation for most small-scale shepherds is to go the professional route, if a reputable shearer is available where you live. On the other hand, Paula and Patrick preferred shearing their own sheep (and those of their neighbors) with hand shears or blades. If you decide to hire a professional, try to be there to help on shearing day — you'll learn a lot, as most shearers are shepherds themselves, and the job will go smoother with extra hands to sort, catch, and move the sheep.

If there aren't any shearers available where you live, or the shearers won't come to your farm because you have too few sheep, or you just really want to try it yourself, then you'll have to learn how to shear. Shearing lessons are often offered in early spring for 1 or 2 days at a nominal fee through county Extension offices and other organizations. They usually limit their instruction to electric shearing, but what you learn is valuable regardless of whether you use electric or hand shears.

Electric clippers, although quicker to use than hand shears, are quite expensive and more apt to result in cut hands and cut sheep, especially when used by an inexperienced shearer. For small flock owners who want to shear their own sheep, hand shears have a number of advantages; they:

- Provide an inexpensive way to get started and can be ordered from any sheep-supply catalog.
- Require no electricity, so you can shear anywhere.
- Are easy and quick to sharpen with just a hand stone.
- Are lightweight and easy to carry with you.

Don't shave the sheep too closely, in order to minimize loss of body heat in cold or rain

Having a set of blades around is a good idea. They come in handy for the occasional trim job that comes up — like trimming excessive wool from around the udder of a new mom. "Rigged" blades have a leather strap taped onto the left handle (for right-handed use), and a rubber stop is taped to the top of the right handle at the base of the blade. These hand shears are more comfortable to use, and the strap prevents them from being kicked out of your hand.

Sharpening Blades

To sharpen, reverse the normal position of the blades, crossing them over each other. Using a medium sharpening stone, grind the stone along the *existing* bevel of each blade with long strokes. Do not sharpen the "inside" surface of either blade. If there are any slightly rough edges when you've finished sharpening, run the stone flatwise along the inside surface of the back (not the edge) to remove the edge burrs. For touch-up sharpening while shearing, close the shears firmly so that each cutting edge protrudes beyond the back of the other blade. Using the fine side of a small ax stone, follow the existing bevel of each blade.

Preparing to Shear

Sheep should be gathered up to 12 hours before you plan to shear and put in a handling facility that minimizes stress on you and them. The pen that holds them should be clean and dry. If the sheep are wet, don't shear them. With a bigger flock, it pays to break them down into groups that have similar fleeces, for example, by breed, staple length, age, and so on.

Shearing can be tough on your back. It's a good idea to do some stretching exercises before you start. You may also want to wear a back-support belt, available from pharmacies.

How to Shear

The real "trick" in shearing isn't learning the pattern of the shearing strokes, which lessens the time involved in removing the wool, but in immobilizing sheep by the various holds that give them no leverage to struggle. A helpless sheep is a quiet sheep. Rendering sheep helpless cannot be done by the use of force alone, for forcible holding makes them struggle more. Try to stay relaxed while you work.

Note both the holds on the sheep, often by use of the shearer's foot or knee, and also the pattern of shearing in the illustrations.

Shearing cuts heal quickly, but use an antibacterial spray to help prevent infections, which may spread to the lymph glands or result in flystrike. Commercial shearers don't normally do this, but if you're there to help, then you can pay attention to these cuts.

Shearing in Twenty Steps

1. Slip your left thumb into the sheep's mouth, in back of the incisor teeth, and place your other hand on sheep's right hip.

2. Bend the sheep's head sharply over her right shoulder, and swing the sheep toward you.

3. Lower the sheep to the ground as you step back. From this position, you can lower her flat on the ground or set her up on her rump for foot trimming.

4. Start by shearing brisket, and then sheer up into the left shoulder area. Place one knee behind the sheep's back, and your other foot in front.

5. *Here, the sheep is on her left side. Trim the top of her head, then hold one ear and shear down the cheek and side of the neck as far as the shoulder, into the opening you made for the brisket.*

6. *Place the sheep on her rump, resting against your legs. Shear down the shoulder while she is in this position.*

7. *With the sheep in this position, hold her head, as shown, and shear down the left side.*

8. *Hold her left front leg up toward her neck, and from this position shear her side and belly.*

9. *With only a minor shift in the position of the sheep, you are now ready to shear the back flank.*

10. *By pressing down on the back flank, the leg will be straightened out, making it easier to sheer.*

11. *From this position, the sheep is shorn along her backbone and a few inches beyond, if possible.*

12. *By holding up the left leg, it is possible to trim the area around the crotch.*

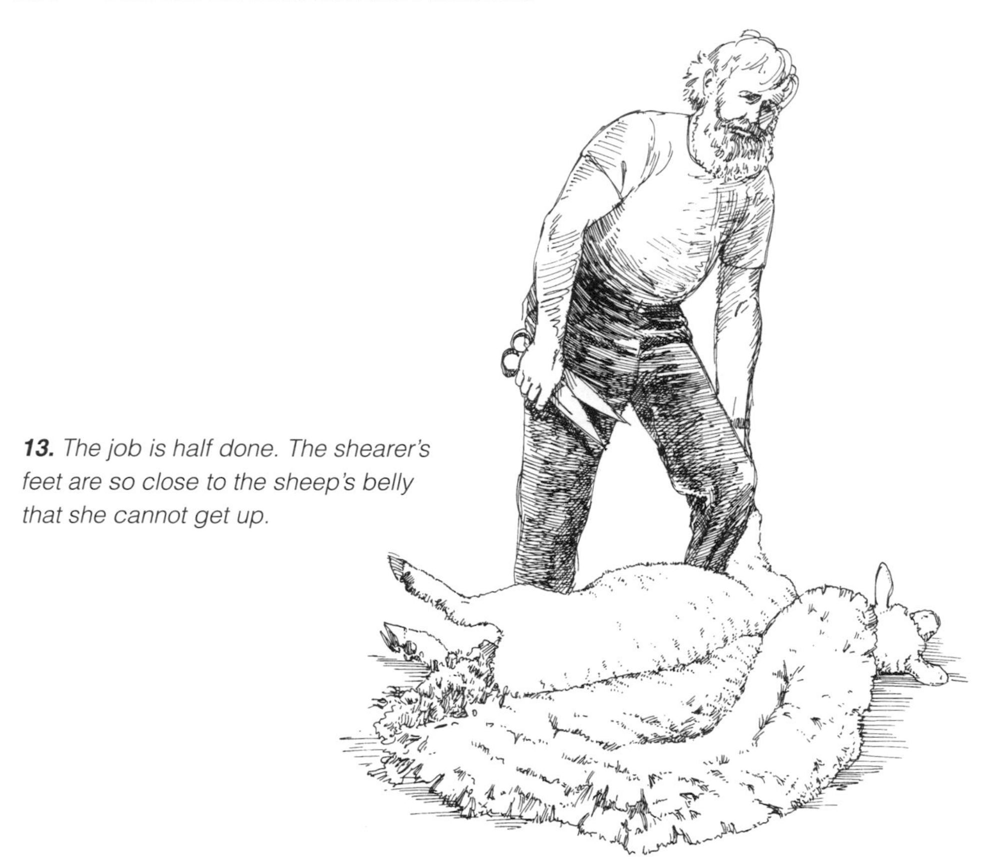

13. *The job is half done. The shearer's feet are so close to the sheep's belly that she cannot get up.*

14. *Holding one ear, start down the right side of the neck. Hold the ear firmly but not tightly — you don't want to hurt her.*

15. *Hold the sheep with your left hand under her chin and around her neck, and shear the right shoulder.*

16. *Pull the sheep up against you to expose her right side, so that you can shear down that side.*

17. *Shifting position, as shown, shear farther down the side and the rump.*

18. *Shifting position again, finish the right flank and shear the sheep's rear end.*

19. *Shift position, holding up the rear leg, and shear the right side of the crotch.*

20. *The job is done, and within a minute the sheep is back on her feet and eating grass.*

Shearing Suggestions

Shearing is something you learn with practice, and over time you'll develop techniques that work well for you, but these suggestions should help you get started.

- Shear as early as the weather permits so shearing nicks will heal before fly season. Ewes can be sheared (gently) before lambing; this makes it easier to help the ewe if necessary and removes dirty wool tags that the lamb might suck on.
- Never shear when the wool is wet or damp. Damp wool is very hard to dry for sacking and storing. It is also combustible and can mildew.
- Pen the sheep in the afternoon prior to shearing so they will not be full of feed when sheared. A covered holding pen with a slatted floor is ideal.
- Shear on a clean tarp, shaken out after each sheep, or on a wood floor that can be swept off. A 4x4-foot (1.2x1.2 m) piece of plywood works well.
- Shear fleece in one piece, but don't trim the wool from the legs or the hooves onto the fleece.
- Remove dung tags, and do not tie them in with the fleece.
- Avoid making second cuts — that is, going twice over the same place to tidy up on overlapping your strokes.
- Roll fleece properly and tie with paper twine if you're selling to a wool dealer or in a wool pool. (See Wool Sales to Hand-Spinners, opposite, for the proper technique for rolling fleeces. If you plan to market wool to hand-spinners, storage techniques are discussed there, too.)
- Skirting the fleece (removing a strip about 3 inches [7.6 cm] wide from the edges of the shorn fleece) is proper, especially if you're selling to spinners. A slatted skirting table makes this easy and allows any second cuts to drop off if the fleece is thrown onto the table with the sheared side down.
- Be sure you shear black sheep and white sheep separately, sweeping off the floor between each. Do not combine white fleece with dark fleece or vice versa.
- For spinning wool, expect top dollar for quality (clean fleeces without manure tags, skirtings, or vegetation).
- For lower quality fleeces, charge lower prices and explain the reason for the price to the customer. These fleeces may be quite adequate for quilt batting, rug yarn, or felting.

Local Shearing Services

Learning to shear your own sheep can sometimes lead to a part-time seasonal income because shearers are scarce in many areas, and some sheep raisers have to wait until the heat of the summer before they can hire one. For flocks of only four or six sheep, professional shearers may not want to spend the time to travel some distance for the small fee that could be charged. Another reason a commercial shearer would not want to do a small number of sheep is that facilities are seldom ideal — often there is no good method or arrangement for catching the sheep and no electricity for his shearing equipment.

When you shear with hand shears, which are so convenient for a small number of sheep, you don't have to worry about electricity. Shearing in your own vicinity obviously eliminates distant travel, and you can make an agreement ahead of time that the owner will have the sheep penned when you arrive. You can either charge for your service, or trade shearing services for wool. If you're charging for your shearing skills, then the person for whom you work expects you to shear the fleece carefully, especially avoiding second cuts.

When you shear as a sideline job, you can expand your service to include trimming the hooves and worming the sheep, but this should be negotiated for a separate fee. The wool that you get can be combined with your clip to provide income or, if you are a spinner, the best could be selected out for your spinning projects.

Wool in Bedding

Batting for wool-filled quilts, comforters, and pillows can provide good income. Though the quantity needed for mattresses and futons is probably beyond the capacity of a very small flock, when these items are made with wool they are comfortable, warm, and lucrative. Despite its supposed advantage of washability, polyester-filled quilts pale in comparison with wool ones — they are not warm enough in winter and are too hot in the summer. Some home quilt makers make their wool-stuffed quilts washable by fashioning the quilt into a big pillowslip like a duvet cover that's removable for washing, leaving the wool batting quilted between two cotton muslin sheets. Three of the 6-foot-long (1.8 m) batts from one of Patrick and Paula's Cottage Industry Carders fills a standard-size quilt.

Wool Sales to Hand-Spinners

The great interest in hand-spinning has created a specialty market for good fleeces. Selling fleeces to hand-spinners in and of itself probably isn't going to create a big cash flow from a flock of sheep, so add other options, like selling breeding stock or locker lambs, to increase your income.

By keeping the fleeces clean, relatively free of grain, hay, burrs, and other vegetation; shearing your sheep carefully (minimizing second cuts); and handling them properly after shearing, you will have a product that is valuable for handcraft use.

After shearing, set aside your best fleeces, which must be absolutely dry for storage. With nice fleeces to sell, it's unlikely that you'll have to hold them very long before they are sold, so for short-term storage, they can be stored in a plastic bag. For longer-term storage, gently place an unrolled fleece into an empty paper feed bag, one fleece to a bag, or lay it out into a large shallow box. You can shake out much of the junk and second cuts before bagging to make the fleece more valuable.

The fleeces that are usually in demand from hand-spinners are those with unusual natural coloring or those from some of the long-wool breeds that are highly lustrous and easy to spin. For your own hand-spinning or for marketing to hand-spinners, consider some of the more exotic breeds, such as Shetland, Icelandic, Cotswold, and Cormo. Even Finnsheep, although not noted for their wool production, have a soft, fine wool that is valuable for blending purposes.

Marketing Your Fleeces: Details Matter

To market your fleeces, find out where the nearest craft classes are given, and let it be known that you have fleeces to sell. Some folks are successfully marketing quality fleeces over the Internet. However you decide to do your marketing, bear in mind that the proof of your success will be repeat customers. This depends not as much on the breed of sheep or the type of wool as it does on the condition of the fleece. If the wool is poorly sheared; full of burrs, seeds, and other vegetable matter that has to be picked out by hand; or if you have not discarded the heavy dung tags, then you may sell someone a fleece once, but you're not likely to sell them another one. The real secret of successful selling to hand-spinners is to offer only your best fleeces, generously skirted, so that you get a good reputation.

Money in Colored Fleeces

Did you know that black lambs have black tongues? The poet Virgil (70–19 B.C.) advised sheep breeders to choose rams without pigment in their tongues if they wanted white fleeces. In many breeds, a dark (or partly dark), varicolored tongue indicates a white sheep with recessive dark genes.

The "black sheep" of the sheep family is the odd dark lamb that crops up occasionally in almost any white breed as the result of recessive genes. In large herds, a black sheep is undesirable. Its fleece must be handled and sacked separately. Even in the flock, its black fibers may rub off on fleeces of white sheep causing the white wool to be discounted in price because of the special problems caused later in the manufacturing processes.

For handcraft use, the picture is different. Many weavers and knitters are spinning yarn for their own use, and a considerable number are spinning for sale. This has created a special market for colored fleeces in natural shades varying from buff to red, or gray to black. As early as 1974, *Shepherd* magazine wrote that "the unwelcome black sheep has suddenly become respectable, with its wool bringing up to several times the price of white wool." But, not just any black fleece — there are so many people now raising dark sheep that it takes a prime fleece, especially clean, to bring top price. There is also the competition from imported fleeces. Australia and New Zealand now have extensive herds of colored sheep and export tons of their wool into North America.

Backcrossing for Black

Once you get a black ram of a suitable wool type, use him to breed a small flock of white ewes. Geneticists say that the offspring will be "white, but carriers of the black gene." However, in practice, we have had people use one of our dark rams on their white ewes and more often than not they got dark lambs.

The first generation of this cross is called the "first filial generation," or F1. If the F1 ewes are bred back to the original black ram, their father, this is called backcrossing. It produces a generation of F2 lambs, and in theory there should be as many black lambs as white ones, with all carrying the recessive black gene. This amount of inbreeding is not likely result in many birth defects, but it is risky to continue breeding with the same ram to succeeding generations (like the F2 offspring). By the time you get a good number of your ram's granddaughters (the black F2 ewes) into your breeding flock, you'd be well advised to sell him and get a different one that is not related to your sheep.

The fleece of most black sheep tends to lighten from year to year. This may be disappointing at first, but in the long run it becomes an advantage because it gives a greater variation in color from a relatively small flock. So, in shopping for a black lamb, remember that however black she is at birth, she probably won't stay that black but will lighten every year. Don't consider the degree of darkness as the main factor in your selection. Look at the body type and wool grade, which does not change and will probably be inherited by her offspring. Refresh your memory about what to look for and what to avoid in earlier chapters, and place health above other criteria when selecting your sheep.

With one or two black sheep to introduce the black genes, you can develop a flock of dark sheep in a few years if this is your goal. The easiest way to achieve this goal is by the use of a good black ram of a nice, spinnable wool breed. If there are spinning classes nearby, ask the teacher what breed of wool is most favored in your area.

Processed Wool

There are still a handful of wool processors in North America who take your raw fleece and process it to finished yarns for you and others that turn out finished products, such as blankets. These products can then be sold. If you have a larger flock, you can do wool processing and custom carding for your own fleeces, and sell your services, too! If you want to go into home processing on a larger scale, Paula's book, *Wool as a Cottage Industry* is a must-read. Patrick and Paula manufacture wool-processing equipment (wool pickers and drum carders) for hand-spinners and for small cottage industry (see Resources).

Cleaning Fleeces for Processing

Most fleeces need to be washed before they are processed. The exception is the occasional really clean lamb fleece, which may be processed and spun first and washed later. If you want some grease to remain, use a natural soap. If your goal is to be grease free, use regular detergent and a degreasing dish soap. Although I've always washed fleeces one at a time in a washtub, I've heard of folks who wash their fleeces in the washing machine and can clean several at a time that way.

1. Regardless of whether you hand wash or machine wash your fleece, fill the container with hot water and add 1 cup (236.6 mL) of laundry detergent and ½ cup (118.3 mL) degreasing dish detergent.
2. Stir to mix well before pushing the fleeces down in. (If using the washing machine method, *do not* run the full cycle, just let the wool soak — you don't want to agitate the fleece.)

3. Every so often, *gently* squeeze the fleeces, but don't twist.
4. Continue soaking for at least 1 hour.
5. After the wool has soaked adequately, loosen all the dirt by emptying the water, lifting the fleece from your container, and patting it dry. (In the washing machine, set to spin dry.)
6. Rinse the container, and refill with warm water.
7. Place the wool back in to soak for about 15 minutes.
8. Repeat the draining and drying step in cool — but not cold — water, adding 1 cup (236.6 mL) of white vinegar.
9. Soak for about 5 minutes in this solution.
10. Keep repeating the draining–drying–soaking step until the drained water is clear.
11. Lay the wool out on old towels, and then roll it up to squeeze out as much water as possible; if you're using a washing machine, the wool can be spun dry the last time.

Paula's Wool-Washing Method

Paula, who washes quite a bit of wool, favors using a 40-gallon (152 L) laundry tub, whose cube shape holds the heat for a good soaking. If water cools down, it deposits gumminess back on the wool. A very hot soaking, with lots of soap or detergent, can clean most wools with one wash, followed by one (or two if really needed) hot rinses. While some amount of dirt may remain, it dries to a fine powder that falls off in the teasing, carding, spinning, and washing of the yarn (to set the twist).

Using the spin cycle of an old washing machine not connected to water will speed removal of wash water and rinse water, shortening the drying period for the wool. This method keeps much of the mess out of the house and doesn't clog up the plumbing, as washing wool in the bathtub would.

Drying Fleeces

The fleece has to be laid out to dry in a way that provides good airflow all the way around it. Nailing 2x4s together and then stapling them on chicken wire can create an easy drying fame. Sweater drying racks, which are sold at hardware stores, also work well. If you're planning to set the fleece outdoors to dry, make sure that it's secured so that it doesn't blow into the dirt and ruin your day's work. It's not a good idea to dry wool outside on very windy days, unless you loosely tie another piece of chicken wire on top of the frame.

After the wool is *completely dry*, you can store it, ship it to a custom processor, or process it yourself. If you want to store it, remember that paper or burlap bags work better than plastic bags. Seal the ends of the bag so that bugs can't get in and ruin your fleece.

Carding

Carding is typically the next step in the processing of wool for yarn. The carding process "teases" the fibers apart, removes short fibers, and sets the fibers to lie in the same direction. At commercial processing facilities, carding is done on large machines, but for home use and 4-H demonstrations, hand-cards can be used. For anyone with two or more sheep, a drum carder is faster, more efficient, and easier on the hands.

As you card, you can pick out any odd pieces of vegetation that survived the washing, but if there was a large amount in the fleece, you may have to "pick" it before carding. Those little burrs, seeds, and pieces of chaff will ruin your finished work. (And you'll quickly realize why maintaining a clean fleece is so important in the first place.)

During carding, you can blend various fibers together, yielding interesting colors and textures. The illustrations below shows the basics of carding with hand-carders. Once your wool is prepared, it can be spun into yarn or used in other handcraft projects.

Extended Cottage Industry Carding Machine with Bump-Winder, produced by Patrick Green Carders, Ltd. (Sardis, BC, Canada)

Hand-Carding in Four Steps

1. Hand-cards work well for occasional small carding jobs, and they're not very expensive if you're just starting out. Spread the wool on the left hand-card, with the shorn ends at the top of the card.

2. Take the right hand-card and lay it in the center of the left hand-card, with the handles in opposite directions, and draw the right hand-card away from you. Repeat this step several times, until the fibers begin to align themselves.

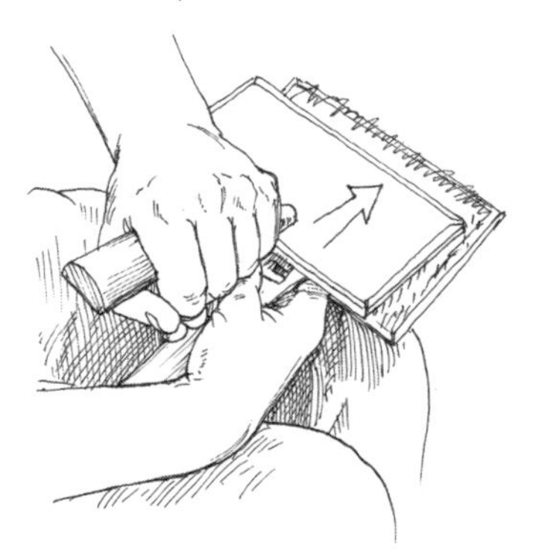

3. When the fibers are well aligned, lay the right hand-card on your knee, and with the handles in the same direction, brush away from yourself. This deposits the wool on the right hand-card. Switch paddles and repeat this step several times.

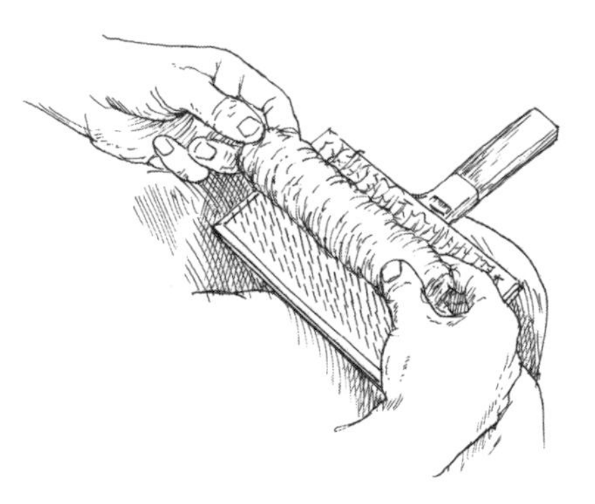

4. Roll short or medium wools off the card, or fold over longer wools. The fibers are ready to spin.

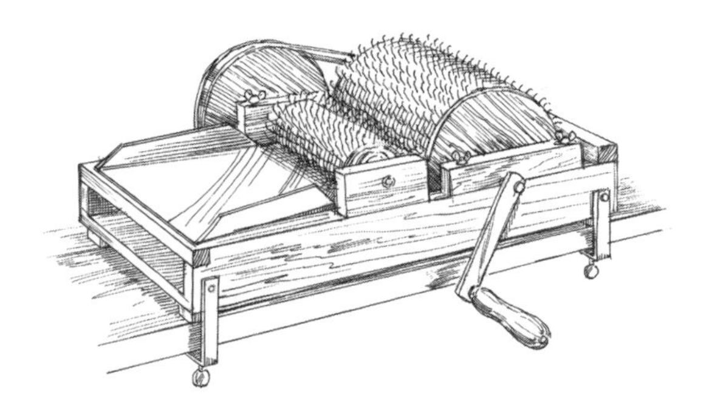

A hand-operated drum carder is ▶ convenient and will save your wrists from carpal tunnel syndrome if you plan to do lots of carding.

Handcraft Uses of Fleece

There are several types of handcraft projects that can be created and marketed. Some are from spun yarn and others are from carded wool. Weaving, knitting, and crocheting are typical uses for yarn. A few possibilities for wool include making it into furry stuffed animals, using it for batting in quilts, or "tying" it for fishing flies. Australian Locker hooking also uses fleece locks, or yarn.

Handspun Yarn

Spinning your own wool into yarn is one way to compound its value per ounce. You can spin it for your own use in knitting, or you can practice on it for home use and go on to sell it to other knitters or weavers when you get good enough. You need to be a knitter in order to knowledgeably sell yarn to other knitters, so you can advise them of needle size, quantity of yarn, and what size yarn to spin for a specific project. The same goes for weaving yarn — unless you can advise your customer, you may be selling them a problem and not know it. (There is a list of spinning literature and sources for equipment in the resources.)

Australian Locker Hooking

Australian locker hooking, using unspun fleece, is a new version of the older craft of locker hooking that used commercial rug yarn. The "locker hook" has a large crochet hook on one end and an eye on the other end so that unspun wool may be hooked into rug canvas and locked in with a binder yarn carried by the locker hook. This technique offers to nonspinners a way of using their wool in an attractive and profitable manner to create rugs, wall hangings, saddle blankets, and heavy garments. (For instruction booklets, see Resources.)

Odd Uses of Wool

Even unprocessed fleece is useful and can be used for the following:

- Cleaning up oil spills around the shop — wool is even being put to that use commercially in large scale oil spills, like that of the Exxon Valdez. Wool can absorb up to 30 times its weight in oil!
- Mulching a garden. Wool works like black plastic, reducing the number of weeds while allowing water, air, and nutrients to pass through to the ground.
- Insulating buildings where fire code restrictions aren't a big concern — like barns or storage sheds. Because wool is flammable,

it is prohibited from use in most construction, but researchers are trying to find a way to create a wool-based insulation that meets fire code.

SHEPHERD STORY

Like Lisa Merian, Donna Herrick, a shepherd from rural Vermont, concentrates on producing high-quality fleeces, but unlike Lisa Merian (see page 279), Donna makes regular use of sheep coats as part of her approach. "My sheep wear their coats 365 days of the year," she told me.

The coats add labor for mending and maintenance, but Donna thinks that she gets higher fleece production and better-quality fleece through using them.

"If you're looking at full-time use of coats, you'll need three or four sizes per sheep because you have to change coats as the wool grows out, though you may not need that many for each sheep. Different size sheep can share coats. My flock includes 38 animals, and have about one hundred coats all together."

As well as processing wool from her own flock, Donna also runs a custom-processing service. She cleans and cards wool for clients she's developed by attending shows and advertising in magazines like *Spin-Off,* but like many small business people, she's learned that word-of-mouth is the best approach to developing a clientele: "I now have more than a dozen clients who all came to me by word-of-mouth from one woman who was really impressed with my wool."

Donna's advice to beginners is this: "Start out small. Sheep are fun and wonderful, but there are hard times. We began with only three sheep and built up our flock. It helps to find a good county Extension agent and a good veterinarian at first."

As for starting a sheep-related wool enterprise, Donna again recommends starting small. "I ran the business for a number of years out of the house. Now it's grown to where we've built a separate shop.

"If I were starting out today, I would just attend lots of shows where potential clients could see the quality of my product. Advertising is expensive, and until you have a clientele, ads don't really pay for themselves."

Meat and Milk

There is generally more money in meat and milk from sheep than in wool. You'll find that if you're willing to make the extra effort of marketing directly to the consumer, you'll get a much better return than you would get from selling to the sale barn, a "lamb pool," or other conventional market. You can give your customers a better buy than they would get at the meat market and pocket some of the money that currently goes to middlemen. Some shepherds have begun milking sheep and making specialty cheeses. Although this is not an option for the part-time, "hobby" shepherd, if your dream is to make a full-time living off of your sheep operation, cheese offers the possibility of a reasonably high return.

Federal and state laws and regulations place definite restrictions on slaughter-and-sell practices. Some are designed to deter rustling; most are designed to enforce sanitation and ensure food safety.

There are two types of butchering facilities, a custom packer and an inspected packer. *Custom packers* butcher, cut, and wrap, but each package of frozen meat must be stamped "not for sale." This is because federal regulations require that each and every carcass be inspected by a federally designated inspector if meat is going to be sold by the package. An inspected packer has a federally designated inspector (either a federal or state employee) on the premises to inspect each animal carcass. Each package of meat coming out of an inspected packing house will be labeled with a USDA approved label. However, this restriction on custom packers need not stop you from legally selling locker lambs (see next section), which are sold to the customer as a live animal, so it's essentially the customer who is engaging the custom packer.

Many small-scale commercial shepherds have met all the regulatory requirements to be able to sell meat directly to the consumer, either by the pound or by the cut. Again, this isn't for someone with only a dozen sheep, but if you're interested in making a living off a small, commercial flock, this can greatly increase your return.

In general, lamb consumers are among the higher-income groups of the population, so this should determine where to advertise, if advertising is needed to sell your meat (or cheese). Always remember the consumer preference for leaner meat, and do not overfatten the lambs in the last month prior to slaughter. If you're going to market meat, it's important to have a good idea of how much you'll get back for different cuts.

There are also real opportunities for selling lamb to certain ethnic markets. A friend of mine who lives in Red Lake Falls, Minnesota, has developed a dedicated clientele for her lamb among Arab community members in nearby

Grand Forks, North Dakota. Research ethnic markets in cities near where you live; lamb is central to the diets of most Arab cultures, as well as Greek and Jewish cultures.

Relative Percentages of Various Cuts

Type of Cut	Relative Percentage of Return
Loin	8% of hanging carcass weight, or 4.4% of live weight
Rib (rack)	7% of hanging carcass weight, or 3.9% of live weight
Leg (boned/rolled)	24% hanging carcass weight, or 13.2% of live weight
Shoulder (boned)	20% hanging carcass weight, or 11.0% of live weight
Ground lamb	10% hanging carcass weight, or 5.5% of live weight
Stew meat	7% of hanging carcass weight, or 3.9% of live weight
Bone, waste, etc.	24% hanging carcass weight, or 13.2% of live weight

So, for example, if you take a 115-pound (52 kg) lamb into be butchered, you'll get about 5 pounds (2.3 kg) of loin cuts (115 x 0.044, or 52 x 0.044).

Locker Lambs

The locker-lamb business legally requires you to sell to your customer in advance, deliver the lamb to the slaughterhouse, and give the slaughterhouse your customer's name. They will notify you, not the customer, of the cutting weight as you direct them. Collect the price per pound on that weight from your customer, who then picks up the meat from the slaughterhouse all cut, wrapped, and frozen, and pays them for the cut-and-wrap charges.

Taking orders in advance is a good idea. By taking advance orders you can plan to deliver lambs about the time the summer pasture starts to dry up. Fast growth of your lambs on good pasture assures that they will be ready for marketing by then, and fast growth is associated with tenderness.

Young lamb is naturally expected to be tender, but several factors, one at a time or combined, can reduce this tenderness:

- Stress imposed on animals prior to slaughter, such as rough handling when catching and loading.
- Slow growth rate; this is a good reason to feed your lambs grain in a creep-feeder if pastures aren't in top form.
- Drying out in slow freezing; most cut-and-wrap facilities do the freezing faster than it could be done in your home freezer.

- Length of time in freezer storage; 1 year is the maximum that lamb should be stored.
- Lamb carcasses can generally be cut and wrapped quickly after slaughter, but if you are butchering a yearling or mutton animal, they must hang at least 1 week in a chilling room prior to cutting and wrapping.

Organic Lambs

With the current trend to health consciousness, there is a very special market for organically grown lambs. This need can be met only by the small grower. Big feeders and producers have more of a disease problem than is ordinarily present on a small farm, so they may use medicated feed as a preventive measure, even when disease is not present.

To market meat as organic, it must be certified by a state agency or a delegated organic certification organization. (See the resources for contact information.) The process can be expensive and time consuming, so it's only really valuable for the small-scale commercial flock. If you are selling the lambs from a really small flock, your customers probably know you, so they shouldn't need to see the certified label to feel comfortable with you or your product.

To become certified, you must be able to provide evidence that all crops raised on your farm for feed are free from most "chemical" pesticides, herbicides, and fertilizers. You must also present evidence that all purchased feeds were certified organic. The use of medications, including wormers, is severely limited if you're seeking a certified label.

Lambs for Easter

Creep-feed your lambs, and try to have some of them ready for sale by Easter. The eating of lamb is part of the religious festivities in the Greek Orthodox tradition, among others. If you have lambs born early (first half of January or before) and do not have them promised, you might tell the nearest Greek Orthodox church of their availability or advertise if there is a Greek newspaper in your area. The size preferred in the Northeast is about 35 to 40 pounds (15.4 kg–18.1 kg) liveweight; in the West, the ideal size is a little larger. Lambs sold at that size are called "milk-fed." The term "hot-house lamb" is sometimes applied to the early January lambs that are sold at Easter and sometimes to the fall lambs that were born out of season and raised mostly indoors for sale in the early spring.

Mutton

Selling an aging ewe or an extra ram is not as easy as selling a lamb, which is expected to be more delicate and tender, because mutton has a rather bad image in this country. Many people (even those who have never tasted it) say they don't like it and expect it to be tough and strong in taste. Prejudices are hard to overcome, so consider saving the mutton for your own locker. You'll be pleasantly surprised to know that there are many uses for mutton, so you'll be able to use whatever "culls" you have and enjoy doing it.

Mutton is known to be very digestible, which makes it a good meat for people who have various gastrointestinal difficulties. Animals that are raised on grass have a high concentration of conjugated linoleic acid, a naturally occurring chemical that researchers are recognizing as being a good antioxidant, anticancer, and anti–heart-disease agent!

The leg of mutton can be smoked, like a ham and most of the rest can be trimmed and boned for use as ground meat. Some good recipes for using ground mutton can be found at the end of this chapter. These recipes make do-ahead meals, casseroles, and quick-fix recipes that are easy to prepare and can be frozen for later. You will find that the money you would be paid for your culls is far less than their value in your freezer.

Another use for mutton meat is to have it made into sausage. Ask your meat cutter or locker owner about custom sausage, or try making your own.

Mutton for Pet Food

Mutton from a really old ram or ewe is often best made into pet food. Use it ground and completely cooked to feed to your own dogs or cats. Selling this premium dog and cat food can also become a profitable sideline. With a secondhand commercial grinder and adequate refrigeration and freezer space, you can take special orders from dog and cat owners. Unlike food for human consumption, individual packages need not have been prepared and labeled at a federally inspected packing plant. David Schafer, a farmer from Kansas who has jumped all the legal hoops to be able to sell his labeled and inspected meat directly to consumers, also sells some of the less desirable cuts this way. As he points out in a recent article in *The Stockman Grass Farmer* magazine, farmers should learn more about "companion animal" feeds. He considers the book *Foods Pets Die For: Shocking Facts about Pet Food*, by Ann Martin to be a good source (see Resources).

(See the resources for home sausage-making supplies.) Hot Italian sausage, made from mutton, is one of my personal favorites. Some sausage recipes can also be found at the end of this chapter.

Cutting Instructions for Lamb and Mutton

You can either take your sheep to a custom packing plant to be slaughtered and butchered or do it yourself. If you want to do it yourself, get a hold of *Basic Butchering of Livestock and Game*, by John Mettler, DVM. He provides excellent slaughter and cutting instructions, with lots of illustrations to help along the way. Your county Extension agent may also have a booklet available on the topic.

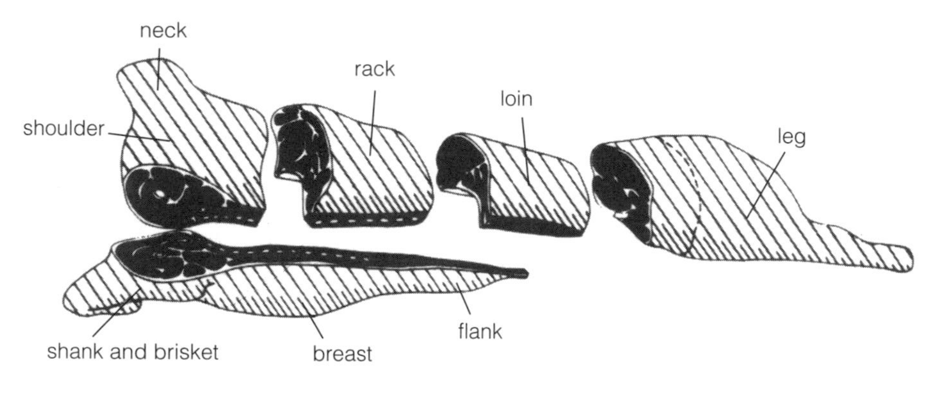

There are several correct ways to break a lamb carcass, and no one method can be considered best. However, for many purposes, the method shown is ideal. (From Lamb Cutting Manual, *American Lamb Council and National Livestock and Meat Board)*

If you're going to work with packers, you will have to give them some directions. To get the maximum use and enjoyment from your sheep, give these instructions:

- Cut off the lower part of hind legs for soup bones.
- For mutton, have both hind legs smoked for "hams."
- For lambs, the hind legs can be left whole, as in the traditional French-style "leg of lamb," or cut into sirloin roasts or steaks, and leg chops or steaks.
- The loin, from either mutton or lamb, can be cut as tenderloin into boneless cutlets or as a loin roast.
- Package riblets, spareribs, and breast meat into 2-pound (0.9 kg) packages. Riblets, which are sometimes also referred to as short ribs, are almost inedible when prepared by most cooking methods, but

when prepared in a pressure-cooker for about 45 minutes, with an inch (2.5 cm) of water in the bottom to start, they are a real delicacy. You can substitute barbecue sauce, curry sauce, or your favorite sauce or marinade for the water. These parts are hard to remove from the bone, but will fall off it easily after being cooked in the pressure cooker. From lambs, the spareribs and breast can be can be barbecued or braised. For mutton, these cuts are pretty much waste products, though spare meat can be trimmed for your companion animal food.

- For mutton, have the rest boned, the fat trimmed, and ground. Double wrap in 1-pound (0.5 kg) packages, and try some of the muttonburger recipes in this chapter. Do not be surprised when the ground mutton seems a lot juicier than other ground meats. Older animals' tissue can bind large amounts of water.
- For lamb, the rack, or rib area, can be cut into that favorite, "lamb chops," or left as a rack roast. The shoulder can be cut into roasts or chops, and the neck and shank can be used as soup bones. Stew meat, or ground lamb, can also come from these "front" cuts.
- If you want kabob meat, make sure to have it cut from the sirloin, or loin.

Tenderstretch

Texas A&M University developed a method of carcass hanging that improves the tenderness of most of the larger and important muscles of the loin and round (most of the steaks and roasts). This procedure is called "tenderstretch" and consists of suspending the carcass from the aitchbone within an hour after slaughter. The trolley hook should be sterilized before inserting in the aitchbone on the kill floor. This method does not require any change in equipment in small slaughterhouses and is suitable for farms.

How Tenderstretch Works

Tenderstetch prevents shortening of the muscle fibers as the carcass passes into rigor mortis. Shortly after death, the muscles are soft and pliable and are very tender if cooked rapidly. But after rigor mortis sets in, the shortened muscles become fixed and rigid. It takes 7 to 14 days at temperatures between 28 and 34°F (–2.2 and 1.1°C) before the muscles lose their rigidity and become pliable again.

With tenderstretching right after slaughter, meat is as tender after 24 hours of chilling as if it were aged for a full week, and further aging further improves the tenderness. Thus, with a little extra effort and no additional cost, the tenderness of many important cuts of the animal is greatly improved. It does not produce the mushy overtenderness that sometimes results with enzyme-tenderized meat.

Cheese from Sheep Milk

Another profitable project to consider is gourmet cheese made from sheep milk. Americans import sheep cheese, yet sheep dairy farming is in its infancy in North America. Farmers who sell cheese made from sheep milk are selling their product for as much as $15.00 per pound (0.5 kg) in the United States, which "ain't too shabby" in the current food economy!

Sheep milk, being high in solids, yields about twice the amount of cheese as cow milk. Per 100 pounds (45 kg) of milk, sheep milk gives about 20 pounds (9 kg) of cheese, goat milk yields 14 pounds (6.4 kg), and cow milk produces 10 pounds (4.5 kg).

A common practice among European producers is to permit the lambs to nurse exclusively for 30 days and then wean them and milk the ewes. However, weaning lambs at 2 or 3 days after birth and feeding them artificially would result in greater total yield of milk. Twice-a-day milking results in the most milk, but another option is to milk once a day in the morning and allow the lambs to nurse later in the day. Good nutrition (high protein) is an absolute necessity for a high volume of milk.

On-farm cheese making is possible, but there are many regulations involving food processing that have to be observed. The factor that makes cheese production from sheep milk a more suitable cottage industry than making cheese from cow milk is that sheep milk can be frozen for thawing and making cheese later with no loss of quality. So, it can be stockpiled and frozen until there is an adequate amount for a cheese project.

Pelts

The pelts of meat lambs can be another source of income. Prime, No. 1 pelts, are from lambs that have been sheared 4 to 6 weeks before slaughter. This shearing before slaughter provides a pelt that is perfect for use as "shearling" lining for slippers or jackets. If you're selling your lambs through conventional markets, you'll be paid extra for your lambs if they have shearling-quality coats — sometimes quite a bit extra, depending on the market. If you are butchering lambs yourself (either literally yourself or at a custom packer), then these pelts provide an excellent by-product. And if you create a product from the pelt — say shearling slippers — you can really increase your return.

Ticks can also ruin a pelt for tanning, which is another reason to keep your sheep free of them. The dark bumps caused by tick bites are called "cockle" in pelts or leather. If you are using tanned shearling pelts to make jackets lined with the wool, the outer surface can be sanded to produce a beautiful suede finish. However, cockle defects seriously impair the softness and appearance of the leather.

Shearing nicks and skinning cuts also show up in the pelt and seriously diminish its value. A good shearer shouldn't leave too many nicks to begin with, and if you're going to be saving the pelts for shearling, make sure the shearer knows it. Shearers should be paid a bonus if they shear without nicks. Skinning should also be done carefully to avoid cutting the hide. If you're working with a custom packer, again, offer a bonus for unblemished skins. Ask your packer to hang the skins over a railing, skin side up, so that they cool quickly. The bonuses you pay for the shearer's or skinner's extra attention don't have to be huge, but if the people who are doing the work know that they're in for an extra $5 or $10 for a job well done, they'll be a little more careful.

Packers who handle a lot of sheep realize a small income from the sale of pelts from the animals being processed, but you can ask for your pelts back. On the other hand, packers who handle few sheep may demand that you pick up

Proper Handling of Live Sheep: The Key to High-Quality Pelts

Timely shearing isn't the only important facet of obtaining a high-quality pelt. If there are more than two cuts in the middle of a skin, tanneries suggest that it be discarded. They also note that many a skin is ruined by improper handling of the sheep while it's alive: skin separation — when the epidermis is pulled away from the skin, a defect caused by picking the animal up by the wool — and stained wool, which is caused by using oil-based markers on sheep, are both handling problems. Skin separation may take as long as 2 months to heal; the pelt is worthless if the healing isn't completed. Several marking crayons, or paints, are prepared especially for use on sheep. These crayons are made with wax or lanolin, instead of oil, so they're "scourable." If you use any markers, make sure they're labeled for wool.

not only the pelt, but the offal as well (trimmed fat, intestines, and so on). This is because the companies that haul off cattle by-products for further processing won't accept sheep by-products mixed in with the load. Pick up the offal as soon as possible. The best way to dispose of it is to bury it in a large compost pile if the pile has lots of dry, carbonaceous materials, such as dried leaves, straw, sawdust, or shredded paper.

Machine-washable pelts (which are prepared using special tanning techniques) are popular as bedpads for people who are bedridden because the pelts distribute pressure evenly, dissipate moisture, do not wrinkle or chafe, and prevent ulcers and bedsores. They are also marvelous for babies.

You can do your own tanning by purchasing supplies through catalog. To decide whether you should do your tanning at home or have it done at a tannery, estimate the cost of your materials and the value of your time if you have little to spare. Weigh this against the cost of shipping and tannery fees to have your pelt done by a commercial tanner.

When trying any tanning process for the first time, be cautious and do only one pelt. After you have done it once, you may see ways to do a better job the second time, or you may prefer to try another process to see if it is easier and more satisfactory.

Once you have perfected your system of tanning and have done it a few times, you should find a ready market in local craft shops or decorator shops. To get a better price, you can sell directly to your customers, or design and produce wearables or furnishings from the tanned pelts. While the tanning chemical is dangerous to handle and must be used with care, the results can be worth the trouble.

Care of Pelts

Whether you're planning to tan your pelts at home or send them to a commercial tanner, some care needs to be taken as soon as you bring them home.

Finding a comfortable position for working is one of the greater challenges of this job. We attach a piece of 8-inch (20 cm) PVC pipe that is about 4 feet long (1.2 m) between two sawhorses and work sitting on a kitchen stool. The pipe provides a good contact surface to scrape against. The edges of the hide may have to be done with you lying on the ground, but with this approach, at least you're not down on your knees, bent over, the whole time you work.

If you are not going to tan the skin the day it comes off the sheep, you should salt it heavily to preserve it for later tanning. As soon as you get the pelt home, rub common pickling salt (don't use products that are labeled "deicer") into the flesh side. Use 5 pounds (2.3 kg) or more of granular salt on big skins and 2 pounds (0.9 kg) or more on lambs. Do a thorough job, being sure to salt the edges well. Spread the pelt out to dry, flesh side up.

Salt draws the moisture out of the skin. If you are salting several hides, stack them leather side up and raised off the floor on boards after salting. Well-fleshed, salted hides can be stored for long periods, but if you didn't do a good job of fleshing and salting, you're likely to have a real mess on your hands, so take your time with these steps.

In 3 or 4 days, hides will be ready to ship to a tannery. With tanning prices known in advance, you can just pack the pelts, salted and folded, inside a feed bag in a carton. Attach a note with your return address and phone number, and indicate whether you want natural or washable tanning (washable tanning costs more).

Fleshing Out the Pelt

Fleshing is dirty work, but the sooner you get this job done, the better.

1. Scrape the flesh side with a heavy, very sharp knife to remove the meat, tissue, fat and grease. Make every effort to remove as much of this stuff as you can without injuring the skin or exposing the hair roots.
2. Scrape off all tough membranes and inner muscular fleshy coat. The cleaner you get the skin, the less chance you have of bug damage or rot.

If for some reason you can't flesh out the pelt the day you bring it home, then salt it well (see page 320). You can scrape the salt and the flesh off the pelt later, though it's more pleasant work if you do it as soon as possible. Plus, if there's a lot of fat attached to the skin, the salt can "melt" the fat into the hide, resulting in ruined patches of skin.

Home Tanning

There are many ways to do tanning at home; two of the most practical are discussed here. Some approaches are more dangerous than others. Neither of these methods results in washable pelt. The acid must be handled carefully and neutralized well so that it does not remain on the skin and damage it.

Preparing the Pelt for Tanning

Whether your pelt is fresh or salted, it needs to be washed before tanning. Do this in the following manner:

1. If the pelt has been salted, soak it overnight in a large tub of cold water containing 1 cup (236.6 mL) of laundry detergent and 1 cup (236.6 mL) of pine-oil disinfectant and then rinse it in cold water in the morning. (If the pelt is fresh, you can skip this presoak step.)
2. Remove this water by spinning the pelt in the spin cycle of your washer.
3. Next is the wash cycle, which is easily done in the washing machine. Use a short cycle with cool or lukewarm water and detergent.
4. Rinse.
5. Spin out the water, again using the spin cycle.

All of the fat, blood, and dirt should be removed from the pelt by now, and you can proceed with your choice of tanning processes. And pleasantly enough, the washing machine doesn't come out any worse for having done this duty.

Salt–Acid Tanning

For the salt–acid tanning solution, use a plastic drum or plastic garbage can. Do *not* use a metal container. For best results, the solution should remain at about room temperature — between 65 and 75°F (18.3 and 23.9°C).

Use your choice of only *one* of the acids, with the water and salt, for tanning. A choice of acids is given so that you can use the one most easily obtained in your area. Whichever acid you use, measure it carefully and store it in a safe place. If you are measuring liquid acid, use a glass or plastic cup, not metal. Adding acid to water is *dangerous!* Add it *slowly*, letting the acid enter at the edge of the water. Rinse the measuring cup in the solution, and stir the mixture with a wooden paddle.

Immerse the pelt in the tanning solution, push it down with the wooden paddle, and stir slowly. Leave the pelt in the solution for at least 5 days (the pelt should be left up to 2 weeks if the temperature of the solution does not get over 75°F [23.9°C]). Keep the pelt submerged, and stir it gently from time to time.

Salt–Acid Tanning Solution

For each 1 gallon (3.8 L) of clear 70-°F (21.1°C) water, use 1 pound (0.45 kg) pickling and canning salt and one of the following acids in the amount specified: 1 ounce (29.6 mL) concentrated sulfuric acid, 4 ounces (118.3 mL) new battery acid, or ½ cup (118.3 mL) sodium bisulfate dry crystals or 2 ounces (59.1 mL) oxalic acid crystals.

To neutralize the tanning solution, follow these steps:

1. Remove the pelt, and spin out the tanning solution in the spin cycle of your washer.
2. Rinse the pelt in clear water twice, then spin out the rinse water.
3. Immerse the pelt in a solution of water and borax, using 1 ounce (29.6 mL) of borax to each gallon (3.8 L) of water.
4. Work the pelt by stirring it frequently for about an hour in this solution, then rinse out in clear water.
5. Spin out the rinse water. This step is necessary to neutralize the acid solution so that it does not remain on the skin and damage it.

When you are finished neutralizing the acid, move down to Drying and Softening the Pelt on page 322.

Baking Soda—Kerosene Method

Although we've never used the baking soda–kerosene method, it sounds good and comes from Paula's friend and long-time veterinarian, Dr. Salsbury. It uses two items that are generally readily available, no matter where you live. Mix baking soda and kerosene until you have a paste that's about the consistency of cake batter before you pour it into the baking pan. This mix takes about 10 pounds (4.5 kg) of baking soda to 1 gallon (3.8 L) of kerosene, which is about enough for one skin.

Apply the paste evenly, about ¼ inch (0.6 cm) thick. Cover the whole pelt, including the edges, well. Leave it alone until the paste has completely dried (1 to 3 weeks, depending on weather). According to Dr. Salsbury, the end result of the process is that the water and kerosene evaporate, and the oils from the skin are absorbed by the baking soda.

Scrape the paste off the pelt. If any areas still appear greasy, reapply some paste to those areas, and let it dry out again.

Drying and Softening the Pelt

Tack the pelt out flat, flesh side up. If you used the salt–acid method, the pelt is wet; if the baking soda–kerosene method was used, it's dry. With either method, the pelt is now cured, but when it's dried, it's quite stiff. (In fact, the words "stiff as a board" come to mind. Ken always jokes that this is where I'm supposed to jump in and chew the hide but there are easier ways to end up with a nice, soft pelt.)

1. Apply a thin coat of neatsfoot oil (a product that's used for waterproofing boots, and is available in most hardware stores) to the flesh side. While the oil is soaking in, which takes from 8 to 10 hours in a warm room, you can dry out the wool side if necessary, using a fan or hair dryer. Then apply a thin coat of tanning oil or leather dressing on the flesh side.
2. When the tanning oil has soaked in, allow the pelt to dry until it starts showing light-colored places. Remove it from the frame, and start the softening process. Stretch the skin in all directions, and flesh side down, work it over the board to soften the skin as it finishes drying.
3. You can sandpaper the flesh side when dry to make it smooth. Comb out the wool with the coarse teeth of a metal dog comb and finish with finer teeth. If the wool seems too fuzzy and dried out, you can rub a hair dressing (such as a hot oil treatment conditioning product) on your hands, rub them lightly through the wool, then brush it gently. Repeat if necessary.

The Live-Animal Business

Selling or leasing breeding stock or selling club lambs (to 4-H groups and other youth programs for showing) may provide an opportunity for profit, particularly if you've invested in some purebred and/or registered animals. But even if you have grade animals that you've bred for certain superior traits, they may be quite marketable as breeding stock or show lambs for nonbreed sanctioned shows.

There's also a unique aspect of the live-animal business that may be worth investigating: sheep as weed control or fire suppression work crews! Shepherds, with a flock of sheep, a portable electric fencing system, and a portable solar fence charger, are finding good money in selling the services of their four-legged mowing crews. In many western states, where overly mature grass becomes a fire hazard, individuals and government entities are hiring shepherds to keep the grass young and vegetative, thereby reducing the fire hazard that the overly mature grass poses.

Ram Rental

Providing breeding services to people who have just a few sheep for "lawnmowers" and do not want to keep their own ram is a little business all in itself. We've been on both ends of ram rental over the years and have found it to be a convenient and reasonable method of making a little extra money off a ram we already owned or for getting some new blood into our sheep without having to go out and purchase another ram. If you have a ram you're willing to rent, look for folks in the area with four or six sheep; microflock owners typically do not want to be bothered keeping a ram and would much rather pay for the use of yours. The rental, for money or for a choice of one of the lambs, can help pay the keep of an extra ram that you might want to keep for yourself to give yourself more breeding options.

Purebred or Specialty Breeding Stock

When raising purebred and/or registered sheep on a small scale, you should try to realize some profits from the sale of breeding stock, though for most breeders in the class, those sales won't offset the extra expenses incurred for maintaining registered stock. Purebred and registered sheep cost more initially, and receipts for the sale of wool and meat aren't substantially higher than what you'd receive with less expensive breeding stock. Maintaining a registered flock involves extra expenses — both in time and money — for record keeping, registration fees, and advertising. You need experience with sheep or good planning skills to improve the flock or even just to keep it from deteriorating. Beginners are often advised to start with less expensive sheep to minimize the loss that may result from inexperience.

After a year or two of raising no-breed-name sheep, it is much easier to decide on the breed that offers you the most potential for profit, given your particular interests. Knowing how much time you can spend with them helps to decide whether the most prolific breeds, which require more attention at lambing time, would be suitable for your situation.

If you are buying purebreds and plan to sell them, try to select a breed that would appeal to a market with which you are familiar, if possible, as well as one suited to your area. Some unusual breeds are in great demand for noncommercial raising, with good sales for breeding stock. Some breeds thrive at high altitudes, some do well in heat, and others prefer cooler climates. Some graze well on rolling hills, and some are more at home on flat meadows. Some breeds can tolerate abundant rainfall; others would suffer with hoof problems and fleece rot if there was too much rain.

Club Lambs

Club lambs are those that are heading for youth program participants — especially 4-H members. They look for carcass characteristics, as opposed to specific breed characteristics, so club lambs don't necessarily have to come from registered flocks. If you are raising lambs that show good carcasses, then you might want to get the word out to Extension personnel.

Club lambs can increase your income because they sell a little higher than market lambs at the same weight (at weaning, usually about 30 to 40 pounds (13.6 to 18.1 kg). But remember, this generally isn't a very big market (go to your county fair and see how many kids are actually showing lambs to give you an idea of the potential market in your area), so don't expect to make your living off club lamb sales.

Mowing Services

This is definitely becoming a more viable income stream for owners of various-sized flocks. It also reduces costs for many shepherds by providing feed that they not only don't have to buy, but get paid to take away! Such shepherds provide a service and are fully responsible for setting up fences and caring for their sheep. Mowing enterprises can run the gamut from taking half a dozen sheep around town for mowing people's yards to large-scale projects. In Canada, California, and New Hampshire, there are big operations on land owned by public utility companies, cities, forests, parks, and private timber companies. For example, Public Service of New Hampshire is paying for sheep to maintain the area under its power lines; they figure sheep are cheaper than humans crews and "greener" than chemical sprays. (See the story on pages 142–143.)

Odds and Ends

There are a few more ideas for marketing that are worth mentioning. The ones you choose depend on where you live, your personality, and your interests.

Manure

Sheep manure is a potential source of income because it can be sold or used in your own garden. It not only stimulates the crop growth, but also adds valuable humus to the soil, which chemical fertilizers do not. You don't have to be modest about proclaiming the superiority of sheep manure over that of other animals, as the accompanying USDA chart will show:

Chemical Content of Sheep Manure

Type of Manure	Pounds per Ton of Manure		
	Nitrogen	Phosphorus	Potash
Sheep	20	9	17
Horse	11	6	13
Cow	9	6	8

Because sheep make use of ingested sulfur compounds to produce wool, their manure does not have the unpleasant-smelling sulfides found in cow manure. It is also in separate pellets or in pellets that hold together in a clump, and thus is less messy in the garden. Sheep manure doesn't even need aging. If you gather it for your own garden, take it first from paths and places where it does not help to fertilize the pasture. Since it contains many of the valuable elements taken from the soil by the plants eaten by the sheep, it is convenient that they spread a lot of it on the pasture. Its pelleted form causes it to fall in the grass instead of lying on top where it might smother the vegetation.

Clean out the barn twice a year, in spring and fall — the wasted hay and bedding left on the barn floor makes great fertilizer because it has absorbed much of the manure and contains valuable nutrients. Being inside, these nutrients are undamaged by rain and sunshine and are just waiting to be reclaimed. Spread a thick mulch of this on a portion of the garden, and don't even dig it in — just set out tomato, zucchini, and cabbage plants in holes in the mulch. By only mulching half of the garden each year, you always have one heavily mulched side for setting out plants and another half to dig up and plant seeds.

Homemade Soap

Homemade soap is one of the "good things" of life — and a profitable small item to add to any product line of sheep-related merchandise. You can make a lot of soap with the fat from lamb or mutton that has been trimmed for locker packaging. Have the slaughterhouse save all the fat trimmings. Some places will grind them for you, which makes the rendering easier. The first step in soap making involves preparing the tallow.

1. **Render the tallow.** Cut up chunks of lamb or mutton fat (tallow), put it in a large kettle, and cook it slowly over low heat. It will take several

hours for a large batch, but don't rush or you'll risk burning it. When the tallow is pretty well melted down, strain it through a cloth.

2. **Purify the tallow.** Boil the fat that you've rendered with about twice its volume of water. Strain it, and set it aside to cool. The clean fat will rise into a solid block. When it has cooled and hardened, remove from the water, turn upside down, cut in wedges, and scrape off the residue of impurities from the bottom. This purified tallow keeps for several weeks in the refrigerator.

Sophia Block's Lamb Tallow Soap Recipe

1. Measure 6 pounds (2.7 kg) of clean purified tallow. Heat it slowly in a large enamel pan to between 100 and 110°F (37.8 and 43.3°C).

2. Put 2½ pints (5 cups) (1.2 L) of water in a smaller enamel pan. Put the pan on a protected surface. Stand back, and slowly pour in one newly opened can of lye (you must use lye, not a chemical drain opener). Turn your face away to avoid breathing the caustic fumes. The lye will heat up the water. Allow it to cool to 98–100°F (36.7–37.8°C). Use a candy thermometer, suspended from the side of the pan and not touching the bottom of the pan.

3. When the lye is at the proper temperature, pour it into a half-gallon (1.9 L) (that is, magnum) liquor bottle by using an agate funnel. Now put the opening of this bottle on the rim of the pot of tallow, and pour the lye mixture very slowly in a thin stream. At the same time, slowly and gently stir the fat and the lye together slowly and gently. It's easier if you have a helper to pour in the lye. The tallow should be at the right temperature (100–110°F [37.8–43.3°C]) and the lye poured into it in a very thin stream. Stirring must be done slowly, very gently, and steadily. If the lye is poured in too fast or the stirring is not slow and gentle, the soap will separate or curdle and you will ruin the whole batch. Stir slowly for 20 minutes, and then pour into containers prepared in the manner discussed in the next section.

Soap Containers

Agate photo-development pans are ideal soap containers. Or use wooden boxes lined with brown paper or with clean cotton cloth, wetted down with water and wrung out. Have the paper or cloth folded out over the outside edge, to make the soap easy to remove when you are ready. You can use cardboard boxes, lined with plastic wrap, which is turned back over the outside edges and stapled to hold it in place while you are pouring the soap.

Pour the soap into these prepared containers and cover the soap with a board or heavy cardboard and then with a blanket. Covering keeps the soap from cooling too fast. Allow it to cool and harden for a day or two in a warm place, away from drafts. The soap will begin to lose its sheen as it hardens. After 2 or 3 days before it gets too hard, you can remove it from the boxes. Cut it into separate bars to age for several weeks, or months, before use. It can be cut neatly with a fine, taut wire wrapped around it and pulled tight. Age these bars unwrapped, with air circulating around them, for several weeks. Any liquid that appears on or in the soap is free lye, and you should discard the soap or reprocess it.

Soap Variations

Mutton tallow soap is often called saddle soap because it cleans and preserves leather so well. It can be used equally well as a bath, laundry, or dishwashing soap, but by making a few variations it can be even more suited to different uses.

Perfumed soap. Add oil of lemon, oil of lavender, or other oil perfumes (not any containing alcohol), or boil leaves of rose geranium and use this "tea" as part of the cold water used with the lye. Reserve part of the lye-dissolving quantity of water, boil the perfuming leaves in it, and add it to the dissolved lye when it has cooled a little. Since soap absorbs odors, it can be perfumed easily after it is in bars (and aged at the same time) by wrapping it in tissue that has been wet with perfume and dried out.

Green soap. Can be made with vegetable coloring obtained by pounding out a few drops of juice from beet tops, or use the vegetable coloring sold for baking.

Mint soap. Use 1 cup (236.6 mL) less water to dissolve the lye. Use this cup of water to make a very strong tea from fresh mint leaves. Add this back to the dissolved lye mixture before adding it to the tallow. Check temperature of lye liquid after adding the mint.

Deodorant soap without chemicals. You can use up to 2 ounces (59.1 mL) of vitamin E oil in your soap recipe, adding it to the mixture after stirring in the lye. It has a mild deodorizing quality, which will prevent any slightly bacony odor if you have used bacon fat along with your tallow; vitamin E is also an antioxidant.

Honey complexion soap. Add 1 ounce (29.6 mL) of honey, and stir it slowly into the soap after adding the lye and before pouring the mixture into the molds.

Laundry soap. To make laundry soap flakes or powder, let the soap age for 3 or 4 days. Grate it on a vegetable grater. Dry the flakes slowly in the oven set at warm, about 150°F (65.6°C), stirring occasionally. The soap can be pulverized when very dry or just left in flakes.

Dishwashing jelly soap. Shave 1 pound (0.5 kg) of hard soap, and boil it slowly with 1 gallon (3.8 L) of water until it is dissolved. Put it into covered containers. A handful dissolves quickly in hot dishwater. For many soft-soap and hard-soap recipes and variations, see the soap- (and candle-) making books in the resources.

Mutton Tallow Candles

Candles are another good use for the lamb or mutton fat. While not quite as practical as soap, candles are a fun way to use excess fat to give as gifts or to sell.

Candle Wicking

Prepared wicking can be purchased (see the resources), but it is simple to make your own from cotton string. You can make a good soaking solution from 8 tablespoons (118.3 mL) of borax dissolved with 4 tablespoons (59.1 mL) of salt in 1 quart water (0.95 L). Soak the wicking in this solution for 2 or 3 hours, then hang out to dry. Some old-time candle makers soaked the wicking in apple cider vinegar or turpentine.

Mutton Tallow

Cut up chunks of mutton or lamb fat, put it in a large kettle, and fry it slowly over low heat as you would for soap. Skim the fat as it rises to the top. Stir occasionally — do not rush the process or you'll burn the fat. A large batch takes several hours to render. When the melting is pretty well complete, strain it through a cloth.

Purifying

In a large kettle, dissolve 5 pounds (2.3 kg) of alum in 10 quarts (9.46 L) of water by simmering. Add the tallow, stir, and simmer about an hour, skimming the fat. This not only purifies the tallow, but also makes a slightly harder texture for use in candles. Cool the tallow, and when you can touch it comfortably, strain it through a cloth and set it aside to cool completely and harden. Scrape off the impure layer on the bottom.

Purified tallow can be stored in a cool place for a week or so until you are ready to make candles. It can also be refrigerated or frozen.

Tallow burns with a less pleasant smell than wax or paraffin. Adding a few drops of pine oil or some other scent after the tallow is melted and before dipping or molding the candles can perfume it.

Candle Dipping

Melt the purified tallow, and pour it into a wide-mouth jar or container placed in hot water to keep it liquid. Next to this container, have another one filled with very cold water, standing in a pan of ice to keep it cold. Since tallow candles have a tendency to droop in hot weather, don't make your candles too long.

Cut a wick about 6 inches (15.2 cm) longer than you want the candle to be, and tie one end of the wick to a small stick. If your containers are large enough, you can tie on several wicks and dip these all at once.

Dip the wick first into the hot tallow. Withdraw it, and let it air harden for a minute. Then dip it in and out of the ice water, which hardens it. Let it drip thoroughly. Keep repeating this process. To make a tapered candle, do not dip all the way to the top each time. Since each single dip into the tallow deposits such a thin layer on the candle, it takes a lot of dippings.

Molded Candles

It is quicker to mold candles. For candle molds, use plastic or paper cups or cut-down milk cartons. They can be sprayed with a nonstick baking spray (the lecithin-based type) to keep the candles from sticking to the mold or just brushed with cooking oil. Metal molds should be both oiled and chilled before you pour in the tallow. There are silicone-type preparations that are also used for "mold release." As with dipped candles, a shorter and wider shape is best when using tallow, which is not as firm as a wax.

Since the bottom of the mold will be the top of the candle, the wick should be threaded out through the bottom and protrude about an inch (2.5 cm). This is easily done when using paper or plastic containers for molds. If you can't make a hole in the bottom of your mold, leave a little coil of extra wick in the bottom that you can pull out when the candle is removed. If you have a wick sticking out the bottom of the mold, knot it there so you can pull it straight and tight while pouring in the tallow. It could be fastened at the top to a wire or stick that rests on top of the mold to keep it straight and centered in the candle until it hardens.

Colored Candles

Stir in 2 teaspoons (9.8 mL) of powdered household dye, such as Rit or Diamond Dye, for each pound (0.5 kg) of tallow, and mix well into the liquid tallow.

Sheep Carpentry

This book has plans for building various pieces of sheep equipment, and there are plan-services booklets available (see the resources). There is always a need for useful equipment, and you could find a ready sale for duplicates of the pieces you make for your own use.

Teaching and Writing

Whatever you do well can always be a source of income by teaching or writing about it. A well-organized small sheep operation can offer farm lecture-tours for a fee. “Sheep lore,” on-farm classes can be a day class or a weekend bed-and-breakfast offering. Wool handcrafts, such as spinning and weaving or locker hooking, can be the meal for your millstone: Teaching classes at a community college or giving private lessons or writing articles for a magazine all provide extra income and may create a market for your fleeces. The only limitation here is your own imagination, inventiveness, and promotion of your availability.

Agritourism

Speaking of bed and breakfasts and on-farm tours or classes, “agritourism” is a growing niche that producers are taking advantage of. With the population losing its connection to the farm, people are looking for ways to reconnect. Offer them a chance to come to your farm and hold a lamb; they’ll pay dearly for the experience. If you’re considering agritourism, check with your insurance agent to determine whether you need additional liability coverage.

Recipes

If you are marketing meat, then learning to cook it yourself in all types of recipes is important. Unless you’re dealing with an ethnic clientele, chances are your customers’ knowledge of how to cook sheep products is limited to lamb chops, shish kabob, and the occasional leg of lamb. You need to be able to educate them about all the ways they can prepare and serve your meat.

There are a bunch of recipes here to get you started, but if you’re looking for more inspiration, check out Jill Stanford Warren’s book, *Lamb Country Cooking: Lamb with All the Trimmings* (see the resources). Jill not only presents some great lamb recipes, but also offers serving suggestions, menus, accompaniments, and other good advice.

Smoked Leg of Mutton "Ham"

1 leg of mutton
Cold water

GLAZE

½ cup (118.3 mL) brown sugar, firmly packed
1 teaspoon (4.9 mL) prepared mustard
1 cup (236.7 mL) orange or pineapple juice
Cloves

Soak mutton in cold water for 1 hour. Dry with paper towels, and wrap securely in a large piece of aluminum foil. Seal edges well, and place in a baking dish. Bake at 350°F (176.7°C) for 30 minutes to the pound (0.5 kg).

Mix together brown sugar, mustard, and fruit juice. Place the precooked leg in a baking dish. Score outer covering with knife, and pour juice mixture over it. Stud with cloves. Bake the leg for an additional 40 minutes, basting often with pan juices. Serve hot or cold.

Note: Simmering may be preferred for the first stage. Soak mutton as above. Plunge into large pan with warm water. Bring to a boil, and simmer for 30 minutes per pound (0.5 kg) or until tender. Allow leg to cool in the liquid. Drain and refrigerate, covered (do not freeze) until needed, then bake with glaze.

In our experience, the yearling or 2-year-old cut is very tender and tasty. Old, old ewes are tasty and but not very tender. So, cut the real old ham into several pieces that will fit into your pressure cooker. Do the 1-hour soak, pressure cook for 15 minutes at 15 pounds of pressure then bake the "ham." It does not need to be baked for long baking and will be both tasty and tender. Grind up leftovers for ham hash. Cook split peas with the bone.

Mutton ham recipe is from the Australian Meat Board, as printed in Shepherd magazine *in 1973.*

Breakfast Sausage

Serves 5–6

1 pound (0.5 kg) lean ground lamb or mutton
⅛ teaspoon (0.6 mL) coarsely ground black pepper
½ teaspoon (2.4 mL) salt (or more)
¼ teaspoon (1.2 mL) powdered marjoram
¼ teaspoon (1.2 mL) powdered thyme
¼ teaspoon (or more) (1.2 mL) powdered sage
¼ teaspoon (1.2 mL) savory seasoning

Mix all ingredients together thoroughly. Cover bowl, and place in refrigerator overnight. To use, shape into patties about ½-inch (1.3 cm) thick. Cook over moderate heat in a heavy skillet until brown. Turn. Brown other side, lower heat to cook through. If you like your sausage a little more moist, you can add about 2 tablespoons (29.6 mL) water and cover the skillet when you lower the heat to cook. For a larger quantity for freezing, add a little ice water and mix in with sausage, so it doesn't crumble when you defrost and cut it into slices.

This recipe was printed in Shepherd *magazine in April 1972.*

Hasty Hash

Serves 4

1 pound (0.5 kg) ground lamb or mutton
1 tablespoon (14.8 mL) vegetable oil
1 small onion, chopped
½ teaspoon (2.4 mL) salt
⅛ teaspoon (0.61 mL) garlic powder
½ teaspoon (2.4 mL) freshly ground black pepper
4 tablespoons (59.2 mL) soy sauce
2 cups (473.2 mL) raw potato, shredded (or defrosted frozen hashbrowns)

Sauté meat with oil until pink color leaves; add onions and sauté until onions are transparent. Separate meat with a fork, as it cooks. Stir in salt, garlic powder, pepper, and soy sauce. Mix. Layer potatoes on top of meat, cover pan, and cook on medium-low heat for 20 minutes, stirring gently from time to time. Uncover and turn heat up a little. Stir and cook until potatoes are beginning to get brown. Good with catsup.

Vi's Tamale Pie

Meat mixture makes two tamale pies; freeze half for later use.

3 pounds (1.4 kg) ground mutton or lamb
1 green pepper, chopped
1 large onion, chopped
1 15-ounce (443.6 mL) can tomatoes
1 15-ounce (443.6 mL) can tomato sauce
1 6-ounce (177.4 mL) can tomato paste (optional)
1 cup (236.6 mL) sliced ripe olives (or more)
2 teaspoons (9.8 mL) sugar
1 teaspoon (4.9 mL) salt, or to taste
1 tablespoon (14.8 mL) chili powder
1 tablespoon (14.8 mL) powdered cumin
Cayenne pepper (optional)
Cornmeal crust (see below)

Sauté meat, green pepper, and onions together until meat loses its red color. Add tomatoes, tomato sauce, tomato paste, olives, sugar, and salt to taste. Add chili powder and cumin seasoning. Simmer 15 minutes, and then taste for seasoning. Add more chili powder if desired. Add some cayenne pepper if you want it hotter. Divide meat mixture, freezing half for later use. Continue with Cornmeal Crust below.

Cornmeal Crust

Serves 6

1 cup (236.6 mL) yellow cornmeal
1 teaspoon (4.9 mL) salt
½ cup (118.3 mL) shredded cheddar cheese
1 cup (236.6 mL) cold water
1 cup (236.6 mL) boiling water

Mix cornmeal with 1 cup (236.6 mL) cold water, then stir it into 1 cup (236.6 mL) boiling water. Cook slowly, stirring constantly, until thick. (If cooking in microwave, stir every 30 seconds until thick.) Spread cornmeal in shallow baking dish, reserving ½ (118.3 mL) cup to decorate the top. Spread meat mixture over cornmeal. Decorate edge with small spoons of cornmeal. Bake 30 minutes at 350°F (176.7°C). Spread shredded cheese on top, bake 5 minutes more. Let stand for 10 minutes before serving.

Garden Meat Loaf Squares

Serves 8 (or it can serve 4; is good reheated, so just save the rest for another meal.)

2 tablespoons (29.6 mL) vegetable oil
⅔ cup (157.7 mL) chopped onions
1 cup (236.6 mL) fresh string beans, cut small, or drained canned beans
½ cup (118.3 mL) green pepper, chopped
1 cup (236.6 mL) celery, chopped
2 pounds (0.9 kg) ground lamb or mutton
1 cup (236.6 mL) bread crumbs
2 teaspoons (9.8 mL) salt
½ teaspoon (2.4 mL) freshly ground black pepper
1 tablespoon (14.8 mL) Worcestershire sauce
1 teaspoon (4.9 mL) soy sauce
1 egg, beaten lightly
⅔ cup (157.7 mL) tomato juice
Garnish, catsup or chili sauce

Sauté onions, beans, pepper, and celery in oil until tender. Mix meat, bread crumbs, salt, Worcestershire sauce, soy sauce, egg, and tomato juice. Mix in the vegetables.

Press mixture into a 9 x 13 x 2 inch (22.9 x 33 x 5.1 cm) pan. Bake 30 minutes at 350°F (176.7°C). Spread top with thin layer of catsup or chili sauce, bake 5 minutes more. Cut into squares to serve.

Sloppy Joes

Serves 8

1 pound (0.4 kg) ground mutton or lamb
1 tablespoon (14.8 mL) dried onion flakes or ½ small onion, chopped
1 teaspoon (4.9 mL) garlic salt
¼ teaspoon (1.2 mL) curry powder
¼ teaspoon (1.2 mL) ginger
½ teaspoon (2.4 mL) coarsely ground black pepper
1 can (6 ounces) (177.4 mL) tomato paste
½ cup (118.3 mL) water
2 tablespoons (29.6 mL) brown sugar
3 tablespoons (44.4 mL) lemon juice
1 tablespoon (14.8 mL) soy sauce
2 teaspoons (9.8 mL) chopped dried or fresh parsley
1 teaspoon (4.9 mL) Worcestershire sauce
8 sliced burger buns, toasted

Brown ground meat and onions, pour off extra fat. Season. Combine tomato paste and rest of ingredients, except buns, and add to meat mixture. Bring to a boil. Serve meat mixture on buns, ¼ cup (59.1 mL) of mixture to each bun.

Anna's Casserole

Serves 4

1¼ pounds (0.6 kg) lean ground mutton
1 onion, chopped, or 2 tablespoons (29.6 mL) dried onion flakes, divided
1 teaspoon (4.9 mL) lamb seasoning salt, divided
1 10-ounce (295.7 mL) package frozen peas, defrosted
2 cups (473.2 mL) celery, thinly and diagonally sliced
½ teaspoon (2.4 mL) freshly ground black pepper
1 10¾-ounce (318 mL) can cream of chicken or cream of mushroom soup
1 ⅞-ounce (55.5 mL) package crushed barbecue potato chips
Paprika

Sauté ground mutton and half of the onion in skillet until lightly browned, seasoning with half of the lamb salt, breaking apart with a fork as it cooks. Drain fat. Spoon meat into medium-size loaf pan. Scatter defrosted peas over the meat. Layer the celery on top of the peas. Mix pepper and the rest of the onion and seasoning salt with the can of soup, spread on top. Put crushed potato chips on it, and sprinkle well with paprika. Bake about 30 minutes at 375°F (190.6°C).

Oregon Lamb or Mutton

Serves 4–6

3 tablespoons (44.4 mL) all-purpose flour
1 teaspoon (4.9 mL) dry mustard
1 teaspoon (4.9 mL) salt
½ teaspoon (2.4 mL) coarsely ground black pepper
4 lamb shanks, split *or* 3 to 4 pounds (1.4 to 1.8 kg) lamb neck, sliced *or* 2 to 3 pounds (0.9 to 1.4 kg) lamb steaks, chops, *or* shoulder roast
Vegetable oil for browning
1 10½-ounce (310.5 mL) can consommé
1 10½-ounce (310.5 mL) can cream of mushroom soup
1 tablespoon (14.8 mL) Worcestershire sauce
1 tablespoon (14.8 mL) Kitchen Bouquet
⅛ teaspoon (0.6 mL) garlic powder
½ teaspoon (2.4 mL) curry powder
½ cup (118.3 mL) white wine
4–6 servings rice, noodles, or potatoes, cooked

Put flour, mustard, salt, and pepper in a paper bag. Add lamb, and shake to coat. Brown lamb in hot oil in nonstick skillet and place in slow-cooking pot. Combine all remaining ingredients except rice, and add to pot; sauce should cover about ¾ of the meat. Cook 1 hour on high, then cook on low for 6 to 8 hours, or until lamb is extremely well done. Serve at once, or pour off all liquid and chill it until the fat can be skimmed off. Pour remaining liquid back over meat and reheat. Serve on rice.

Recipe from My Secret Cookbook *by Paula Simmons, Pacific Search Press, 1979.*

TWELVE

SHOWING SHEEP

Some folks are interested in the world of showing sheep, either for themselves or for their children. If you think showing might be for you, just keep in mind that it requires a serious commitment. Care of show sheep begins months before the show with grooming and training. There are extra costs associated with showing, such as entry fees, transportation, additional vaccinations, and supplies for *fitting* (or grooming) your show sheep. At the same time, showing can be rewarding, especially for children, who learn responsibility by showing their animals.

This chapter is a brief introduction on how to show sheep — it isn't intended to be comprehensive. *The Banner* is a general sheep magazine that's really aimed at those who show sheep. (See the resources.)

Kinds of Shows

Shows may be shows only or they may be a combined show and sale. They fall into two main categories:

- Registered breed shows, which are sponsored by the breed's association or some other national sheep organizations. These shows are often held in conjunction with state fairs and special expositions, like the North American International Livestock Exposition or the National Western Stock Show. All animals must have current registration papers, and those papers must be up to date and accurate.
- There are some national shows that are not breed specific, but most "open-class" shows are local and are often sponsored through the

Cooperative Extension Service and county fair system. Some of these open shows are for adults, but most are for youth programs, like 4-H and Future Farmers of America. In these shows, animals are generally not required to have registration papers from a particular breed association, though if one or two breeds are very popular in the area, there may be "classes" in which registered animals compete. These breed classes may or may not be sanctioned by the breed's association.

Each show has its own rules and regulations. These rules tell you everything from when entry fees and forms must be submitted, to what health certificates you need to bring, to what criteria the judges use. A copy of the rules should be acquired and studied well ahead of time so your sheep aren't disqualified for some infraction. If you're planning on showing in more than one show, get the rules for each — don't assume that they're all the same.

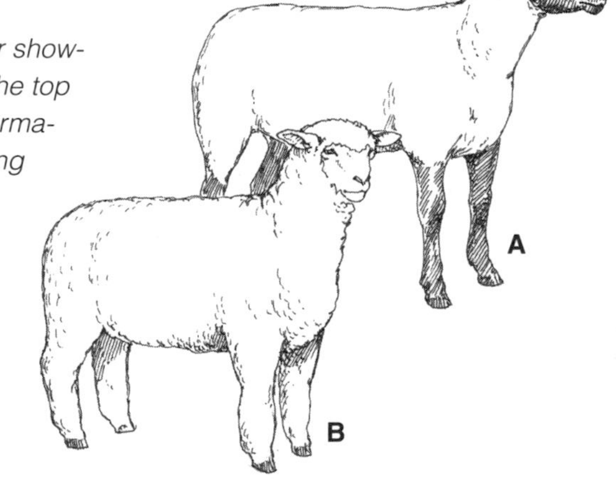

*When buying or raising sheep for showing, conformation is important. The top sheep (**A**) shows excellent conformation for a meat breed, with a strong straight back, whereas the lower sheep (**B**) shows the correct conformation for a wool breed, with a slightly sloping back.*

Show Classes

In shows, there are generally various classes of sheep competing against each other. The classes that have been established for a given show may relate to the age, sex, production type, or breed of sheep, or they may relate to the person doing the showing. Open classes are those that are available for various breeds and to any competitor within the appropriate category — for example, any woman can compete in an open "ladies lead" class or any child can compete in an open "market lamb" class. A closed class is limited to participants who meet certain requirements, for instance, a 4-H show is usually only open to children who've participated throughout the year on a sheep project through a local 4-H group.

SHEPHERD STORY

Rod Crome's parents ran a diverse livestock farm in southern Illinois and his father was a well-known and well-respected showman and breeder of sheep. Rod followed in his father's footsteps — quite literally — from the time he was 3 years old, accompanying his father to shows throughout the United States.

Rod's own career in showing began with 4-H, and by the time he was in high school, he was showing in open classes at state fairs around the United States. Rod still spends time in the ring, but now he's often there as a judge. He has judged at national shows and state fairs throughout the country.

"When I'm judging, I know the kids are nervous, but what they need to think about is that the show is really about the sheep! The judge spends 90 percent of his or her time looking at the sheep and studying them and only about 10 percent of the time actually handling the sheep. You'll get the attention of a judge by being proud of your animal, not by being a show-off.

"My dad always said, 'Try to get the most out of your animal, but try to make it look easy.' That takes practice; you have to know your animal really well, so you know when it's looking its best, and you know when you need to reposition it."

Rod says that the best asset a showman can have is "confidence without arrogance," and again, he says that comes with practice.

Some of the characteristics that Rod looks for when judging sheep are:

- The sheep should be true to type for the breed, and meet the breed's characteristics
- The sheep should have a good, long, straight top (the line along the spine, or backbone)
- The sheep should have good, sturdy feet and legs with short pasterns, and the legs should be well set under the four points of the body
- The sheep should be straight and wide at the dock

Training Sheep

So, you have a sheep you want to train to take to shows — let's call her Fluffy. How hard could it be to train a sheep anyway? They're not very big, and you've watched a few competitors. They made it look easy: They walk around the ring with their sheep cooperatively trotting along on a halter, they stand for the judges and their sheep stand still — no sweat. But like anything else you do with critters, you quickly learn that Fluffy has a mind of her own. In this case, she has no burning desire to be a cooperative show sheep. She must be well trained if you expect her to perform in the showring. And even if she's doing pretty well walking around on a halter at home, the first time she's in the spotlight she may become scared or confused and not do what you want her to do.

Training should begin 2 to 3 months before the first show. As with any type of animal training, frequent — but short — sessions are better than infrequent, long sessions. Fifteen to 20 minutes per day, 4 to 5 days per week accomplishes far more than one 2-hour session per week.

You need patience when training any kind of animal, and this is especially true for sheep. If Fluffy doesn't follow your lead, don't yell or scream, drag her off her feet, or hit her. Stand still, and hold her head upright, close to where the lead rope connects to the halter. Once she settles down, try again. When she does well, give her a pat, a word of encouragement, and an occasional treat, like a piece of apple.

Leading on a Halter

The first step is to lead the sheep on a halter. Halter training can actually begin with lambs as young as 1 month old. In fact, if a child is going to do the training, lambs of this age are very good candidates for halter training. Children that are trying to train a lamb for the first time should be carefully supervised, so they don't hurt it. Conversely, a mature sheep should be trained by an adult to protect the child. Halters can be purchased or made out of rope.

Sheep are led from the sheep's left side. Hold the lead rope close to the halter, usually about 1 foot (30 cm) from the sheep's head. Begin walking forward. At first, Fluffy will pull back against the halter and fight the movement.

Once Fluffy begins to get the hang of walking willingly on a halter and lead rope, have your helper work with you to acclimatize her to the noises and sights of a showring. While you walk with Fluffy, your helper can make noise, rustle paper, or shake jackets just at the edge of her flight zone; shine bright flashlights in her face; and create other distractions that are similar to the showring.

Handling Obstinate Students

If Fluffy fights too much, have a second person walk behind her about 6 or 8 inches (15.2 or 20.4 cm) and tickle her butt with a switch (which can be cut out of a thin branch). *The person doing the tickling doesn't need to beat Fluffy!* Just tickle with the switch. If you're forced to work alone, you may be able to this yourself, but it's easier with a second person. Another approach to an obstinate student that won't walk forward (especially if you're alone) is to lay a hand on her dock, or tail, and push her forward. Although this method works, if you have to keep at it for any length of time, it becomes back numbing.

Training to Stand

The second part of the training regimen is teaching Fluffy to stand correctly. During a show, both you and Fluffy are required to stand for the judge's close inspection. Part of what the judge is looking for is the animal's conformation, but he also judges on how well the animal stands in one position and on the correctness of its position.

In many shows, when you hold the sheep in the ring, you're expected to be able to do so without the assistance of a halter. (The exception is young exhibitors, who may be allowed to continue using the halter throughout judging.) But when you first begin practicing, it may be a good idea to leave the halter on, so you have control — just in case Fluffy decides to take a hike.

The correct standing position for Fluffy is with her head held high and her legs planted firmly on the ground, placed squarely at the four corners of her body. She shouldn't be overstretched in any direction or all scrunched up. The correct position for you is typically squatting though some shows allow you to stand. Shows in which younger children can participate generally allow them to stand throughout.

You hold Fluffy, by facing her on her left side and cupping your left hand under her chin. Your thumb and fingers exert slight pressure in the indentation behind her teeth. Control improves if you cup your right hand behind her head — just behind her ears, though once you're in the ring, the judge scores higher if you maintain control with just one hand.

During a show, exhibitors stay low while the judge views their sheep. The exhibitor always faces the judge. When the judge comes for a close-up inspection, the exhibitor stays low but moves around the sheep as the judge moves around it, helping the judge to clearly observe the sheep from all angles.

Once Fluffy has learned to stand well, you can begin switching sides. Start out on her left side, then slide around in front of her, so you're holding her from her right side. In the ring, the judge will first look at one side, then the other side, and then front and back. You should remain facing the judge, which requires your being able to switch sides as he moves around.

Training Sheep for Handling during a Show

Judges come up and handle sheep during a show. They feel along the backbone and rump to feel how muscular or fat the sheep is. Teaching Fluffy not to be upset when strange people handle her just requires doing it — have someone she doesn't know come out and feel her body while she's standing still.

Sheep that tighten their muscles when the judge is handling them score better because muscles feel firmer. Getting Fluffy to tighten her muscles isn't too hard once she's trained to stand for handling. From a position in front of Fluffy, place your hands on either side of her neck and squeeze gently but firmly. This forces her head down and toward you, and she'll naturally tighten up.

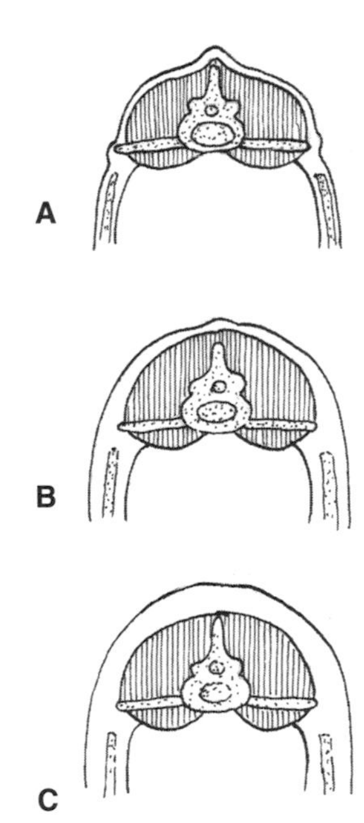

Judges typically run their hands down the sheep's spine and hips to feel for condition. The illustrations provide a cross-section view of the spine and hips. **A.** *This sheep is too thin, with too much backbone protruding.* **B.** *This sheep is a well-conditioned animal, with just enough meat covering the backbone.* **C.** *This sheep has too much fat covering the backbone.*

Fitting

In sheep-show lingo, *fitting* is the name given to preshow preparation: trimming feet, shearing, washing, and carding. Fitting doesn't start the day before the show — like training, it starts months before with early shearings, which result in the fleece being an optimum length for the show.

When early shearings actually begin depends on the class of sheep: For breeding animals, shearing is typically done the first time between 2 and 3 months before the show; market lambs are sheared about 2 weeks before. Putting a blanket or a coat on the sheep after shearing helps keep the fleece cleaner, making your later work a little easier, but it isn't absolutely necessary.

The fitting work that takes place just before the show is most easily accomplished with the aid of a fitting stand. A fitting stand is a wooden and metal contraption that holds the sheep still in a standing position while you work on it. You occasionally have to lean over a sheep in a fitting stand, but most of your work can be done from a standing position. Animals should never be left unattended in a fitting stand (or when tied off) because they may hurt themselves. Have the supplies you'll need at hand before you start, or have a helper who can retrieve that forgotten currycomb!

One or 2 days before the show, shear Fluffy's belly and inside her legs. Then she'll need a bath and a rinse. This is where your fitting stand comes in handy, but if you don't have one you can tie her off with her halter.

The shading indicates wool that is typically trimmed off when fitting a sheep for show. The goal in fitting is to have a nice consistent fleece length over the animal's whole body.

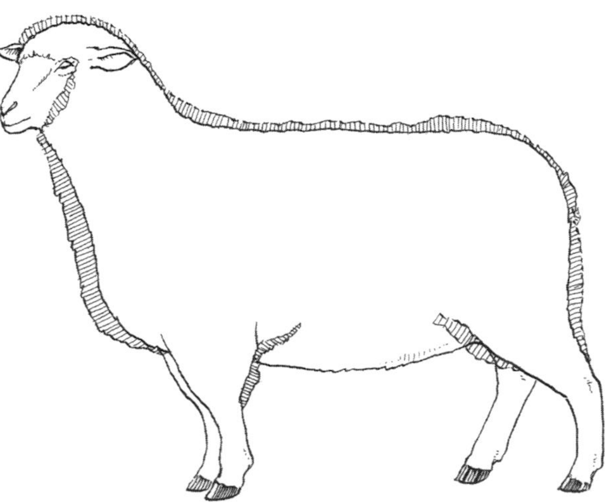

Warm water is preferable for washing, and as you're working around Fluffy's head, try to keep both the soap and the water out of her eyes and ears. (Sheep hate it as much as people do.) Use a hose with fairly light water pressure to wet her down.

Place a small cube of mild hand soap — about 1 inch (2.5 cm) square — in an old empty dish soap or squirt jar (this is part of the reuse, recycle, and reduce program), and fill with warm water. Shake well, then let it set for an hour or so. Shake it again to mix the soap and water, and begin applying evenly over the fleece with the squirt top.

Rub in the soap mixture well. An old terrycloth towel cut into a square works nicely. Make sure you get at those hard-to-hit spots, like the belly and insides of the legs.

Now it's time to rinse. Get all of the soap out of Fluffy's fleece or it will dry in a weird, patchy manner that won't win you any awards. Again, light water pressure is sufficient for rinsing.

When you're done rinsing, remove the excess water with a clean, round currycomb. Let her air dry until Fluffy's fleece is almost dry, and then place a show blanket on her to keep her from getting debris or dirt in her fleece.

She can be let out into a freshly bedded stall at this point to wait for her final fitting, which is usually done within an hour or so before you take her into the ring.

Final Fitting

The final fitting consists of trimming and carding, and carding and trimming. This is like Fluffy's day at the beauty salon.

Since market lambs were sheared 2 weeks before, they should still be pretty slick skinned. But if there are some spots where the shearer left a little more than other spots, use your hand shears, or blades, to clip them even.

Preshow Carding

After trimming, use a number 2 or number 3 hand-card to fluff out that fleece (use a number 5 on the head). When carding, work from back to front again, lifting and straightening the fibers. You can't overdo it. After you've carded the first time, (no, you're not done, silly) do it again. The carding should make any long spots stand out for more trimming. The whole trimming and carding process may be repeated three or four times.

Some market lambs are shown with a "poodle-cut." If you plan to show your lambs this way (and the show allows it), then tell the shearer not to clip the hips, rump, or dock in his earlier shearing. During final fitting, poodle cuts should be trimmed to about 1 inch (2.5 cm) long.

Breeding sheep usually have a nice regrowth of fleece since they were sheared 2 or 3 months ago. They should be trimmed so that their fleece is about 1 inch (2.5 cm) all the way around.

SHEPHERD STORY

Rebecca Roberts has been showing sheep since the ripe old age of seven and, as a result, is now a young woman with many great experiences. "My mom and dad both showed sheep when they were kids, so when I was seven, my dad asked me if I wanted to show sheep. He showed me a sheep magazine to decide what kind of sheep I wanted to show, and I picked out Suffolks, because I thought they were the prettiest sheep in the magazine.

"I showed in open classes until I was nine, than in open shows and 4-H. When I got to high school, I started showing in FFA [Future Farmers of America], too."

Showing has given Rebecca the opportunity to meet different people, to travel to many different places, and to learn life skills that she probably wouldn't have learned without her sheep project. In 1998–1999, Rebecca was the National Ambassador for the National Junior Suffolk Sheep Association. The experience of promoting her breed was really fun and educational. She traveled to events all over the country and considers a visit to the Iowa State Fair as the highlight of her year: "The Iowa State Fair is really big, but everybody was so nice, and welcomed me so warmly," she told me.

Now, thinking ahead toward college, she hopes to study agriculture and help promote the sheep industry in the future. Her sheep project will help her with that quest — she's already received one scholarship through 4-H, and she plans to look into more.

Rebecca says that even though it's really rewarding, it's also lots of work. She spends an hour a day doing the regular chores, such as feeding, but there's other work involved with showing. Her training schedule is year-round, since her show schedule is spread out throughout the year. Then there's paperwork, which needs to be done for maintaining her sheep's registrations and for signing up for shows.

Trimming is done most easily by working from back to front. Take little snips at a time, and be patient. Keep your blades clean while you work by dipping them in a bucket of hot soapy water as needed (the soap should be a strong dish detergent with a degreasing agent).

Showring Strategies

Watch some earlier classes of sheep being shown if you can. It will give you a chance to scope out the ring itself — you'll want to avoid any low or damp spots in the ring if you can. Watch how the judge works.

Be on time when your class is called to enter the ring. Move your animal into the line, leaving 2 to 3 feet (61 to 91 cm) between you and the next exhibitor. Get Fluffy standing well, and get into your squat next to her.

Follow any directions the judge gives, and if he or she has you move around, then get Fluffy resquared and squat down again, quickly, after he's placed you in a new position. Judges often ask about your sheep's breed, age, or some questions about your care and training. Be prepared to answer them by practicing before the show.

Always be polite to other exhibitors. And good luck!

LAMBING CALENDAR

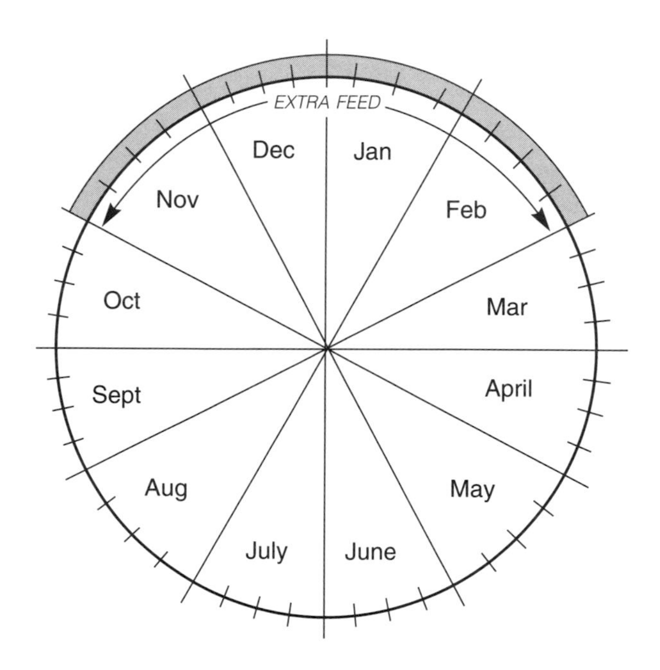

Template for personal use

A lambing calendar needs to be flexible so it can be adjusted to suit your specific needs. This circular calendar is designed with flexibility in mind. The template above is for your personal use; the sample calendar on the right is filled out according to the instructions beneath it. Note that in the example calendar, the shepherd has established mid-April as the desired lambing date. Once you've determined your desired lambing date, follow Steps 2 through 9.

How to Use the Template

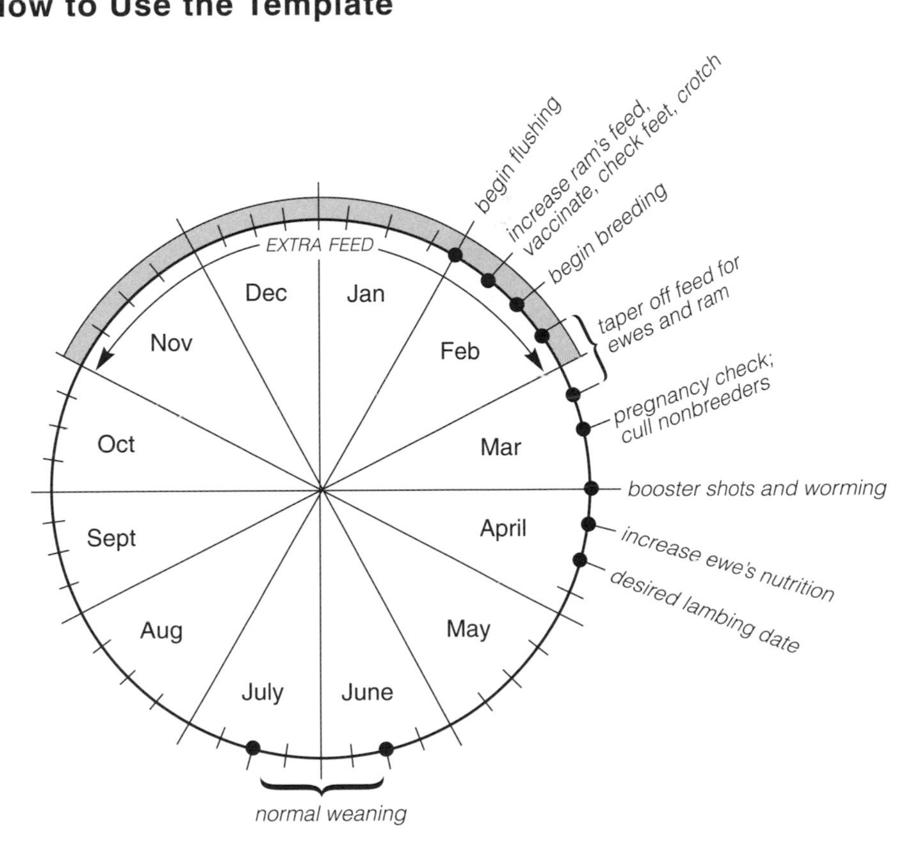

1. Record desired lambing date.
2. Count back (counter clockwise) 1 week, and note "increase ewe's nutrition."
3. Count back 1 week more and note "booster shots," if needed in your vaccination program, "and worming."
4. Count back 2 weeks more and note "pregnancy check; cull nonbreeders."
5. Count back 1 week more and then 2 weeks more. Bracket this period, and note "taper off feed for ewes and ram." You'll taper off the extra "flushing" feed at this time.
6. Count back 1 week more and note "begin breeding."
7. Count back 1 week more and note "increase ram's feed, vaccinate, check feet, crotch."
8. Count back 1 week more and note "begin flushing."
9. Count forward 8 to 12 weeks from desired lambing date. Note "weaning."

Feed Requirements for Sheep

For Growing Sheep:

Live Weight in Pounds	50	75	100	125
Dry matter lb	2.2	3.5	4.0	4.6
Crude protein %	12.0	11.0	9.5	8.0
Crude protein lb	0.26	0.39	0.38	0.37
TDN %	55	58	62	62
TDN lb	1.21	2.03	2.48	2.85
Energy Mcal	1.14	1.18	1.27	1.27
Calcium %	0.23	0.21	0.19	0.18
Phosphorus %	0.21	0.18	0.18	0.16

For Breeding Sheep:

	First Two-Thirds of Gestation	Last Third of Gestation	First 10 Weeks of Lactation	Last 14 Weeks of Lactation	Rams at Moderate Work
Dry matter lb per 100 lb of body weight	2.5	3.5	4.2	3.5	3.5
Crude protein %	8.0	8.2	8.4	8.2	7.6
Crude protein lb per 100 lb of body weight	0.20	0.29	0.35	0.29	0.27
TDN %	50	52	58	52	55
TDN lb per 100 lb of body weight	1.25	1.82	2.44	1.82	1.93
Energy Mcal per lb of feed	1.0	1.1	1.2	1.1	1.2
Calcium %	0.24	0.23	0.28	0.25	0.18
Phosphorus %	0.19	0.17	0.21	0.19	0.16

TDN = total daily nutrients

Note: Don't feed sheep feed formulas or mineral mixtures that are not specifically recommended for them. The amounts of some trace minerals, such as copper, that are in feed for other classes of livestock may be toxic to sheep.

Composition of Common Feedstuffs

Feed, Common Name	Description	Typical % Dry Matter (DM)	Crude Protein %, DM Basis	Crude Fiber %, DM Basis	Calcium %, DM Basis	Phosphorus %, DM Basis	Total Digestible Nutrients %, DM Basis	Digestible Energy, Mcal/lb
Forages								
Alfalfa	Fresh, vegetative	21	20.0	23	2.19	0.33	57–61	1.01–1.22
Alfalfa	Hay, early-bloom	90	18.0	23	1.41	0.22	55–60	1.00–1.31
Alfalfa	Hay, mature	91	13.0	38	1.13	0.18	50–55	0.90–1.10
Alfalfa	Silage	38	15.5	30	1.30	0.27	55–58	1.06–1.17
Bermuda grass	Fresh, vegetative	34	12.0	26	0.53	0.21	50–60	0.82–1.32
Bermuda grass	Hay	90	6.0	31	0.43	0.20	45–49	0.94–1.10
Bird's-foot trefoil	Fresh, vegetative	24	21.0	25	1.91	0.22	63–66	0.99–1.50
Bluegrass	Fresh, vegetative	31	17.4	25	0.50	0.44	56–72	0.92–1.40
Brome	Fresh, vegetative	34	18.0	24	0.50	0.30	68–80	0.90–1.26
Brome	Hay	89	10.0	37	0.30	0.35	54–55	0.99–1.29
Clover, red	Fresh, vegetative	20	19.4	23	2.26	0.38	57–69	0.92–1.39
Clover, red	Hay	89	16.0	29	1.53	0.25	49–60	0.91–1.37
Clover, crimson	Fresh, vegetative	87	18.4	30	1.40	0.20	49–57	0.92–1.39
Clover, ladino	Fresh, vegetative	19	27.2	14	1.93	0.35	56–68	1.13–1.57
Fescue	Fresh, vegetative	28	22.1	21	0.53	0.38	70–73	0.79–1.24
Fescue	Hay	92	9.5	37	0.30	0.26	48–62	0.82–1.24
Oat	Hay	92	4.4	40	0.24	0.06	40–47	0.81–1.22
Orchard grass	Fresh, vegetative	23	18.4	25	0.58	0.54	55–72	0.93–1.34
Orchard grass	Hay	91	8.4	34	0.26	0.30	45–54	0.86–1.38
Redtop	Fresh, vegetative	29	11.6	27	0.46	0.29	60–65	0.84–1.24
Redtop	Hay	94	11.7	31	0.63	0.35	54–57	0.90–1.15

Composition of Common Feedstuffs *(continued)*

Feed, Common Name	Description	Typical % Dry Matter (DM)	Crude Protein %, DM Basis	Crude Fiber %, DM Basis	Calcium %, DM Basis	Phosphorus %, DM Basis	Total Digestible Nutrients %, DM Basis	Digestible Energy, Mcal/lb
Forages								
Reed canary	Fresh, vegetative	23	17.0	24	0.36	0.33	47–75	0.91–1.10
Ryegrass, annual	Fresh, vegetative	25	14.5	24	0.65	0.41	50–60	0.79–1.24
Ryegrass, annual	Hay	88	11.4	29	0.62	0.34	52–57	0.70–1.12
Ryegrass, perennial	Fresh, vegetative	27	10.4	23	0.55	0.27	60–68	0.80–1.35
Ryegrass, perennial	Hay	86	8.6	30	0.62	0.32	45–60	0.80–1.20
Sudan grass	Fresh, vegetative	18	16.8	23	0.43	0.41	63–70	0.83–1.40
Sudan grass	Hay	91	8.0	36	0.55	0.30	55–56	0.87–1.12
Timothy	Fresh, vegetative	26	18.0	32	0.39	0.32	61–72	0.76–1.34
Timothy	Hay	89	9.1	31	0.48	0.22	45–60	0.78–1.31
Vetch	Fresh, vegetative	22	20.8	28	1.36	0.34	55–57	1.02–1.23
Vetch	Hay	89	20.8	31	1.18	0.32	67–72	0.91–1.10
Wheatgrass, crested	Fresh, vegetative	28	21.5	22	0.46	0.34	70–75	0.95–1.26
Wheatgrass, crested	Hay	93	12.4	33	0.33	0.21	50–53	0.85–1.11

Feed, Common Name	Description	Typical % Dry Matter (DM)	Crude Protein %, DM Basis	Crude Fiber %, DM Basis	Calcium %, DM Basis	Phos-phorus %, DM Basis	Total Digestible Nutrients %, DM Basis	Digestible Energy, Mcal/lb
Other Feeds								
Barley	Grain	88	13.5	6	0.05	0.38	80–84	1.34–1.75
Beet pulp	Dried with molasses	92	9.0	13	0.56	0.08	68–70	2.99–3.07
Brewer's grain	Dehydrated	92	30.0	14	0.33	0.55	64–68	1.14–1.60
Corn	Shell (grain)	86	9.0	2	0.03	0.27	78–79	3.45–3.48
Corn ears	Ground	87	9.0	9	0.07	0.28	74–83	1.36–1.70
Corn	Distiller's grains	94	23.0	12	0.11	0.43	70–86	1.25–1.75
Corn	Silage	33	8.1	24	0.24	0.22	66–71	1.32–1.42
Cotton	Seed hulls	91	4.1	48	0.15	0.09	33–42	0.65–0.97
Cotton	Seed meal	93	44.3	13	0.21	1.16	75–78	0.97–1.72
Cotton	Seeds	92	24.0	21	0.16	0.75	90–96	1.05–1.57
Oats	Grain	89	13.0	12	0.07	0.38	76–77	1.29–1.54
Rye	Grain	88	11.3	2	0.07	0.34	71–78	3.15–3.43
Soybean	Meal	89	50.0	7	0.33	0.71	82–86	1.22–1.71
Soybean	Seeds	92	43.0	6	0.27	0.65	56–64	1.67–1.88
Sunflower	Seeds, no hulls	93	47.0	11	0.53	0.50	61–68	2.67–3.01
Turnip	Roots, fresh	10	1.0	1	0.03	—	7–8	0.32–0.37
Wheat	Middlings	87	16.0	3	0.08	0.50	73–78	3.21–3.45

Note: The ranges for total digestible nutrients and digestible energy that are available in a feedstuff vary by species. As a rule of thumb, monogastric animals are at the low end of these ranges and ruminants are at the highest end of these ranges.

Resources

Books

Campbell, Stu. *Let it Rot*. Pownal, VT: Storey Books, 1998.
Great information about preparing and using compost.

Carroll, Ricki and Robert. *Cheesemaking Made Easy*. Pownal, VT: Storey Books, 1996.
A good starting point if you want to try making your own sheep's-milk cheese.

Cavitch, Susan Miller. *The Natural Soap Book: A Comprehensive Guide with Recipes, Techniques & Know-How*. Pownal, VT: Storey Books, 1995.
Basic soapmaking instructions and specialty techniques like marbling, layering, and making transparent and liquid soaps.

Damerow, Gail. *Fences for Pasture & Garden*. Pownal, VT: Storey Books, 1992.
This book is packed full of great advice and techniques for building all types of fencing.

Ekarius, Carol. *Small-Scale Livestock Farming*. Pownal, VT: Storey Books, 1999.
If you want to learn more about managed grazing, marketing, and general animal husbandry, read this book.

Fogt, Bruce. *Lessons from a Stock Dog*. Sidney, OH: Working Border Collie, Inc., 1996.
For any novice herding dog trainer, from the publisher of Working Border Collie Magazine.

Haynes, Bruce. *Keeping Livestock Healthy*. Pownal, VT: Storey Books, 1994.
Anyone raising livestock should have a copy of this book.

Hirning, H. J., et al. *Sheep Housing and Equipment Handbook*. Ames, IA: MidWest Plan Service, 1994.
Excellent plans for sheep buildings and all sorts of equipment from Midwestern Agricultural Colleges. For more information, contact MidWest Plan Service, Agricultural and Biosystems Engineering, 122 Davidson Hall, Iowa State University, Ames, IA 50011.

Martin, Ann. *Food Pets Die For: Shocking Facts about Pet Food.* Troutdale, OR: New Sage Press, 1997.
Good info if you want to "market" pet food.

Oppenheimer, Betty. *The Candlemaker's Companion: A Comprehensive Guide to Rolling, Pouring, Dipping, and Decorating Your Own Candles.* Pownal, VT: Storey Books, 1997.
A great book for candlemakers of all abilities.

Reavis, Charles. Home Sausage Making. Pownal, VT: Storey Books, 1981.
Sausage is a great way to use mutton!

Schroedter, Peter. *More Sheep, More Grass, More Money.* Manitoba, Canada: Ramshead Publishing, 1997.
A great little book about how this Canadian shepherd turned his operation into a profitable enterprise by using spring lambing on pasture.

Simmons, Paula. *Spinning and Weaving with Wool* (rev. ed). Unicom Books & Crafts, 1991.
The best book about spinning and weaving for beginner and experienced spinners and weavers.

——. *Turning Wool Into a Cottage Industry.* Pownal, VT: Storey Books, 1991.
If you think you want to start a business marketing fleece, yarn, or finished wool products, this is a must-have book.

Simmons, Paula, and Darrel Salsbury. *Your Sheep.* Pownal, VT: Storey Books, 1992.
A great book for kids who are getting their first sheep.

Spaulding, C. E. *A Veterinary Guide for Animal Owners.* Emmaus, PA: Rodale Press, 1996.
Another must-have for animal owners.

Warren, Jill Stanford. *Lamb Country Cooking.* Lake Oswego, OR: Culinary Arts, Ltd., 1995.
Lots of good lamb recipes, as well as serving hints, and ideas for using leftovers.

Magazines

The Banner (Cuba, IL). Phone: 309-785-5058.
This is especially suited for those folks who are interested in registered sheep and in showing sheep.

Black Sheep Newsletter (Scappose, OR). Phone: 503-621-3063.
Geared toward anyone who is interested in raising naturally colored sheep.

Country Side & Small Stock Journal (Withee, WI). Phone: 800-551-5691.
A good general homesteading type of magazine, with lots of info on sheep and goats.

Handwoven and/or *Spin-Off* (Loveland, CO: Interweave Press).
Phone: 970-669-7672.
The magazines for spinning and weaving.

Sheep Canada (Edmonton, Alberta, Canada). Phone: 780-414-2043.
Contains lots of information on sheep shows and registered sheep in Canada.

Sheep Industry News (Englewood, CO: American Sheep Industry Association). Phone: 303-771-8200.
News about what's happening in the sheep industry, including legislative information.

sheep! magazine (Lake Mills, WI). Phone: 920-648-8285.
This is the best all-around magazine for shepherds with small and medium-sized flocks. Lots of informative articles and good ads.

The Shepherd (New Washington, OH). Phone: 419-492-2364.
Aimed at larger scale, commercial shepherds; contains lots of information on sheep research coming out of USDA and the agricultural colleges, as well as information on laws and regulations that impact the sheep industry.

The Shepherd's Journal (Mossleigh, Alberta, Canada).
Phone: 403-534-2185.
An excellent general sheep magazine aimed at Canadian shepherds, but with information that is applicable to shepherds in the United States as well.

Stockman Grass Farmer (Ridgeland, MS). Phone: 800-748-9808.
Excellent source of information about managed grazing. If we could afford just one agricultural magazine, this would be the one we'd keep.

The Working Border Collie (Sidney, OH). Phone: 937-492-2215.
Excellent for anyone who wants to train and work with herding dogs.

General Information

American Kennel Club
5580 Centerview Drive, Suite 200
Raleigh, NC 27606
919-233-3600
www.akc.org

American Herding Breed Association
277 Central Avenue
Seekonk, MA 02771
508-761-4078, or
www.primenet.com/~joell/ahba/main.htm

U.S. Border Collie Handlers Association
2915 Anderson Lane
Crawford, TX 76638
254-486-2500
www.bordercollie.org/usbcha.htm
These three organizations can provide information on herding dogs. AKC can also provide contact information for guardian-dog breeder's groups.

American Livestock Breeds Conservancy
21 Hillsboro Street
Pittsboro, NC 27312
919-542-5704
www.albc-usa.org
If you are interested in raising heritage breeds, these folks can help you locate other shepherds who work with the breed that interests you.

Forage Information System, Oregon State University
forages.orst.edu/
Oregon State has the best site on the Web for learning about forage plants of all types. The site also lists plant tissue–testing labs.

Oklahoma State University, Department of Animal Science
www.ansi.okstate.edu/breeds
Oklahoma State has put together a really good Web page with information about livestock breeds from around the world.

Appropriate Technology Transfer for Rural America (ATTRA)
P.O. Box 3657
Fayetteville, AR 72702
800-346-9140
www.attra.org
ATTRA is a great resource and can provide information and answer questions. Their services are free.

Organic Trade Association
P.O. Box 547
Greenfield, MA 01302
413-774-7511
www.ota.com
The Organic Trade Association can help you identify an organic certifying agency that is working in your state.

United States Department of Agriculture
14th & Independence Avenue SW
Washington, DC 20250
202-720-2791
www.usda.gov
For general information.

USDA Animal and Plant Health Inspection Service
U.S. Department of Agriculture
12th & Indepedence Avenue SW
Washington, DC 20250
www.aphis.usda.gov
For information on wildlife services and scrapie certification.

Commercial Providers

All Kinds of Supplies

NASCO Farm & Ranch
901 Janesville Avenue
Fort Atkinson, WI 53538
800-558-9595
www.enasco.com
This is sort of like the Sears Catalog of agriculture. It has just about anything you might need, but can't find elsewhere.

Artificial Insemination

Elite Genetics
605 Rossville Road
Waukon, IA 52172
319-568-4551
elitegenetics.com
These are the leaders in AI for the sheep industry.

Cottage Industry Wool Supplies

Patrick Green Carders
48793 Chilliwack Lake Road
Chilliwack, British Columbia, Canada V4Z1A6
604-858-6020
Paula and Patrick's company, which makes all the equipment you might need if you plan to go into the Cottage Wool business. Even if you aren't interested in commercial preparation of wool, they sell an excellent drum carder.

Fencing and Sheep Supplies

Premier
2031 300th Street
Washington, IA 52353
800-282-6631
Ask for both Premier's fencing catalog and shepherd's catalog. A wide selection of specialized products, including ram masks.

Natural Wormers and Test Kits

Farmstead Health
P.O. Box 985
Hillsborough, NC 27278
919-643-0300
www.farmsteadhealth.com
Linda Phillips and Susan Glandin offer herbal parasite control and do-it-yourself test kits.

Sheep Coats

Powell Sheep Company
P.O. Box 183
Ramona, CA 92065
760-789-1758
Powell offers 12 sizes of sheep coats.

Sheep Supplies

Sheepman Supply Company
P.O. Box A8102 Liberty Road
Frederick, MD 21702

MidStates Livestock Supplies
125 East 10th Avenue
South Hutchinson, KS 67505
800-835-9665

Sausage Making Supplies

ABS Supply
38 Scarlett Road
Toronto, Ontario, Canada M6N4K2
416-769-0907
www.sausagemaker.net

Eldon's Jerky and Sausage Supply
HC75 Box 113 A2
Kooskia, ID 83539
208-926-4949
800-352-9453 (to order)
www.eldonsausage.com
These companies supply equipment, sausage casings, and seasonings.

Glossary

Abscess. A localized collection of pus, generally caused by an infected wound, sting, or a "splinter" that's encapsulated under the skin.

Acclimatization. Becoming accustomed to a new environment.

Acute infection. An infection or disease that has rapid onset and pronounced signs and symptoms.

Additive. An ingredient or substance added to a feed mixture, generally in small quantities. May be added for nutritional reasons, such as vitamins or minerals, or for medicinal purposes, such as antibiotics.

Afterbirth. The placenta and membranes that are passed from the ewe's body after giving birth to a lamb or lambs.

AI. Abbreviation of *artificial insemination*.

Anemia. A deficiency in the oxygen-carrying capacity of blood. Can be caused by loss of blood or by certain disease conditions, but in sheep is most often caused by loss of blood due to blood-feeding worms.

Anestrus. The nonbreeding season; females that are not in heat.

Anthelmintic. A drug that kills or expels intestinal worms.

Antibiotic. A medicine that inhibits the growth of, or kills, bacteria. Antibiotics have no effect on viruses, fungi, or worms.

Antibody. A protein substance developed in the body to fight a specific antigen.

Antigen. A "foreign invader," to which the body's immune system recognizes as such. Usually a bacteria or a virus.

Antiseptic. A chemical used to reduce or kill bacteria.

Artificial insemination. The introduction of semen into the reproductive system of a ewe for the purposes of impregnating her. Must be done by a trained technician or veterinarian.

Ash. The mineral matter of feed.

Bacteria. A single-celled microorganism. Some bacteria are beneficial, and necessary for good health — for example, the bacteria that regularly live in the rumen — and others cause disease.

Bag. The ewe's udder, or mammary glands.

Balanced ration. A feed ration that supplies all the required nutrients for the animal's needs at the time.

Band. Used by ranchers running thousands of sheep on a range to signify one group. Like a flock, but much larger. Also, a small rubber device that looks like a little donut that is used for castrating and docking.

Black wool. Any wool containing black, or dark, fibers.

Bloat. A disorder characterized by an abnormal accumulation of gas in the rumen. Bloat is often fatal if not caught and treated quickly.

Blood grading. The degree of fineness of wool. Measured as fraction. Originally, the fraction indicated the portion of Merino blood in the animal.

Body condition scoring. A system of assessing the condition a sheep is in, with scores ranging from 1 to 5, with 1 being thin and 5 being fat.

Bolus. Regurgitated food that is being chewed, or has been chewed and is ready to be swallowed (see Cud); a large pill for animals.

Breech. The buttocks; a birth in which the fetus is presented "rear" first.

Breed. A like group of animals that have been bred to exhibit certain definable, inherited traits; the mating of animals.

Bright wool. Light, clean wool.

Broken mouth. Old ewe or ram that have lost teeth. Usually begins around 4 to 5 years old.

Browse. Woody or brushy plants that can be eaten.

Buck. Mature male (see Ram).

Bummer. A lamb that has to be bottle fed by the shepherd. Usually an orphan, though sometimes a lamb whose mother doesn't enough milk for multiple lambs.

Burdizzo. A tool used to castrate lambs by severing the cord without breaking the skin.

Bushel. A unit of capacity approximately equal to 1.25 cu ft (37.5 cm^3).

Carcass. The dressed body of a meat animal from which internal organs and offal have been removed.

Carding. An operation that converts loose, clean wool into continuous, untwisted strands. May be done with "hand cards" or a carding machine.

Carpet wool. Coarse, harsh, strong wool suitable for producing carpets.

Carrier. An animal that carries a disease but doesn't show signs of it.

Carrying capacity. The number of animals a piece of land is capable of sustaining for a given period of time.

Castrate. To remove the testicles from a ram so that he is permanently incapable of breeding.

Clean wool. Usually refers to scoured wool, though hand spinners may describe a grease wool that has little or no vegetable contamination as clean wool.

Clip. The total, annual wool production from a flock.

Closed face. A sheep that has heavy wool about the eyes and cheeks.

Club lamb. A lamb raised as a 4-H, FFA, or other club, project.

Colic. An abdominal condition generally characterized by severe pain.

Colostrum. The first milk produced by a ewe after giving birth. Colostrum contains antibodies from the ewe's immune system, which can be absorbed through the lamb's intestines for about the first 24 hours of life.

Combing. An operation that removes short fibers and leaves long fibers laid out straight and parallel.

Composite. A uniform group of animals created through selective crossbreeding.

Concentrate. A high-energy, low-fiber feed.

Conformation. The shape, proportions, and "design" of the animal.

Count. The fineness to which yarn may be spun; a system of grading wool based on how finely it can be spun.

Creep. An enclosure that allows lambs to enter for supplemental feeding, but prohibits older animals from entering.

Crimp. The "wave" effect in wool fibers.

Crossbred. Animals that are known to have more than one breed in their lineage. Many crossbreds perform well due to hybrid vigor.

Cross-fencing. Fences used to subdivide pastures into smaller paddocks.

Cud. A bolus of regurgitated food.

Cull. Remove a breeding animal from the flock that isn't meeting the needs of breeding animals within the flock, often because of health, age, poor reproductive record, and so on. In the case of rams, they're often culled so they don't breed their own daughters. One farmer's cull animal may make a fine addition to another farmer's flock!

Dam. The female parent.

Deficiency disease. An illness caused by a lack of one or more nutrients. For example, calcium deficiency in heavily-milking ewes (also known as milk fever) can cause stiffness, lameness, bone deformities, and convulsions in the ewe or her lambs.

Degreased wool. Wool that's been cleaned chemically to remove all "grease," or lanolin.

Density. Number of wool fibers per unit area of a sheep's body. Fine-wool breeds have greater density than coarse wool breeds.

Dental pad. The firm upper gum, which lacks teeth.

Dock. To cut the tail off; the remaining portion of the tail of a sheep that has been docked.

Drench. A liquid medicine given orally.

Drift lambing. A pasture-based system of lambing in which ewes that have not yet lambed are moved each day, while ewes with new lambs remain in the pasture where they dropped their lambs.

Dry. A nonlactating ewe; the period between lactations.

Dry matter. The proportion of a given feedstuff that doesn't contain any water. Found in a laboratory by "cooking" a feed sample at about 120°F (48.9°C) to drive off all water molecules.

Drylot. A small enclosure in which animals are confined.

Edema. Swelling due to excess accumulation of fluid in tissue spaces.

Elasticity. The ability of wool fibers to return to original length after being stretched. Good-quality wool has a great deal of elasticity.

Elastrator. A device that is used to apply a heavy rubber band to the tail or scrotum of a lamb for docking or castrating.

Electrolytes. Salts naturally found in an animal's blood. May be administered orally or intravenously during illness.

Emaciated. An animal that is overly thin, often caused by illness.

Eruction. The elimination of gas by belching.

Estrous cycle. The time and physiological events that take place in one heat period of 17 days in a ewe.

Estrus. The time during which a ewe will allow a ram to breed her. Normally, a ewe is in estrus for about 28 hours.

Ewe. Mature female.

Ewe breed. Fine wool, prolific breeds of sheep.

Ewe lamb. Immature female.

Extensive management. A management system that relies on low input methods of production, centered on pastures and grazing.

Facing. Trimming wool from around the face of closed-face sheep (see Wigging).

Fecundity. The ability to produce many offspring, either within 1 year or over a lifetime.

Feeder lambs. Animals under 1 year of age that make good gains if placed on high-input feedstuffs.

Feedstuff. An ingredient or material fed to an animal.

Felting. The interlocking of fibers when rubbed together under conditions of heat, moisture, and pressure.

Fermentation. Microbial decomposition or organic matter in an oxygen-free environment, including the breakdown of food by microorganisms in the sheep's rumen.

Fertility. Ability of an animal to reproduce.

Fetus. An animal in the uterus until birth.

Finish. To fatten animals for slaughter.

Fitting. Preparing an animal for show.

Fleece. The wool from one sheep.

Flock. A group of sheep, also see brood.

Flushing. Feeding ewes additional feed for 2 to 3 weeks prior to breeding.

Forage. Vegetable matter in pasture, hay, or silage.

Free choice. Food available at all times.

Freshening. Giving birth.

Gestation. Time between breeding and lambing. In sheep, between 147 and 153 days.

Grade. Grade animals may be crossbred, or may be purebred, but there definitely are no records of breeding recorded with a breed association.

Grading. Classifying fleeces according to fineness, length, character, and quality.

Graft. Having a ewe accept and mother a lamb that isn't her own.

Grease wool. Wool as it comes from the sheep.

Gummer. An old sheep missing all or most of its teeth.

Halter. A rope or leather head gear used to control or lead an animal.

Hothouse lamb. A lamb born in fall or early winter and butchered at 9 to 16 weeks of age.

Hybrid vigor. The extra vigor, strength, hardiness, and productive capacity that comes from crossbreeding animals.

Immunity. The animal's ability to resist or overcome infection. May be natural or the result of vaccinations.

Inbreeding. The breeding of animals that are closely related.

Infestation. Presence of a large number of parasites or insects.

Intensive management. A production system that relies on high levels of inputs, including harvested feeds and specialize facilities.

Jug. A small pen large enough for just one ewe and her offspring. Generally used for several days after birthing.

Ked. An external parasite that affects sheep. Sometimes referred to as a sheep tick, but actually the organism is a flat, brown, wingless fly.

Lactation. The period during which a ewe is producing milk.

Lamb. Newborn or immature sheep, typically under 1 year of age.

Lanolin. The naturally occurring "grease" that coats wool.

Long wool. Wool that is 12 to 15 inches (30.5–38.1 cm) long, typically from the Lincoln, Leicester, and Cotswold breeds.

Luster. The natural gloss or sheen of a fleece.

Maintenance requirement. Feed ration required to maintain an animal's condition at rest; does not provide adequate nutrition for growth.

Mastitis. Infection of the mammary gland or udder.

Meconium. The first manure passed by a lamb.

Milk letdown. A physiological process that allows milk to be removed from the udder by sucking or mechanical means.

Minor breeds. Those breeds that have fallen from favor in commercial agriculture; consequently their numbers decrease — sometimes to the extent that the breed becomes endangered or extinct.

Mutton. Meat from a mature or aged sheep.

Open face. A sheep that doesn't have much wool around the eyes and cheeks.

***Ovis aries*.** Scientific name for domestic sheep.

Oxytocin. A hormone that controls milk letdown.

Packers. Animals that are destined to go to the butcher. Cull ewes and rams.

Palatable. Acceptable taste and quality for an animal to readily ingest.

Papered. Same as registered.

Pathogen. A disease-causing organism.

Pelt. The skin from a slaughtered sheep that still has the wool on it.

Physiological. Pertaining to the science that deals with the functions of living organisms.

Piebald. An animal that is spotted.

Polled. Naturally hornless, born without horns.

Purebred. Purebred animals have 100 percent of their bloodlines coming from one breed.

Quarantine. Keeping an animal isolated from other animals to prevent spread of infections.

Ram. Mature male (see also Buck).

Ram lamb. Immature male.

Ration. The amount of feed supplied to an animal, or a group of animals, during a specific period.

Registered. Registered animals are purebred, or bred in accordance with the standards of the breed's association or registry.

Replacement. A young animal selected to be kept for the breeding flock.

Retained placenta. A placenta not passed as afterbirth.

Roughage. Course and bulky feed that is high in fiber, like hay or silage.

Ruminant. An animal, such as sheep, goats, or cattle, that have a four-compartment stomach system.

Scours. Diarrhea.

Second cuts. Short lengths of wool resulting from cutting the same spot twice during shearing.

Shearing. The act of clipping wool from a sheep.

Shearling. Pelt from a slaughtered sheep that carries less than 1 inch (2.5 cm) of new wool.

Shrinkage. The amount of weight an animal loses during adverse conditions or transport; the loss of carcass weight during aging; the loss of weight in wool during scouring.

Skirting. The practice of removing the edges of a fleece at shearing.

Stanchion. A device for holding the head of an animal for milking or to perform veterinary procedures.

Stripping. Removing milk from the udder. Usually refers to removing the last of the milk.

Tagging. Cutting dung locks off a sheep.

Tags. Locks of wool contaminated by dung and dirt.

Tallow. The extracted fat from sheep and cattle.

Teaser. A ram or stag that is incapable of breeding ewes but is used to find ewes that are in heat.

Udder. The mammary glands with nipples.

Unsound. An animal that has health problems, poor conformation, and so on.

Weaning. Stopping lambs from allowing to suckle on their dams.

Wether. Castrated or neutered male.

Yearling. Ewe or ram between 1 and 2 years old.

INDEX

Note: Page numbers in *italic* refer to photographs or illustrations; those in **boldface** refer to charts.

B

Other Storey Titles You Will Enjoy

Basic Butchering of Livestock & Game by John J. Mettler Jr., DVM. Provides clear, concise, step-by-step information for individuals interested in slaughtering their own meat. 208 pages. Paperback. ISBN 0-88266-391-7.

Building Small Barns, Sheds & Shelters by Monte Burch. Covers tools, materials, foundations, framing, sheathing, wiring, plumbing, and finish work for barns, woodsheds, garages, fencing, and animal housing. 248 pages. Paperback. ISBN 0-88266-245-7.

The Canning, Freezing, Curing & Smoking of Meat, Fish & Game by Wilbur F. Eastman Jr. Safe, step-by-step instructions for preparing and storing fresh meat, plus recipes and instructions for building smokehouses. 208 pages. Paperback. ISBN 0-88266-045-4.

Fences for Pasture & Garden by Gail Damerow. The complete guide to choosing, planning, and building today's best fences: wire, rail, electric, high-tension, temporary, woven, and snow. 160 pages. Paperback. ISBN 0-88266-753-X.

How to Build Small Barns & Outbuildings by Monte Burch. This book takes the mystery out of small-scale construction. Projects are offered with complete plans and instructions. 288 pages. Paperback. ISBN 0-88266-773-4.

Keeping Livestock Healthy: A Veterinary Guide to Horses, Cattle, Pigs, Goats & Sheep by N. Bruce Haynes, DVM. A complete guide to preventing disease through good nutrition, proper housing, and appropriate care. 352 pages. Paperback. ISBN 0-88266-884-6.

Making Your Small Farm Profitable by Ron Macher. This practical, step-by-step guide to operating a small farm in the new millennium examines 20 alternative farming enterprises. Readers will learn how to target niche markets and sustain a farm's biological and economic health. 288 pages. Paperback. ISBN 1-58017-161-3.

Raising Animals by the Moon by Louise Riotte. Practical advice on breeding, birthing, weaning, and raising animals in harmony with nature. 176 pages. Paperback. ISBN 1-58017-068-4.

Small-Scale Livestock Farming: A Grass-Based Approach for Health, Sustainability, and Profit by Carol Ekarius. Helps you determine what you want from your farming life; choose suitable livestock; understand housing, fencing, and feeding needs of livestock; learn about caring for your animals' health; investigate conventional and cutting-edge market strategies; create a complete financial and biological farm plan; and make decisions that are good for you, for your animals, and for your land. 192 pages. Paperback. ISBN 1-58017-162-1.

Turning Wool into a Cottage Industry by Paula Simmons. The complete, no-nonsense, up-to-date information source for anyone operating a wool business at home. Includes merchandising tips and descriptions of 18 entrepreneurs. 192 pages. Paperback. ISBN 0-88266-685-1.